D1084308

VOLUME FOURTY THREE

EXPERIMENTAL METHODS IN THE PHYSICAL SCIENCES

RADIOMETRIC TEMPERATURE MEASUREMENTS

II. Applications

EXPERIMENTAL METHODS IN THE PHYSICAL SCIENCES

Thomas Lucatorto and Albert C. Parr, *Editors in Chief*

Founding Editors
L. MARTON
C. MARTON

VOLUME FOURTY THREE

EXPERIMENTAL METHODS IN THE PHYSICAL SCIENCES

RADIOMETRIC TEMPERATURE MEASUREMENTS

II. Applications

Editors

Zhuomin M. Zhang
George W. Woodruff School of Mechanical Engineering
Georgia Institute of Technology
Atlanta, GA 30332-0405, USA

Benjamin K. Tsai
Optical Technology Division, National Institute of Standards and Technology, Gaithersburg, MD 20899-8441, USA

Graham Machin
Engineering Measurement Division
National Physical Laboratory
Teddington, Middlesex, TW11 0LW, UK

ELSEVIER

Amsterdam • Boston • Heidelberg • London • New York • Oxford
Paris • San Diego • San Francisco • Singapore • Sydney • Tokyo
Academic Press is an imprint of Elsevier

Academic Press is an imprint of Elsevier
Linacre House, Jordan Hill, Oxford OX2 8DP, UK
84 Theobald's Road, London WC1X 8RR, UK
Radarweg 29, PO Box 211, 1000 AE Amsterdam, The Netherlands
30 Corporate Drive, Suite 400, Burlington, MA 01803, USA
525 B Street, Suite 1900, San Diego, CA 92101-4495, USA

Notice
No responsibility is assumed by the publisher for any injury and/or damage to persons
or property as a matter of products liability, negligence or otherwise, or from any use
or operation of any methods, products, instructions or ideas contained in the material
herein. Because of rapid advances in the medical sciences, in particular, independent
verification of diagnoses and drug dosages should be made

Library of Congress Cataloging-in-Publication Data
Application submitted

British Library Cataloguing in Publication Data
A catalogue record for this book is available from the British Library

ISBN: 978-0-12-375091-4
ISSN: 1079-4042

For information on all Academic Press publications
visit our web site at elsevierdirect.com

Printed and bound in United States of America

10 11 12 13 14   10 9 8 7 6 5 4 3 2 1

In memory of David P. DeWitt — a pioneer, teacher, and friend

CONTENTS

List of Contributors   *xi*

Volumes in Series   *xiii*

Preface   *xvii*

1. Industrial Applications of Radiation Thermometry   **1**
Jörg Hollandt, Jürgen Hartmann, Ortwin Struß and Reno Gärtner

   1. Introduction   2
   2. Industrial Radiation Thermometers and Line-Scanners   4
   3. Specifications of Radiation Thermometers   24
   4. Applications of Radiation Thermometry in Industrial
      Production Areas   28
   5. Summary   54
   Acknowledgement   54
   References   54

2. Experimental Characterization of Blackbody
Radiation Sources   **57**
Sergey N. Mekhontsev, Alexander V. Prokhorov and
Leonard M. Hanssen

   1. Introduction   57
   2. Reflectometric Determination of the Effective Emissivity
      of Blackbody Radiators   60
   3. Radiometric Measurement of Blackbody Sources   74
   4. Measurement of Input Parameters for the Calculation
      of Blackbody Radiator Characteristics   98
   5. Conclusions   129
   References   130

3. Radiation Thermometry in the Semiconductor Industry   **137**
Bruce E. Adams, Charles W. Schietinger and Kenneth G. Kreider

   1. Introduction   138
   2. Basics of Optical Fiber Thermometry (OFT)   143
   3. Temperature Measurements in the Semiconductor Industry 152
   4. Applications   159

5. Calibration   164
6. *In situ* Calibration of Radiation Thermometers in RTP Tools   170
7. Emissivity Methods   189
8. Problems to Solve   208
9. Summary   208
Acknowledgments   209
References   209

4. **Thermometry in Steel Production**   **217**
   Tohru Iuchi, Yoshiro Yamada, Masato Sugiura and Akira Torao

   1. Introduction   217
   2. Review of the Steel Production Process   219
   3. Characteristics of Radiometric Temperature
      Measurement in Steel Processes   224
   4. Applications   228
   5. Summary   270
   References   271

5. **Thermal Imaging in Firefighting and Thermography
   Applications**   **279**
   Francine Amon and Colin Pearson

   1. Introduction   279
   2. Thermography Engineering Applications   287
   3. Thermal Imaging in Firefighting   316
   4. Standards   327
   5. Summary   328
   Acknowledgments   328
   References   328

6. **Remote Sensing of the Earth's Surface Temperature**   **333**
   Peter J. Minnett and Ian J. Barton

   1. Introduction   334
   2. Statement of the Problem   337
   3. Remote Sensing of Surface Temperature   345
   4. Spacecraft Radiometers   354
   5. Validation of Surface Temperature Retrievals   361
   6. Residual Uncertainties   367
   7. Applications of Remotely Sensed Surface
      Temperatures   374
   8. Atmospheric Profiles   380
   9. Future Missions   381
   10. Conclusions and Outlook   383
   References   384

7. Infrared and Microwave Medical Thermometry   **393**
E. Francis. J. Ring, Jürgen Hartmann, Kurt Ammer, Rod Thomas,
David Land and Jeff W. Hand

  1. Introduction   394
  2. Infrared Ear Thermometers for Clinical Thermometry   394
  3. Infrared Thermal Imaging in Medicine   400
  4. Pulsed Photo-Thermal Radiometry (PPTR)   411
  5. Microwave Radiometry for Medical Applications   427
  6. Summary   442
  References   442

Appendix A: Fundamental and Other Physical Constants   **449**

Subject Index   **451**

LIST OF CONTRIBUTORS

Numbers in parenthesis indicate the pages on which the author's contributions begins.

Bruce E. Adams (137)
Applied Materials, Santa Clara, CA, USA

Kurt Ammer (393)
Institute for Physical Medicine and Rehabilitation, Hanusch Krankenhaus, Heinrich Collinstrasse 30, Vienna A1140, Austria

Francine Amon (279)
Building and Fire Research Laboratory, National Institute of Standards and Technology, Gaithersburg, MD 20899, USA

Ian J. Barton (333)
Marine and Atmospheric Research, Commonwealth Scientific and Industrial Research Organisation, Hobart, Tasmania 7001, Australia

Reno Gärtner (1)
Raytek GmbH, Berlin, Germany

Leonard M. Hanssen (57)
National Institute of Standards and Technology, 100 Bureau Drive, Gaithersburg, MD 20899-8442, USA

Jeff W. Hand (393)
Radiological Sciences Unit, Hammersmith Hospital, DuCane Road, London W12 0HS, UK

Jürgen Hartmann (1, 393)
Physikalisch-Technische Bundesanstalt, Braunschweig and Berlin, Germany

Jörg Hollandt (1)
Physikalisch-Technische Bundesanstalt, Braunschweig and Berlin, Germany

Tohru Iuchi (217)
Department of Mechanical Engineering, Toyo University, 2100 Kujirai, Kawagoe Saitama 350-8585, Japan

Kenneth G. Kreider (137)
Process Measurements Division, National Institute of Standards and Technology, Gaithersburg, MD, USA

David Land (393)
Department of Physics and Astronomy, University of Glasgow, Glasgow G12 8QQ, UK

Sergey N. Mekhontsev (57)
National Institute of Standards and Technology, 100 Bureau Drive, Gaithersburg, MD 20899-8442, USA

Peter J. Minnett (333)
Meteorology and Physical Oceanography, Rosenstiel School of Marine and Atmospheric Science, University of Miami, 4600 Rickenbacker Causeway, Miami, FL 33149-1098, USA

Colin Pearson (279)
BSRIA Design and Facilities Innovation, Old Bracknell Lane West, Brackness, Berkshire, RG12 7AH, UK

Alexander V. Prokhorov (57)
National Institute of Standards and Technology, 100 Bureau Drive, Gaithersburg, MD 20899-8442, USA

E. Francis J. Ring (393)
Medical Imaging Research Unit, Faculty of Advanced Technology, University of Glamorgan, Pontypridd CF37 1DL, UK

Charles W. Schietinger (137)
Lopez Island, Washington, DC, USA

Ortwin Struß (1)
Heitronics Infrared GmbH, Wiesbaden, Germany

Masato Sugiura (217)
Environment & Process Technology Center, Nippon Steel Corporation, 20-1 Shintomi, Futtsu, Chiba 293-8511, Japan

Rod Thomas (393)
Faculty of Applied Design and Engineering, Swansea Metropolitan University, Mount Pleasant, Swansea, SA1 6ED, UK

Akira Torao (217)
New Products Planning & Development Department, JFE Advantech Co. Ltd., 3-48 Takahata-cho, Nishinomiya, Hyogo 663-8202, Japan

Yoshiro Yamada (217)
National Metrology Institute of Japan, National Institute of Advanced Industrial Science and Technology, 1-1-1 Umezono, Tsukuba, Ibaraki 305-8563, Japan

Volumes in Series
Experimental Methods in the Physical Sciences
(formerly Methods of Experimental Physics)

Editors-in-Chief
Thomas Lucatorto and Albert C. Parr

Volume 1. Classical Methods
Edited by lmmanuel Estermann

Volume 2. Electronic Methods, Second Edition (in two parts)
Edited by E. Bleuler and R. O. Haxby

Volume 3. Molecular Physics, Second Edition (in two parts)
Edited by Dudley Williams

Volume 4. Atomic and Electron Physics - Part A:
Atomic Sources and Detectors; Part B: Free Atoms
Edited by Vernon W. Hughes and Howard L. Schultz

Volume 5. Nuclear Physics (in two parts)
Edited by Luke C. L. Yuan and Chien-Shiung Wu

Volume 6. Solid State Physics - Part A: Preparation, Structure,
Mechanical and Thermal Properties; Part B: Electrical,
Magnetic and Optical Properties
Edited by K. Lark-Horovitz and Vivian A. Johnson

Volume 7. Atomic and Electron Physics - Atomic Interactions
(in two parts)
Edited by Benjamin Bederson and Wade L. Fite

Volume 8. Problems and Solutions for Students
Edited by L. Marton and W. F. Hornyak

Volume 9. Plasma Physics (in two parts)
Edited by Hans R. Griem and Ralph H. Lovberg

Volume 10. Physical Principles of Far-Infrared Radiation
Edited by L. C. Robinson

Volume 11. Solid State Physics
Edited by R. V. Coleman

Volume 12. Astrophysics – Part A: Optical and Infrared Astronomy
Edited by N. Carleton
Part B: Radio Telescopes; Part C: Radio Observations
Edited by M. L. Meeks

Volume 13. Spectroscopy (in two parts)
Edited by Dudley Williams

Volume 14. Vacuum Physics and Technology
Edited by G. L. Weissler and R. W. Carlson

Volume 15. Quantum Electronics (in two parts)
Edited by C. L. Tang

Volume 16. Polymers – Part A: Molecular Structure and
Dynamics; Part B: Crystal Structure and Morphology;
Part C: Physical Properties
Edited by R. A. Fava

Volume 17. Accelerators in Atomic Physics
Edited by P. Richard

Volume 18. Fluid Dynamics (in two parts)
Edited by R. J. Emrich

Volume 19. Ultrasonics
Edited by Peter D. Edmonds

Volume 20. Biophysics
Edited by Gerald Ehrenstein and Harold Lecar

Volume 21. Solid State Physics: Nuclear Methods
Edited by J. N. Mundy, S. J. Rothman, M. J. Fluss, and
L. C. Smedskjaer

Volume 22. Solid State Physics: Surfaces
Edited by Robert L. Park and Max G. Lagally

Volume 23. Neutron Scattering (in three parts)
Edited by K. Skold and D. L. Price

Volume 24. Geophysics – Part A: Laboratory Measurements;
Part B: Field Measurements
Edited by C. G. Sammis and T. L. Henyey

Volume 25. Geometrical and Instrumental Optics
Edited by Daniel Malacara

Volume 26. Physical Optics and Light Measurements
Edited by Daniel Malacara

Volume 27. Scanning Tunneling Microscopy
Edited by Joseph Stroscio and William Kaiser

Volume 28. Statistical Methods for Physical Science
Edited by John L. Stanford and Stephen B. Vardaman

Volume 29. Atomic, Molecular, and Optical Physics –
Part A: Charged Particles; Part B: Atoms and Molecules;
Part C: Electromagnetic Radiation
Edited by F. B. Dunning and Randall G. Hulet

Volume 30. Laser Ablation and Desorption
Edited by John C. Miller and Richard F. Haglund, Jr.

Volume 31. Vacuum Ultraviolet Spectroscopy I
Edited by J. A. R. Samson and D. L. Ederer

Volume 32. Vacuum Ultraviolet Spectroscopy II
Edited by J. A. R. Samson and D. L. Ederer

Volume 33. Cumulative Author Index and Tables of Contents,
Volumes 1-32

Volume 34. Cumulative Subject Index

Volume 35. Methods in the Physics of Porous Media
Edited by Po-zen Wong

Volume 36. Magnetic Imaging and its Applications to Materials
Edited by Marc De Graef and Yimei Zhu

Volume 37. Characterization of Amorphous and Crystalline Rough
Surface: Principles and Applications
Edited by Yi Ping Zhao, Gwo-Ching Wang, and Toh-Ming Lu

Volume 38. Advances in Surface Science
Edited by Hari Singh Nalwa

Volume 39. Modern Acoustical Techniques for the Measurement
of Mechanical Properties
Edited by Moises Levy, Henry E. Bass, and Richard Stern

Volume 40. Cavity-Enhanced Spectroscopies
Edited by Roger D. van Zee and J. Patrick Looney

Volume 41. Optical Radiometry
Edited by A. C. Parr, R. U. Datla, and J. L. Gardner

Volume 42. Radiometric Temperature Measurements. I. Fundamentals
Edited by Z. M. Zhang, B. K. Tsai, and G. Machin

Volume 43. Radiometric Temperature Measurements. II. Applications
Edited by Z. M. Zhang, B. K. Tsai, and G. Machin

PREFACE

Temperature measurement and control has played and continues to play a vital role in many scientific and technological advances. Radiometric temperature measurement, that is the measurement of temperature based on thermal radiative emission, has a long history from fundamental studies of Planckian emission, to many industrial applications including iron and steel production and materials and chemical processing, to playing a fundamental role in the realization and dissemination of successive international temperature scales. Radiation thermometry is attractive in many challenging temperature measurement situations because it is a noncontact, nonintrusive, and fast technique.

Thermal radiation is governed by the fundamental physical laws established over one hundred years ago by Kirchhoff, Stefan, Boltzmann, Wien, and, in particular, Planck. These laws directly link emitted blackbody radiation, total or spectrally resolved, to the thermodynamic temperature of the emitting source. Actual practical measurements by radiation thermometry, however, are prone to a number of uncertainties associated with, for example, surface emissivity and environmental effects such as absorption by dust or smoke and reflected ambient radiation. While a number of books have been published on thermometry, in general, no comprehensive book devoted to radiometric temperature measurement has been published since the publication in 1988 of *Theory and Practice of Radiation Thermometry*, edited by D.P. DeWitt and G.D. Nutter.

In recent years, there have been tremendous developments in instrumentation. For instance, infrared focal plane arrays can now produce images with a spatial resolution of order $10\,\mu\text{m}$ with a temperature resolution of $0.01\,\text{K}$. While the expert in the field can keep abreast of these rapidly advancing techniques through the information presented at periodic international temperature symposia and through the technical literature, it is very difficult for a newcomer to find a definitive up-to-date summary of the practice of radiation thermometry. This book aims at filling that gap by covering basic theory, measurement fundamentals, standards and calibration, and summaries of current practice of radiation thermometry in different technical fields at a level accessible to the newcomer but also comprehensive enough to provide the information needed to understand and bring to bear the latest technique to a particular radiometric temperature measurement problem.

This two-volume set on *Radiometric Temperature Measurements* (I. Fundamentals and II. Applications) is written for those who will apply radiation thermometers in industrial practice, who will use thermometers in scientific research, who design and develop thermometers for instrument manufacturers, and who will design the thermometers to address particular measurement challenges. These volumes are more than a practice guide. We hope that by presenting the fundamental principles and pointing out the pitfalls in applying radiation thermometry in various settings, our readers will gain knowledge in: (1) the proper selection of the type of thermometer; (2) the best practice in using radiation thermometers; (3) awareness of the uncertainty sources and subsequent appropriate procedure to reduce the overall measurement uncertainty; and (4) understanding of the calibration chain and its current limitations. We have also added a large number of references at the end of each chapter as a source for those seeking a deeper or more detailed understanding.

The author(s) of each chapter were chosen from a group of international scientists who are experts in the field and specialist(s) on the subject matter covered in the chapter. It is intended that together the two volumes will form a comprehensive summary of the current practice of radiation thermometry. The first volume concentrates on the fundamental aspects, while the second volume mainly focuses on the industrial and practical applications. In the fundamental volume, Chapter 1 provides a historical overview of radiation thermometry, explains the basic fundamentals and commonly used terms, and lists the various types of radiation thermometers. The concepts of temperature, its scale realization, calibration, traceability, measurement, uncertainty analysis, and future approaches, are extensively elaborated in Chapter 2. The basic theory on blackbody radiation, radiative properties, and the electromagnetic wave theory are discussed in Chapter 3. Chapter 4 focuses on the design and characterization of radiation thermometers. Chapter 5 addresses the theoretical and computational characterization of isothermal and nonisothermal blackbody cavities by analytical and Monte Carlo methods. In Chapter 6, radiance sources used for calibration such as fixed-point blackbodies, variable temperature blackbodies, cryogenic blackbodies, high stability and other tungsten-based lamps are described. Chapter 7 is an overview of some complementary surface temperature measurement techniques, such as thermal reflectance, interferometry, ellipsometry, and photothermal radiometry with application examples.

The volume on applications begins with a review of the state-of-the-art industrial applications of radiation thermometry, including a critique of multiwavelength thermometry (Chapter 1). Chapter 2 describes experimental characterization of blackbody cavities with an extensive survey on the measurement techniques. Chapter 3 focuses on the application of optical fiber thermometry for semiconductor processing, with an emphasis on rapid thermal processing and *in situ* calibration of lightpipe thermometers using thin-film thermocouples. Chapter 4 reviews the state-of-the-art practice of radiation thermometry in the steel industry, highlighting specific manufacturing processes. Chapter 5 deals with thermal imaging in firefighting and other thermographic applications along with standards of measurement and application. Chapter 6 discusses remote sensing of earth and sea surface temperatures and reviews different instruments and their measuring capabilities. Finally, Chapter 7 covers four aspects of clinical radiation thermometry: ear thermometry, medical thermal imaging, medical pulsed photothermal radiometry, and microwave radiometry for clinical applications.

This two-volume set is a tribute to David DeWitt (1934–2005) who has been an inspiration for us and to many others in the radiation thermometry community. In his last eight years, he dedicated his research to temperature measurement and calibration for rapid thermal processing in microelectronics manufacturing industry. He will always be remembered as a leader in the fields of radiation thermometry and heat transfer engineering.

The editors sincerely thank all of the chapter authors for their outstanding contributions and hard work. We also express appreciation to Dr. Tom Lucatorto and Dr. Albert C. Parr, the series editors, for their constant encouragement during this process and their careful review of the chapter contents. Finally, we would like to thank our families for their full support and enduring patience throughout the writing and editing of this book.

Zhuomin M. Zhang
Georgia Institute of Technology

Benjamin K. Tsai
National Institute of Standards and Technology

Graham Machin
National Physical Laboratory

June 2009

CHAPTER 1

INDUSTRIAL APPLICATIONS OF RADIATION THERMOMETRY

Jörg Hollandt[1], Jürgen Hartmann[1], Ortwin Struß[2] *and* Reno Gärtner[3]

Contents

1. Introduction	2
2. Industrial Radiation Thermometers and Line-Scanners	4
2.1. General characteristics of radiation thermometers and influence of environmental conditions	4
2.2. Types of industrial radiation thermometers	9
2.3. Line-scanners	19
3. Specifications of Radiation Thermometers	24
4. Applications of Radiation Thermometry in Industrial Production Areas	28
4.1. Glass industry	28
4.2. Steel industry	36
4.3. Aluminium industry	37
4.4. Plastics industry	43
4.5. Laser welding and laser cutting of metals and plastics	49
4.6. Semiconductor industry	50
4.7. Measurement in the tunnel furnace	51
4.8. Other applications	53
5. Summary	54
Acknowledgement	54
References	54

[1] Physikalisch-Technische Bundesanstalt, Braunschweig and Berlin, Germany
[2] Heitronics Infrared GmbH, Wiesbaden, Germany
[3] Raytek GmbH, Berlin, Germany

Experimental Methods in the Physical Sciences, Volume 43
ISSN 1079-4042, DOI 10.1016/S1079-4042(09)04301-X

1. INTRODUCTION

Temperature is the most important thermodynamic state variable determining, among other things, the speed of chemical reactions, the rate of reproduction of living cells, the degree of efficiency of thermal engines and the emission of thermal radiation. Practically all quantities relevant in industry and research are a function of temperature, and this makes temperature, in addition to time, the most often measured physical quantity. It is clear that good temperature measurement and control for successful industrial processes is required. In addition, because of the globalization of industrial production, simple, precise and yet globally uniform temperature determination is essential. This uniformity is ensured by the implementation of the International Temperature Scale of 1990 (ITS-90) [1] (see Chapter 2 of the companion volume [2] for a detailed description).

Radiation thermometry determines the temperature of an object through measuring its emitted thermal radiance. The technique is a non-contact, quick and non-intrusive surface temperature measurement that, with proper traceable calibration, can be robustly linked to the ITS-90. Among the variety of existing temperature measuring methods, radiation thermometry has a unique role in temperature measurement for industrial production processes. It allows temperature measurement in the following situations which would be very difficult by any other means:

- rapidly moving objects;
- very small objects;
- objects with small heat capacity and/or low thermal conductivity;
- objects with rapidly changing temperature;
- objects for which spatial temperature distributions need to be determined;
- objects at very high temperatures;
- processes where contamination/intrusion has to be avoided.

Although non-contact industrial temperature measurement in steel and glass manufacture and processing had its origin mainly at high temperatures above 700°C, there is at present a rapid expansion occurring in the application of non-contact thermometry in the low-temperature range, particularly exploiting the atmospheric window from 8 μm to 14 μm. For example, non-contact thermometry at lower temperatures has expanded into the non-metal working industry (e.g. plastics processing and production), maintenance and repair (e.g. non-destructive testing and evaluation), heat and refrigeration engineering and food processing, nearly all of which are at temperatures below 300°C. In no particular order, important industrial production and processing areas in which radiation thermometry is used extensively are, for example:

- glass industry;
- steel industry;
- aluminium industry;
- plastics industry;
- semiconductor industry;
- asphalt, cement and chalk industry;
- paper industry;
- printing and surface-coating industry;
- lacquer curing and lacquer drying;
- laser welding and laser cutting;
- food industry;
- waste incineration, hot gases and flames;
- energy generation and power plant operation.

The detector of thermal radiation is *the* essential component of a non-contact thermometer. It can be a fast photoelectric semiconductor detector (e.g. based on silicon or indium gallium arsenide (InGaAs)) or for lower temperature applications a thermal detector (e.g. a pyroelectric detector, bolometer or thermopile). Detectors are usually thermally stabilized (or their temperature is monitored and corrected for) to negate changes in responsivity due to drifts in the detector temperature during operation. Furthermore, developments in the semiconductor industry have allowed greatly improved thermal radiation detectors to be developed − leading to enhanced performance radiation thermometers [3].

Depending on the application, *stationary* and *non-stationary* radiation thermometers are in use. For industrial process control, *stationary radiation thermometers* (this includes line-scanners, which give the temperature distribution in one direction) are essentially applied. Non-stationary radiation thermometers are generally *handheld radiation thermometers* and *thermal imagers* which are often used as multi-purpose instruments for irregular temperature measurements. Thermal imagers provide two-dimensional temperature distributions.

The *stationary radiation thermometers* are widely used for *in situ* monitoring and for continuous process control in an industrial environment. This type of device has shown considerable growth in the number of applications, fuelled by improvements in detector performance and thermometer design aspects such as permissible ambient temperature, installation size, field-of-view diameter and cost.

In addition to the stationary radiation thermometers, *handheld radiation thermometers* and *thermal imagers* are increasingly being employed. They are often used as a hot-spot finder in the electrical and electronic industry or, in the case of devices with a sufficiently small measurement uncertainty, for temperature control in the food industry. In the case of thermal imagers, the development of micro-bolometer arrays [4], which are produced as evacuated thin-film bolometers based on vanadium oxide or amorphous silicon, has changed the market. Almost 90% of all modern thermal imagers used in the civilian domain are based on these focal-plane arrays. Such an array generally consist of 12,000−310,000 single bolometers, each having a typical dimension of 25 μm × 25 μm or 35 μm × 35 μm. Devices having 160 × 120 pixels, a temperature resolution of 70 mK and an image frequency of 25 Hz are currently the standard. These specifications are improving rapidly with both higher spatial and temperature resolution. However, it should be noted that presently in industrial process control, they only find niche applications. Having said that, thermal imagers along with infrared line-scanners *are* increasingly being used in industrial process control, where in addition to the temperature, the spatial distribution/ uniformity is a crucial quality criterion.

In this chapter, the principal design and current types of industrial radiation thermometers in common use are presented. Some of the selection criteria and essential specifications necessary for the practical implementation of radiation thermometers are discussed. A survey is given of the use of radiation thermometers in important sectors of industry. This will illustrate the broad spectrum of the application fields for radiation thermometry.

2. Industrial Radiation Thermometers and Line-Scanners

2.1. General characteristics of radiation thermometers and influence of environmental conditions

The general layout of a radiation thermometer, including a typical application environment, is shown schematically in Figure 1. An optical system images the measured object onto the field stop in front of the detector. The aperture stop and the field stop define the solid angle and the surface area (measurement field/field-of-view) of the thermal radiation measured by the detector. In the case of most radiation thermometers, a spectral filter is used to limit the wavelength range of the radiation reaching the detector. The spectral filter is often positioned in front of the detector or integrated as the window of the detector housing.

The performance of a radiation thermometer is essentially determined by the quality of its optical system and the detectivity of its radiation detector. High-quality radiation thermometers operate with a fixed focus (which corresponds to the recommended measuring distance) or with a variable focus for adjustment to the measuring distance. From the distance-dependent setting of the focus, a measurement field results in which the

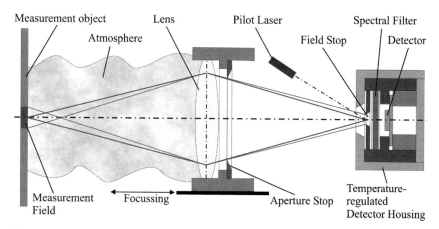

Measurement object   Lens   Pilot Laser   Spectral Filter

Atmosphere   Field Stop   Detector

Measurement   Focussing   Aperture Stop   Temperature-
Field   regulated
                                                                Detector Housing

Figure 1   Schematic view of the assembly of a radiation thermometer.

mean radiation temperature of the measurement object is determined. Industrial radiation thermometers for process temperature measurement operate, as a rule, with high-quality optical imaging systems. These can be made of lenses, mirrors, fibre optics or a combination of these. Depending on the spectral range in which the radiation thermometer operates, optical glass or infrared-transparent materials, such as silicon, germanium, zinc selenide, sapphire or barium fluoride, are employed for lens systems. In the case of low-cost radiation thermometers, optical components made of plastic are often used.

The alignment of radiation thermometers (particularly those used for lower temperature applications) and/or the marking of the measurement field is achieved with the aid of pilot lasers, transparent visors or even integrated digital camera systems. Integrated alignment lasers are often used as focus marking devices, which show the actual dimension of the measurement field with changing measuring distance.

For the measurement of thermal radiation temperatures in the range of about 200°C and above, photoelectric semiconductor detectors based on silicon or InGaAs in conjunction with glass optics are generally used. InGaAs detectors are used to measure temperatures from around 200°C, with silicon detectors being used from 450°C upwards. Semiconductor detectors, such as lead sulphide and lead selenide, are typically used for thermometers to measure object temperatures of 50°C and above. However, for very low-temperature applications down to −100°C, thermal detectors, such as thermopiles or pyroelectric detectors, are generally deployed. It should be noted that by cooling semiconductor detectors, their detectivity can be significantly improved. For example, cooling semiconductor detectors, like indium antimonide or mercury cadmium telluride, enables them to measure very low photon fluxes of

Table 1 Minimum temperature of the thermometer in relation to the ambient temperature and humidity to avoid condensation.

Ambient temperature (°C)	Relative humidity of ambient air						
	2%	4%	10%	20%	30%	50%	70%
30.0	5.0	5.0	5.0	6.0	11.0	19.0	25.0
40.0	5.0	5.0	5.0	13.0	20.0	28.0	34.0
50.0	5.0	5.0	10.0	21.0	28.0	38.0	45.0
60.0	5.0	5.0	18.0	28.0	38.0	47.0	54.0
70.0	5.0	9.0	24.0	38.0	45.0	57.0	n.p.
80.0	5.0	15.0	32.0	45.0	55.0	n.p.	n.p.
90.0	10.0	21.0	38.0	52.0	n.p.	n.p.	n.p.
> 100.0	15.0	27.0	45.0	60.0	n.p.	n.p.	n.p.

(n.p.): Operation not possible, as minimum cooling temperature is above 60°C.

thermal radiation, but they are not significantly used for industrial applications.

With some industrial radiation thermometers, the detector and sometimes also the spectral filter are temperature-stabilized. Often, however, internal reference radiation sources (or simply the measurement and processing of the internal temperature of the detector, and/or of the optical components) ensure a stable reading of measurement values of the radiation temperature within the specified temperature range of the radiation thermometer. Industrial radiation thermometers typically work in an ambient temperature range from −20°C to 60°C without requiring active heating or cooling. For more extreme conditions, custom-made products and the use of cooling jackets allow the use of radiation thermometers even under extreme ambient temperatures to above 300°C. When cooling the radiation thermometer, condensation, particularly on the optical system, but also on the thermometer housing must be avoided. See Table 1 for the minimum temperature for given environmental humidity conditions.

Industrial radiation thermometers generally operate within a limited spectral range, or spectral band. This band is generally adapted to the required temperature range of the thermometer, the transmission characteristic of the atmosphere and, for specific applications, the emission spectrum of the material to be measured.

Dependent on the temperature of the object being measured, different wavelength bands are most appropriate for the measurement. This is due to the peak thermal radiation emission (via Planck's radiation law; see Chapter 3 of Ref. [2]) shifting to longer wavelengths at lower temperatures. Above 450°C, where Si photodiodes are the detector of choice, wavelengths of 0.95 µm or less are most suitable. From 250°C upwards, where the InGaAs detector is most appropriate, the wavelength range from 0.9 µm to 1.7 µm

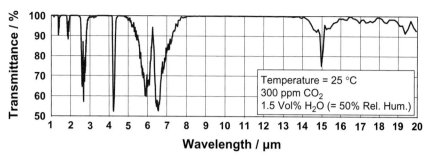

Figure 2 Transmittance of an air path 1 m long with 300 ppm CO_2 concentration, an air temperature of 25°C and a relative air humidity of 50%.

is suitable. Radiation temperatures above 0°C can be measured in an atmospheric window of 2.5−5 µm. For the measurement of radiation temperatures of ambient temperature and below the spectral range of 8−14 µm is used. This spectral range is also one of the choices for simple, multi-purpose radiation thermometers.

In principle, three phenomena have to be considered which may be caused by the atmosphere. Atmospheric *absorption* reduces the spectral radiance of the measured object, while atmospheric *emission* increases the spectral radiance observed by the radiation thermometer. Emission of the atmosphere is in most cases less significant than absorption, as long as objects are observed with a temperature significantly above atmospheric temperature. Atmospheric *scatter* becomes a major problem when a significant amount of small particles like dust, smoke or vapour is in the atmosphere. From among the gases present in air, water and carbon dioxide have a significant influence on the measurement of temperature. Figure 2 shows the transmittance of a 1-m-long air path with 300 ppm CO_2 concentration at a temperature of 25°C and a relative humidity of 50%. As can clearly be seen, there are "spectral windows" in the atmosphere where the transmission is practically unaffected by these molecular absorption bands.

The effect of the atmosphere must be considered, particularly when viewing targets that are at a significant distance from the thermometer. For example, in many industrial processes, dust and high air humidity are unavoidable consequences and these will strongly influence the readout of even well-designed radiation thermometers even at small measurement distances. Some high-quality radiation thermometers allow, if the atmospheric conditions are known, a correction to be made to the measured value of the temperature for the effect of distance, humidity and even carbon dioxide concentration. When using radiation thermometers under difficult and variable atmospheric conditions, measurement errors and contamination of the optical system can be avoided or reduced through judicious mounting of the radiation thermometer as well as through the use of protection tubes and/or air purging of the optical system. Furthermore, the use of fibre optics, which are brought close to the measuring object, or the use of ratio radiation thermometers, at least where the attenuation is wavelength independent, presents possible solutions. An overview on the environmental effects on radiation thermometry is given in Ref. [5].

The emissivity of the measurement object has a great influence on the radiometric measurement of the object's temperature (see Chapter 3 of Ref. [2]). The emissivity of a material is not only dependent on the temperature, wavelength and direction of observation, but also on the form and condition of the surface (e.g. contamination, oxidation, roughness and structure). Consequently, in the course of an industrial process, considerable changes in the emissivity can occur within very short time spans, and it is, in general, impossible to calculate these changes. In only a few very special cases can emissivity be calculated (e.g. reflecting metal surfaces). In nearly all cases, diffuse reflection prevails in practice and one is dependent on *a priori* values of emissivity from literature, from radiation thermometer manufacturers or in extreme cases a direct *in situ* measurement. In the case of a low emissivity and a low object temperature, the uncertainty of the radiation temperature measurement is likely to be large, because in addition to low levels of thermal radiation from the target (due to its intrinsic temperature and emissivity), the ambient thermal radiation reflected by the object has to be taken into account. This precludes, or at least makes very difficult, the measurement of shiny metal surfaces. However, high emissivity, diffusely reflecting materials, for example, construction materials, such as stone, cement or ceramic, have emissivities close to unity and as such the temperatures of such materials are quite straightforward to measure using radiation thermometry. In addition, for materials which appear transparent in the visible such as glass, water and plastic film, as well as for hot gases and flames, characteristic wavelength ranges exist in the infrared, in which the absorptivity and emissivity are large so that with spectrally adapted radiation thermometers, non-contact temperature measurement is possible. In practice, for most real surfaces, it is possible to adjust the thermometer for known emissivity and ambient temperature so that the thermometer corresponds to the "true" temperatures of the measured object. An important general principle to minimize emissivity errors is to measure at as short a wavelength as practicable, if the object to be measured is not surrounded by a furnace with a higher temperature than the object itself (see Section 4.7). An overview on the thermal radiative properties of materials can be found in Refs. [6,7] and methods how to reduce the influence of emissivity on the temperature measurement with radiation thermometers are given in Ref. [8].

Now in principle the output signal, S_m, of a radiation thermometer is a complex function of the dimensions of the aperture and the field stop, the transmission of the optical components, the responsivity of the detector and the performance of the associated signal processing electronics. However, this situation can be greatly simplified if (a) the radiation thermometer is operated in its linear range where the output signal is proportional to the measured radiance L_m and (b) the spectral responsivity of the radiation thermometer is sufficiently narrow so that its output signal is proportional to Planck's law $L_{BB}(\lambda, T)$ at the effective wavelength λ [9,10] of the radiation thermometer when observing a blackbody. Under these assumptions, the output signal and the spectral radiance measured by a radiation thermometer when observing an opaque object are

$$S_m \propto L_m = \varepsilon(\lambda, T_o)L_{BB}(\lambda, T_o) + [1 - \varepsilon(\lambda, T_o)]L_{BB}(\lambda, T_a) - L_{BB}(\lambda, T_d) \quad (1)$$

where $\varepsilon(\lambda, T_o)$ is the spectral emissivity of the observed object, T_o the temperature of the observed object, T_a the temperature of the surrounding (ambient temperature) and T_d the temperature of the detector.

The first part of Equation (1) represents the spectral radiance emitted from the observed object due to the object temperature. The second term is the radiance originating from the surrounding of the measured object and reflected by the observed surface. The last term represents the radiance emitted by the detector. It is assumed that the emissivity of the detector is unity, since it is usually enclosed in a housing at the same temperature.

For low-temperature measurements, the detector radiance must be considered in the radiation budget, but for thermometers which measure temperatures above 200°C the last term in Equation (1) is negligible and the output signal is given by

$$S_m \propto L_m = \varepsilon(\lambda, T_o)L_{BB}(\lambda, T_o) + (1 - \varepsilon(\lambda, T_o))L_{BB}(\lambda, T_a) \quad (2)$$

2.2. Types of industrial radiation thermometers

The various types of radiation thermometers are generally categorized by their respective spectral responsivity.

2.2.1. Spectral radiation thermometers

Spectral radiation thermometers operate in a narrow spectral range with a typical bandwidth ($\Delta\lambda$) of less than 20 nm in the visible, and less than 100 nm in the infrared (Figure 3a). Their waveband is generally so narrow that in industrial applications it is possible to assign to them an effective wavelength independent of temperature (Table 2). As the radiant power striking the detector is low, radiation detectors with a high detectivity must be used. Spectral radiation thermometers are generally used in industrial process control for specific temperature measuring tasks. Their spectral responsivity range is usually selected to match the particular industrial process. One specific example is that by selecting suitable narrow spectral bands, radiation thermometers can be used to measure both the

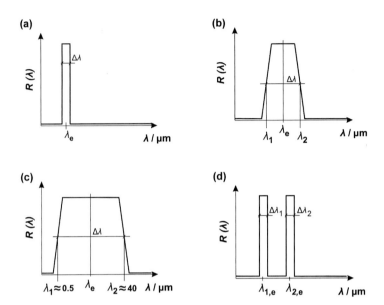

Figure 3 Schematic sequence of the spectral responsivity for various types of radiation thermometers: (a) spectral radiation thermometer, (b) broadband radiation thermometer, (c) total radiation thermometer and (d) ratio radiation thermometer.

surface and below-surface temperature, for example, in glass, plastic or a gas, and also behind an object, for example, behind a window (Figure 4). Besides industrial applications, special spectral radiation thermometers that have a signal strictly proportional to the spectral radiance with very high-temperature resolution and a very small size-of-source effect (SSE; see Section 3) [12] are suitable as metrological transfer radiation thermometers. These can be calibrated with very small uncertainty using blackbody radiators (see Chapter 6 of Ref. [2]) and their reference function can be modelled very precisely. These types of thermometers act as transfer standards in the calibration chain providing the link (traceability) for industrial radiation thermometers to the primary standards of the ITS-90 (see Chapter 2 of Ref. [2]) [11,13].

2.2.2. Broadband radiation thermometers

Many industrial radiation thermometers are *band* or more properly *broadband radiation thermometers*. Strictly speaking, the spectral radiation thermometer is also a band radiation thermometer, but with a very narrow spectral band (see Section 2.2.1). For broadband radiation thermometers, the bandwidth of the spectral filter is considerably larger than for spectral radiation thermometers (Figure 3b) and they can no longer be taken to be quasi-monochromatic in operation. They often use the entire spectral range

Table 2 Effective wavelengths at various radiation temperatures for the spectral ranges of a spectral radiation thermometer at 0.65 μm and typical industrial band radiation thermometers [11].

Temperature (°C)	λ_e (μm)					
	From 0.65 to 0.65	From 0.7 to 1.2	From 1.1 to 1.7	From 2.0 to 2.5	From 4.5 to 5.5	From 8.0 to 14.0
−100	−	−	−	−	5.16	11.43
0	−	−	−	−	5.07	10.71
100	−	−	−	2.34	5.03	10.36
200	−	−	1.57	2.31	5.00	10.16
300	−	−	1.57	2.30	4.99	10.04
400	−	1.13	1.55	2.28	4.97	9.95
500	−	1.11	1.53	2.27	4.96	9.89
600	0.65	1.10	1.51	2.26	4.95	9.84
800	0.65	1.08	1.47	2.25	4.94	9.78
1000	0.65	1.05	1.44	2.24	4.94	9.73
1200	0.65	1.03	1.42	2.23	4.93	9.70
1400	0.65	1.01	1.40	2.23	4.93	9.68
1600	0.65	1.00	1.39	2.23	4.93	9.66
1800	0.65	0.98	1.38	2.22	4.92	9.64
2000	0.65	0.97	1.37	2.22	4.92	9.63
2500	0.65	0.94	1.35	2.22	4.92	9.61
3000	0.65	0.92	1.34	2.21	4.91	9.60

of an atmospheric window. The effective wavelength of a broadband radiation thermometer changes according to the temperature of the object measured. This shift in effective wavelength is caused by a change in the spectral radiance distribution within the bandwidths of the thermometer with temperature according to Planck's law. Table 2 shows the effective wavelengths at various radiation temperatures for the spectral ranges of some typical broadband radiation thermometers. By using the temperature-dependent effective wavelength, the signal to temperature characteristic of a broadband radiation thermometer can simply be described with Equation (1) or (2) according to a spectral radiation thermometer.

The greater bandwidth of such thermometers leads to a higher radiant power on the detector, which for the most part facilitates good resolution of the measured temperature values, and enables the thermometer to function at lower radiance temperatures than spectral radiation thermometers. Their operation in a spectral band also reduces the influence of atmospheric absorption compared to total radiation thermometers (see Section 2.2.3). Because of the avoidance of atmospheric absorption bands, the measurement uncertainties attainable with broadband radiation

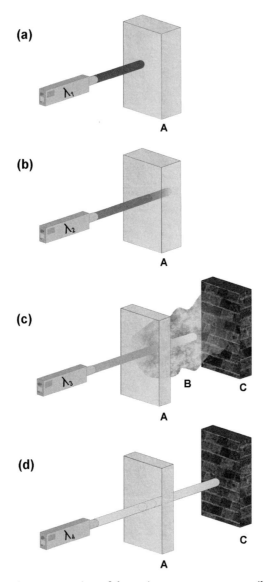

Figure 4   Schematic representation of the various measurement possibilities of a spectral radiation thermometer as a function of its effective wavelength and the absorption and/or emission behaviour of the observed materials. (a) At the effective wavelength λ_1, material A is highly absorbent. The radiation thermometer measures the surface temperature of A. (b) At the effective wavelength λ_2, material A is weakly absorbent. The radiation thermometer measures — if A has sufficient material thickness — a mean temperature within the depths of A. (c) At the effective wavelength λ_3, material A is transparent and gas B is weakly absorbent. The radiation thermometer measures — if there is a sufficiently large gas volume — a mean temperature of the gas in an enclosed volume. (d) At the effective wavelength λ_4, material A is transparent and material C is highly absorbent. The radiation thermometer measures the surface temperature of C. *Note*: The schematic presentation of the various measurement possibilities has been simplified for clarity — that is, in practice the cases (c) and (d) may need the residual absorption of the window A to be taken into account.

thermometers can be kept low, but are usually larger than those obtained with spectral radiation thermometers.

2.2.3. Total radiation thermometers

Total radiation thermometers operate mostly without spectral filters and detect nearly the entire spectrum of the thermal radiation of the measurement object (Figure 3c). The radiant power striking the detector is large due to the very broad bandwidth. However, the calibration and application of total radiation thermometers with low measurement uncertainty can only be performed under very constant atmospheric conditions, either with an inert gas or in a vacuum. The industrial use of total radiation thermometers is impractical due to their sensitivity to variable atmospheric conditions.

2.2.4. Ratio radiation thermometers

In the case of the *ratio radiation thermometer*, the temperature is determined from the ratio of two signals, taken at the same temperature but at different wavelengths. It operates in two, mostly narrow, spectral ranges lying close together (Figure 3d). The radiation temperature of the measurement object is determined by forming the ratio of the spectral radiances in the two spectral ranges. The output characteristics of such an instrument are this ratio as a function of temperature. Ratio radiation thermometers are designed to avoid the problem of unknown emissivity of the measured object and unknown transmittance of the optical path. They are well suited for the temperature measurement of grey bodies, that is, objects whose emissivity does not change with wavelength, if the reflected radiation of the surrounding from the object can be neglected. In this case, from Equation (2) follows for the output signal of a ratio radiation thermometer $S_{m,R}$ working in the two wavelength ranges characterized by λ_1 and λ_2

$$S_{m,R}(T_o) \sim \frac{\varepsilon(\lambda_1, T_o)L_{BB}(\lambda_1, T_o)}{\varepsilon(\lambda_2, T_o)L_{BB}(\lambda_2, T_o)} \tag{3}$$

where $\varepsilon(\lambda, T)$ is the spectral emissivity of the observed object and T_o the temperature of the observed object.

From Equation (3), it is obvious that if the emissivity of the object is the same at the two wavelengths λ_1 and λ_2 (grey conditions: $\varepsilon(\lambda_1) = \varepsilon(\lambda_2)$), the ratio depends only on the temperature of the observed object. The ratio thermometer can also be applied when small emissivity differences between the two wavelengths occur. However, it should be noticed that the ratio thermometer is in general more sensitive to measurement errors, including a non-grey behaviour of the measured object, than a typical band radiation thermometer as it determines the temperature from the shape of the thermal radiation curve, while the band radiation thermometer measures the intensity of the thermal radiation, which is much more strongly changing with the temperature than the shape of the spectral distribution (i.e. the ratio of the signal at the two wavelengths).

In real industrial situations, materials often display, in the case of changing temperatures, changing emissivity ratios rendering the ratio radiation thermometer unsuitable for "emissivity free" thermometry. However, often the great advantage of these devices lies in three other stated factors: measurements with a ratio radiation thermometer can significantly reduce the influence of window contamination and of a transmittance change of the line of sight by either dust or smoke, as these often lead to a wavelength-independent loss of signal (grey absorption). Finally the ratio radiation thermometer can perform temperature measurements for objects that are only partially filling the measurement field.

The importance of the last item, where the target to be measured is small compared to the measurement field of the thermometer, should not be underestimated. The method is often the only "simple" way in which very small and/or moving measurement objects, which do not completely fill the field-of-view of the radiation thermometer, for example, the temperature measurement at the laser focus in material processing with lasers, can be measured. However, this is only possible if the radiance contribution from the fraction of the measurement field which is not filled with the object of interest is considerably smaller than the radiance contribution of the measured object.

In the case that the reflected radiation from the measured object cannot be neglected, the signal of a ratio radiation thermometer is derived from Equation (2) to

$$S_{m,R}(T_o) \sim \frac{\varepsilon(\lambda_1, T_o)L_{BB}(\lambda_1, T_o) + (1 - \varepsilon(\lambda_1, T_o))L_{BB}(\lambda_1, T_a)}{\varepsilon(\lambda_2, T_o)L_{BB}(\lambda_2, T_o) + (1 - \varepsilon(\lambda_2, T_o))L_{BB}(\lambda_2, T_a)} \qquad (4)$$

where $\varepsilon(\lambda, T)$ is the spectral emissivity of the observed object, T_o the temperature of the observed object and T_a the temperature of the surrounding (ambient temperature).

From Equation (4), it follows that if the reflected radiation cannot be neglected even for objects with true grey-body behaviour, the measured ratio is not independent of the emissivity. Therefore, ratio radiation thermometers are only used for measuring temperatures above approximately 150°C, as the influence of the ambient radiation reflected from the object being measured must be kept small. Furthermore, it becomes obvious from Equation (4) that a ratio radiation thermometer cannot be used for temperature measurement of objects in a furnace. As the furnace temperature is generally even higher than the object temperature, a large error in object temperature measurement results (see Section 4.7) [14].

2.2.5. Multi-wavelength radiation thermometers

The extension of the idea of the dual-wavelength ratio radiation thermometer is the *multi-wavelength radiation thermometer*. The temperature is determined from the ratios of three or more signals, taken at the same temperature but at different wavelengths. In this case, it is usually assumed that the problem of the wavelength dependence of the emissivity can be solved by describing the logarithm of the emissivity as a series expansion in wavelength [15]. This means a three-wavelength radiation thermometer can measure true temperature if the logarithm of the emissivity depends linearly on wavelength, a four-wavelength thermometer if the logarithm of emissivity is a quadratic function of wavelength and so on. However, the sensitivity of the multi-wavelength radiation thermometer to measurement errors increases rapidly with the number of parameters in the emissivity model, that is, the number of wavelengths of the thermometer. So in practice the use of such thermometers cannot be regarded as a general solution to the emissivity problem, and very significant temperature errors can result even for materials that differ only slightly from grey-body conditions [16].

A critical analysis of the potential of multi-wavelength radiation thermometry to derive true temperatures is given in Ref. [17].

2.2.6. Typical commercially available radiation thermometers

Tables 3−5 show the characteristics, for example, the spectral responsivity, temperature ranges, the radiation detectors used, as well as typical applications, of commercially available typical radiation thermometers.

Nearly all industrial radiation thermometers operate with digital signal processors for signal processing and parameter setting. This results in a reliable and reproducible operation with good temperature resolution. In addition, addressable interfaces are now routinely available for remote signal transfer and for controlling of the devices.

To obtain correct temperature measurement from radiation thermometers, they must be calibrated against blackbody radiators of known temperature (see Chapter 6 of Ref. [2]). The calibration of the radiation thermometer, usually carried out by the manufacturer, establishes traceability to ITS-90. At the beginning of the traceability chain are the National Metrology Institutes (NMI) which realize and disseminate the ITS-90 (and all the other base SI units) at the highest metrological level [13]. The calibration of radiation thermometers is possible with a small uncertainty and allows worldwide comparable and reproducible temperature measurements with an uncertainty for calibrated industrial radiation thermometers of potentially less than $1°C$ in the temperature range from $-50°C$ to $1,000°C$. For higher temperatures up to $3,000°C$, the uncertainties attainable with commercial radiation thermometers are below

Table 3 Spectral and band radiation thermometers [17].

	Wavelength[a] (μm)						
	1.1 (0.7–1.1)	1.6 (1.45–1.8)	3.5 (2–5)	2.4 (2.0–2.8)	3.9	5 (4.8–5.6)	7.5–8.2
Temperature range[b,c] (°C)	450–3000	250–2500	0–2500	50–2500	75–2500	100–2500	0–2500
Response time[d] (ms)	1	1	1.5	1.5	5	5	5
Types of detectors used[e]	Si	InGaAs, Ge	PbSe, Pyro, TP	PbS, Pyro, TP	Pyro, TP	Pyro, TP	Pyro
Typical applications	Metals, molten glass, ceramics, semiconductors	Metals, ceramics, semiconductors	Metals, ceramics, hot-melt adhesives	Metals, molten metal, non-ferrous heavy metals, molten glass, ceramics, graphite	Glass, by means of flames	Glass surface	Glass surface, ceramics

[a]Indicated are the central wavelength and the typical spectral limits in parentheses.
[b]Indicated is the total temperature range offered. Individual radiation thermometers have, as a rule, a lower measuring temperature range.
[c]The lower temperature range is dependent on the optical system used (measuring field size) and the response time.
[d]Indicated is the shortest possible response time. In the lower temperature range, longer response times are usually necessary for a good temperature resolution.
[e]Si: silicon, Ge: germanium, InGaAs: indium gallium arsenide, PbS: lead sulphide, PbSe: lead selenide, Pyro: pyroelectrical detectors, TP: thermopiles.

Table 4 Spectral and band radiation thermometers [17].

	Wavelength[a] (μm)						
	3.43	6.8	7.95/8.05	8–10	8–14	9.5–11.5	4.3/4.5
Temperature range[b,c] (°C)	75–300	40–400	0–400	0–1000	–100–1000	–50–500	300–2500
Response time[d] (ms)	50	30	30	5	5	5	5
Types of detectors used[e]	PbSe, Pyro	Pyro	Pyro, TP	Pyro, TP	Pyro, TP	Pyro	Pyro
Typical applications	Plastic film, glass, semiconductors (wafers)	Plastic film	Plastic film	Plastics, non-metals, ceramics	Organic materials, lacquers, rubber, oils	Meteorological measurements, environment, large distances	Hot gases, emissions in incineration plants, flames

[a]Indicated are the central wavelength and typical spectral limits in parentheses.
[b]Indicated is the total temperature range offered. Individual radiation thermometers have, as a rule, a lower measuring temperature range.
[c]The lower temperature range is dependent on the optical system used and the response time.
[d]Indicated is the shortest possible response time. In the lower temperature range, longer response times are usually necessary for a good temperature resolution.
[e]Si: silicon, Ge: germanium, InGaAs: indium gallium arsenide, PbS: lead sulphide, PbSe: lead selenide, Pyro: pyroelectrical detectors, TP: thermopile.

Table 5 Ratio radiation thermometers [17].

	Wavelengths[a] (μm)				
	1 (0.85/1.1)	1 (0.95/1.05)	1/1.5	1.5 (1.4/1.75)	2
Temperature range[b,c] (°C)	600–3000	600–3000	500–3000	300–2000	200–1400
Response time[d] (ms)	1	10	2	20	50
Types of detectors used[e]	Si/Si	Si/Si	Si/Ge, Si/InGaAs	InGaAs	Pyro, TP
Typical applications[f]	Metals, molten metal, graphite, molten glass				

[a]Indicated are the central wavelength and the typical spectral limits in parentheses.

[b]Indicated is the total temperature range offered. Individual radiation thermometers have, as a rule, a lower measuring temperature range.

[c]The lower temperature range is dependent on the optical system used and the response time.

[d]Indicated is the shortest possible response time. In the lower temperature range, longer response times are usually necessary for a good temperature resolution.

[e]Si: silicon, Ge: germanium, InGaAs: indium gallium arsenide, PbS: lead sulphide, PbSe: lead selenide, Pyro: pyroelectrical detectors, TP: thermopiles.

[f]In the case of short-wave ratio radiation thermometers, in the case of temperatures under 300°C, an increased "photosensitivity" can occur (disturbances due to sunlight and artificial lighting).

2°C, or 0.05% of temperature. These uncertainties are attainable, however, only with high-quality industrial radiation thermometers and only take into account the calibration uncertainties. In an industrial setting, the uncertainty contribution due to the actual temperature measurement, often dominated as it is by the emissivity of the measurement object, is likely to be much larger — typically at least an order of magnitude.

2.3. Line-scanners

In addition to single-point measurements with radiation thermometers, one- and two-dimensional temperature distributions are increasingly measured on expanded workpieces in the production process with *infrared line-scanners* and *thermal imagers*. Thermal imaging is dealt with in detail in Chapter 5. Here a short description of line-scanners and their application in industrial processes is given. Line-scanners are widely used in the industry complementing single-spot non-contact temperature measurements with radiation thermometers.

Infrared line-scanners enlarge the "temperature scene" visible to a single-spot thermometer by moving the field-of-view of the device over a single-element detector or by directly applying a multiple-element linear detector array in the focal plane of an imaging system. Instruments range from very simple designs where a single-element detector is moved over the scene to more sophisticated designs where a rotating mirror is placed in front of a single-element detector to those that utilize various techniques to project the scene onto a multiple-element detector. Even more sophisticated designs with piezoelectric scanning devices, acoustic−optic and electro-optic scanners will not be discussed as they generally have military rather than civil applications.

Line-scanners available for the industrial market mainly fall into one of the two categories: those that use a single-element detector with a scanning device and those that use a multiple-element detector to capture the thermal radiation energy and convert it into an equivalent voltage or current.

Single-element detector designs utilize continuously or non-continuously rotating or oscillating opto-mechanical components to redirect the measured infrared radiation onto the single-element detector. There are numerous different designs available on today's market, particularly polygon scanners which use multiple facets in order to maximize the field-of-view and the scan efficiency at the lowest possible speed to minimize wear of the bearings. However, the simplest and lowest cost technique makes use of a rotating mirror (one facet). This principle is shown in Figure 5.

Multiple-element detector designs generally use single-line arrays usually with either 128 or 256 detector elements. The single-line array is placed at the focal plane of an optical system which gathers the thermal radiation energy. In the case of a pyroelectric detector array, which requires

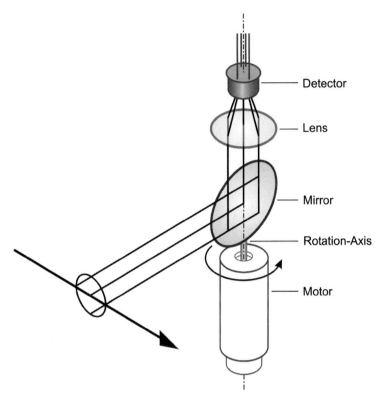

Figure 5   Schematic view of a line-scanner design with a single-element detector.

an alternating signal for proper operation, a mechanical chopper will be placed between the source of thermal radiation and the detector (Figure 6). Otherwise, no additional moving mechanical parts are needed in this design (e.g. micro-bolometer single-line arrays can be used without any moving parts in the line-scanner system).

Moving on from the basic principles towards a more detailed description of various actual line-scanner models, Table 6 can be taken as an illustration of the main features of modern instruments. This table is a generalized list of parameters and not all the stated features are available on any particular device.

Manufacturers of infrared line-scanners are offering a broad spectrum of different models that are designed to meet a wide variety of industrial applications. The achievable performance of actual instruments is strongly influenced by choosing the right instrument for a specific application. Certain applications may demand very high speed as objects are fast moving, whereas other applications can only be served if the instrument is limited in its spectral responsivity by use of optical filter (glass and plastics

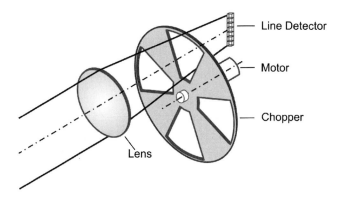

Figure 6   Schematic view of a line-scanner design with a multiple-element detector.

Table 6   Selected parameters and their typical performance of one- and multiple-element detector line-scanner designs.

Parameter	Single-element detector line-scanner	Multiple-element detector line-scanner
Field-of-view (FOV)	Up to 100°	Up to 60°
Moving mechanical parts	Yes (mirror)	No (optionally chopper)
Scan rate	10−150 Hz	Up to 5 kHz
Number of pixel	Up to 1120	Up to 512
Scan efficiency (per 360°)	0.25	n.a.
Spectral responsivity ranging	1−5 µm	1.4−14 µm
Temperature range	20−1600°C	0−1300°C
Optical resolution (typical)	Up to 500:1	Up to 1000:1
Detector type	MCT, PbSe, Si, InGaAs	Pyroelectrical, InGaAs, bolometer
Mean time between failures (MTBF) (typical) (h)	40000	40000

industries are typical examples). In high-temperature applications, like primary or secondary steel processing, short-wavelength single-element infrared line-scanners equipped with diode-type detectors operating in either the 1 µm region (Si-cell) or 1.6 µm region (InGaAs) or 2.4 µm region (extended InGaAs) are the instruments of first choice. Physical limitations, usually defined by the actual layout and set-up of the production line, may demand that the instrument has to be installed far away from the object to be measured. In such cases, a high optical resolution could be the dominating factor, driving the decision to choose a specific model.

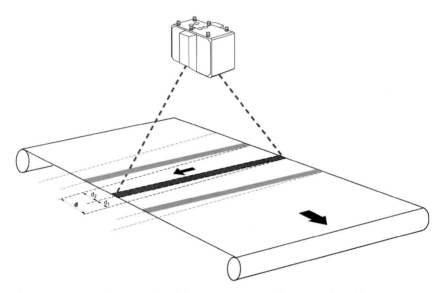

Figure 7   Single-element infrared line-scanner measuring a moving object.

To understand how the main parameters of infrared line-scanners depend on one another, we describe here the construction of a thermogram from multiple scanned lines of an object under observation. It is assumed that the object is generally moving orthogonally with respect to the instrument's scanning axis. In Figure 7, the larger arrow indicates the direction of motion of the object while the smaller arrow shows the direction of the scan of the observed surface temperature. Depending on the scan rate and the speed of the object, the scanned line will be more or less tilted (d_1 in Figure 7). This phenomenon is obviously not present when utilizing multiple-element infrared line-scanners. Figure 7 also shows that subsequent lines will have a certain stand-off distance (d_2 in Figure 7), leaving areas unobserved which might have been of interest for the process. This distance is again determined by the scan rate of the scanner, the speed of the moving object and to some extent the optical resolution of the instrument (measuring field). The effect of stand-off distance between subsequent lines is common to both principles, though is less of an issue in multi-element systems.

Looking more specifically at the shape of one single line, it becomes apparent that the shape of a single measurement spot changes with the angle of incidence. When viewing directly below the scanner, the field-of-view is circular. However, its shape changes towards an ellipse at the edge of the possible field-of-view. Figure 8 depicts this behaviour. Multiple-element line-scanners do not show this behaviour; however, one drawback in using such multiple-element designs is that they might suffer from the effect of a

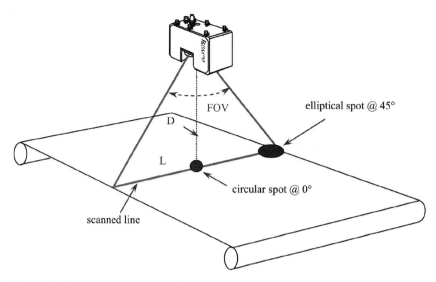

Figure 8   Single-element infrared line-scanner: spot shape versus angle of incidence.

non-homogeneous responsivity along the single-line array detector. This could be due to variability in their manufacture or simply the inability to use the same optical path per detector element. For example, each individual element of the sensor uses a different part of the lens which may have different local optical properties like transmission and of course will have aberration effects as a function of its diameter or thickness that can influence the thermal radiation reaching the detector. The practical implication of the effect of non-homogeneous responsivity is addressed through proper calibration and scheduled re-calibration of the instrument, thus resulting in a homogeneous response. Correcting the optics for spherical aberration would also help but certainly results in more sophisticated and therefore more expensive designs. The effect of chromatic aberration and scatter is common to both single- and multiple-element designs and should be compensated for.

Line-scanners are the instruments of choice wherever an object to be measured is continuously moving and when either a thermogram or more particularly a temperature distribution along a line-shaped field-of-view is needed. Their wide spectral responsivity and corresponding wide temperature range makes them distinct from thermal imagers. Typical applications are wide ranging from glass processing (float line, automotive glass, container glass, etc.), steel production, plastics industries (continuous film and container production, etc.) and cement kiln monitoring to combustion surveying and other quality assurance or safety relevant measurement tasks. For these applications, there is an extensive set of common useful functions possessed by line-scanners beyond their basic operation and essential optical and electrical parameters. These can include digital data acquisition and processing, analogue/digital interfaces, integrated aiming aids, remote control, cooling accessories, lens protection accessories, mechanical support structure that allow swivelling of the instrument and many others. It is clear that line-scanners fulfil an important, indeed unique, measurement function in industry and it is unlikely that they will be superseded by thermal imagers for many years to come.

3. SPECIFICATIONS OF RADIATION THERMOMETERS

It is very important for users of radiation thermometers to have a common basis on which to select a radiation thermometer for a particular application. The most important characteristic of a radiation thermometer is its spectral range. However, there are many other characteristics, or specifications, that a radiation thermometer could take; some of these are listed in Table 7 and an outline of their relevance for practical applications is given. The terms used here are consistent with the recommendations of the International Electrotechnical Commission (IEC) in accordance with the Technical Specification IEC 62942-1 TS [18], in which numerous relevant specifications for industrial radiation thermometers are defined with explanations and examples.

The *spectral responsivity range* of an industrial radiation thermometer is selected for many reasons. For example, besides the required temperature range, it will be selected on the basis of the emissivity of the material to be measured and/or the transmittance of the atmosphere. In addition, the transmittance of all materials (windows, gases, flames, etc.) in the line of sight has to be considered. The spectral responsivity range should be selected so that the emissivity of the surface is as high as possible in order to maximize the characteristic radiation from the measurement object and minimize the reflected ambient radiation. In general, the influence of disturbances and drifts on the measuring signal due to dust and smoke in the line of sight, as well as the emissivity uncertainty, on the temperature readout is lower for shorter wavelengths of operation. Therefore, the spectral range should *above all* be selected to be as short as practicable. Of course, the lowest temperature required to be measured sets the lower wavelength limit. An important exception to this general principle is measuring the temperature of an object in an oven of higher temperature, a situation that occurs often in industry. The thermal radiation emitted by the hot oven walls is reflected from the colder measurement object and this leads to on over-estimate of object temperatures both with the spectral radiation thermometer and especially with the ratio radiation thermometer. In this case, ratio radiation thermometers should not be used, and measurements should be made with a radiation thermometer operating at as

Table 7 Specifications of radiation thermometers.

Specification	Definition
Spectral range or spectral responsivity range	Parameter which gives the lower and upper wavelength limits of the responsivity range over which the radiation thermometer operates
Measuring temperature range	Temperature range for which the radiation thermometer is designed
Measuring distance	Distance or distance range between the radiation thermometer and the target (measured object) for which the radiation thermometer is designed
Field-of-view or measurement field	A usually circular, flat surface of a measured object from which the radiation thermometer receives radiation
Distance ratio	The ratio of the measuring distance to the diameter of the field-of-view, when the target is in focus
Temperature measurement uncertainty	Parameter associated with the result of a temperature measurement that characterizes the dispersion of the values that could reasonably be attributed to the measurand
Noise equivalent temperature difference	Parameter which indicates the contribution of the measurement uncertainty in degrees Celsius, which is due to instrument noise
Temperature parameter	Parameter which gives the additional uncertainty of the measured value depending on the deviation of the temperature of the radiation thermometer from the value for which the technical data are valid after warm-up time and under stable ambient conditions
Size-of-source effect	The difference in the radiance or temperature reading of the radiation thermometer when changing the size of the radiating area of the observed source
Response time	Time interval between the instant of an abrupt change in the value of the input parameter (object temperature or object radiation) and the instant from which the measured value of the radiation thermometer (output parameter) remains within specified limits of its final value
Exposure time	Time interval between an abrupt change in the value of the input parameter (object temperature or object radiation) has to be present, such that the output value of the radiation thermometer reaches a given measurement value

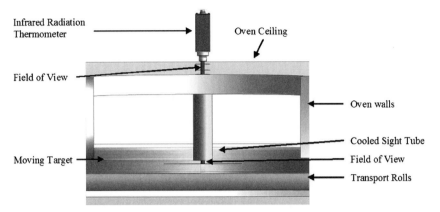

Figure 9 Arrangement for temperature measurement in an oven using a sight tube.

long a wavelength as practicable (see Section 4.7) [19]. A further measure for inhibiting the wall radiation is, in some cases, the use of sight tubes which are close to the measurement object and through which the object is viewed. The open end of the sight tube has to mount as close as possible to the surface of the target. The outer diameter of the sight tube should be typically five times bigger than the field-of-view. The typical installation is vertical to the surface of target. The sight tube provides a cooled surface for the reflected radiation from the background, while the hot walls of the oven are not seen by reflection on the surface of the target (see Figure 9). In the case of large temperature differences between the measurement object and oven, the tube must be cooled to the approximate temperature of the measurement object. However, in some cases the use of sight tubes is not possible, for example, because of condensation of hot gases from the oven on or in the tube or because of an unacceptable heat loss from the oven to the tube.

In the case of a radiation thermometer, the *measuring temperature range, field-of-view, response time* and *noise equivalent temperature difference* are interdependent. Effectively, the higher the temperature measured, the larger the field-of-view and the longer the response time, the more the temperature resolution is improved. Conversely, when the temperature resolution is fixed, the higher the temperature and the larger the measurement field, the quicker the measurement can be made. These interrelationships are not linear, however, and the absolute values are, in addition, also dependent on the spectral responsivity range of the measurement instrument.

The *temperature parameter* describes how well changes in the actual temperature of the radiation thermometer are compensated for when the thermometer has attained a stable temperature again. This is particularly an issue for thermometers that work at lower temperatures, that is, longer wavelengths, since all the individual components of the instrument contribute to the measurement signal. It is possible to design well-compensated low-temperature radiation thermometers that operate reliably in conditions of rapidly changing ambient temperatures and such devices are suitable for use under frequently changing ambient temperatures.

Aberrations, instrument-internal reflections, scattering and diffraction lead to an out-of-focus boundary of the field-of-view diameter of a radiation thermometer. That means that the measurement signal of a radiation thermometer depends to a greater or lesser extent on the size of the observed measurement object and on the thermal radiation sources outside of the nominal field-of-view diameter. This characteristic is described by the *SSE*. By using a high-quality optical system and judicious placement of internal baffles/stops, the SSE can be distinctly reduced. The SSE is an essential parameter, particularly in the use of industrial radiation thermometers which can end up viewing objects with very inhomogeneous temperature distributions.

In the industrial use of radiation thermometers for process monitoring and control, an important condition is that the thermometer has a sufficiently small *response time* compared to the timescale of the process being controlled. In many industrial processes, the workpiece moves on a conveyor belt through the field-of-view of the radiation thermometer. In this case, the *exposure time* of the radiation thermometer must be sufficiently small in comparison to the retention time of the object in the field-of-view.

Other important criteria for users to consider in selecting a radiation thermometer include temperature uncertainty, traceability issues, repeatability and stability, cost and other physical features of the radiation thermometer. A user needs to know and understand the desired or required temperature uncertainty for the specific application in order to purchase the appropriate type of radiation thermometer. After purchasing a radiation thermometer, the traceability of the radiation thermometer has to be determined by accepting the calibration from the manufacturer, calibrating the radiation thermometer in one's own laboratory and/or requesting a secondary calibration laboratory or a NMI to provide the primary calibration. The repeatability of temperature measurements and the stability or drift of the detector and the radiation thermometer should be considered in the selection of the required radiation thermometer. Of course, cost is a very important factor in the quality of the radiation thermometer that can be purchased. Other notable features in a radiation thermometer that should be considered include emissivity settings, focal lengths and measurement distances, operating conditions such as temperature and humidity, display unit and resolution, battery life, power supply, accessories, outputs, software and the radiation thermometer weight and size.

4. Applications of Radiation Thermometry in Industrial Production Areas

Radiation thermometers are used for the measurement and control of temperature in a broad — and continuously expanding — range of industrial settings. Also there is not only an increasing use of general-purpose radiation thermometers, but also a strong increase in solutions especially tailored to the needs of particular industrial requirements. In general, the need for fast and reliable temperature measurement in many areas of industrial production is driven by the requirement for:

- improved product reliability and reproducibility (achieved by on-line temperature control), that is, improved quality control;
- increased productivity due to remote temperature measurements integrated in a computer-controlled production process;
- cost reduction through improved energy efficiency;
- nominally zero waste as the process is run optimally with contingent environmental and economic benefits.

In the following section, some major production areas — which rely strongly on radiation thermometry for temperature measurement — will be described, together with their specific requirements. Only an overview on selected industrial applications of radiation thermometry can be given here. Not included in this chapter are applications, such as combustion, cryogenics, food processing, medical and thermography. A more detailed description of the benefits of radiation thermometry in specific production processes is given in the following chapters of this book or in Refs. [20,21]. Note that as continuous tunnel furnaces are frequently used in many industrial processes, a separate sub-chapter explains the specific problems when using radiation thermometers in such environments, that is, for temperature measurements when the surrounding is hotter than the measured object (see Section 4.7).

4.1. Glass industry

Glass is manufactured from the base materials of silica (in the form of sand), soda ash and limestone. It has no specific melting or freezing point but its fluidity gradually changes from solid to liquid as its temperature increases. This variable viscosity is a very useful property allowing glass to be used to produce many different types of glass products essential for modern life. Temperature is the key parameter in the processes of making the final glass products and non-contact temperature measurements are applied in almost all the process steps of glass production.

It should be noted that there are many different types of glass with different chemical and physical properties. Each can be made by a suitable adjustment of its specific chemical composition. The main types of glass are:

Commercial glass is most of the glass we see around us in our everyday lives in the form of bottles and jars and also, for example, as the flat glass for windows.

Borosilicate glass is a type of glass that is used as ovenware and other heat-resisting ware. It has good chemical durability and thermal shock resistance.

Glass fibre has many applications from roof insulation to medical equipment. Its composition varies depending on its application.

Technical glass is a group of glasses which are adapted for use in the electronics industry. This type of glass is used for soldering glass, metals or ceramics as it melts at the relatively low temperature of 450−550°C, for example, for sealing to tungsten, in the manufacture of incandescent and discharge lamps.

Glass is, over a wide range of the infrared spectrum, a semi-transparent material with a relatively poor thermal conductivity. Therefore, radiation thermometers enable us to determine different temperatures (e.g. the surface temperature or the interior temperature) depending on the effective wavelength of the radiation thermometer used. At short wavelengths (typically below 2.7 μm), it is possible to look through the glass and measure the temperature of objects behind it (Figure 10). Short-wavelength radiation

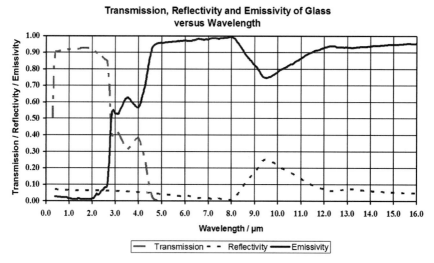

Figure 10 Emissivity (ε), reflectivity (ρ) and transmission (τ) of uncoloured commercial glass at room temperature.

thermometers will, however, also measure the bulk temperature of glass if it is thick enough to have high absorption at the wavelength of measurement. At long wavelengths (typically from 5.0 µm to 8.3 µm), glass exhibits a high emissivity, and it is possible to carry out surface-temperature measurements. However, above 8.3 µm, glass is highly reflecting (up to 20%), which could result in large reflections from the hot surroundings; therefore, for surface temperature measurements on float lines, the operating wavelength of thermometers is restricted to narrow spectral bands of around 5.0 µm and 7.8 µm. At intermediate wavelengths (typically from 2.7 µm to 4.5 µm), where the emissivity of glass is moderate and changes strongly with the wavelength, it is possible to measure the interior or the mean bulk temperatures. The emission of thermal radiation of sheets of glass at various temperatures and of various thicknesses is reviewed in Ref. [22].

Similar to the steel industry, the glass is often in motion and heated to high temperatures. Temperatures of up to 1,600°C are reached for temperature measurement on the bridge wall, which separates the melter basin from the refiner. These measurements are among others intended to control the melt furnace temperature. Typical radiation temperature measurements in glass production are, for example, the measurement of the in-furnace temperature, in order to determine the exit temperatures of the molten glass, also the measurement of the outside furnace wall temperature for confirming the integrity of the furnace liners. Different aspects of the glass production process will be described below with emphasis given to the relevance of non-contact temperature measurement during the process. The following production areas will be covered: flat glass, automotive glass, container glass, tubing glass, glass fibres, optical glass and insulation material, as well as processes of drying and coating, sealing and cutting, bending and tempering.

All production processes have in common the mixing of the raw materials and the heating and melting of the glass. For mass production, the mixing and melting of many tonnes of raw materials is performed. In smaller scale and specialist production, the raw materials are accurately weighed to give accurate composition and hence physical properties to a particular batch of glass.

4.1.1. Flat glass
In flat glass production (Figure 11), radiation thermometers measure temperatures at each processing stage. During these processes, the temperatures change rapidly from 1,500°C to 50°C. At every stage from the furnace to the tin bath, the annealing tunnel (lehr) and, finally, before cutting, the temperature is required. Incorrect or uneven temperatures or those changing too rapidly can lead to an uneven expansion and contraction of the product. Only through close control of the temperatures

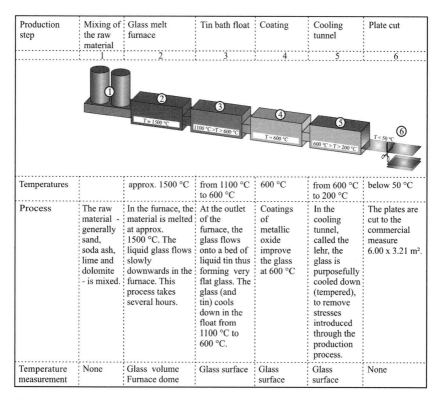

Production step	Mixing of the raw material	Glass melt furnace	Tin bath float	Coating	Cooling tunnel	Plate cut
	1	2	3	4	5	6
Temperatures		approx. 1500 °C	from 1100 °C to 600 °C	600 °C	from 600 °C to 200 °C	below 50 °C
Process	The raw material - generally sand, soda ash, lime and dolomite - is mixed.	In the furnace, the material is melted at approx. 1500 °C. The liquid glass flows slowly downwards in the furnace. This process takes several hours.	At the outlet of the furnace, the glass flows onto a bed of liquid tin thus forming very flat glass. The glass (and tin) cools down in the float from 1100 °C to 600 °C.	Coatings of metallic oxide improve the glass at 600 °C	In the cooling tunnel, called the lehr, the glass is purposefully cooled down (tempered), to remove stresses introduced through the production process.	The plates are cut to the commercial measure 6.00 x 3.21 m².
Temperature measurement	None	Glass volume Furnace dome	Glass surface	Glass surface	Glass surface	None

Figure 11   Schematic view of flat glass production.

can proper annealing of the glass be achieved yielding the desired properties of flatness and post-process workability (e.g. cutting).

From the exit of the furnace, the glass flows onto a bed of liquid tin (step 3 in Figure 11). The temperature of the glass surface is measured at different points in the direction of the glass flow. In addition, a line-scanner is (or a number of spot pyrometers are) used to measure the profile across the direction of the glass flow to maintain a uniform temperature profile, that is, to ensure that there is a controlled temperature gradient in the direction of the glass flow (Figure 12). At the end of the float, when the glass is leaving this part of the process, a device is installed to detect breaks. This is typically a simple radiation thermometer.

In the production line of flat glass, a coating is often applied to the glass surface to give particular characteristics such as tints, reflections or even self-cleaning (step 4 in Figure 11). This coating process is performed in another tunnel facility. Different coatings, either metal or organic, can be applied to the surface at this station. For this process to be successful, it is very important to know the glass temperature before the coating starts and

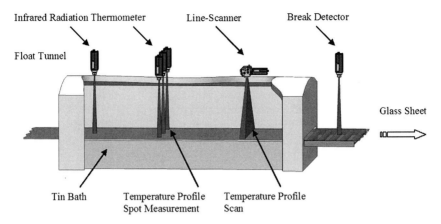

Figure 12 Temperature measurement in the tin bath float section for flat glass production with a break detector at the exit.

therefore one or more infrared radiation thermometers are installed at the entrance of the facility (Figure 13).

In the cooling tunnel (lehr), the coated or uncoated glass is purposefully cooled down in a carefully controlled manner to ensure that there is no breakage of the glass (step 5 in Figure 11). In addition, if the glass has been coated, it is very important to dry the coating at this point. The knowledge of the temperature gradient along the direction of motion as well as the temperature profile perpendicular to this direction is very important for the quality of the final glass sheets. Infrared radiation thermometers as well as line-scanners are installed here, with line-scanners typically installed at the exit of the cooling tunnel (Figure 14).

4.1.2. Container glass

For container (e.g. bottle) glass production, the molten glass flows from the furnace into a distributor station where the liquid glass is distributed to one or more forehearths. The temperature of the glass is made isothermal by heat conductivity and mechanically mixing the glass while it flows through the forehearth. Knowledge of the temperature of the glass in this part of the process is very important and it is generally monitored by noble metal thermocouples or infrared radiation thermometers. At the exit to the forehearth, that is, at the feeder, a silicon detector-based radiation thermometer in combination with a fibre optic cable is often installed to measure the temperature of the glass, and hence determine its viscosity. At the end of the feeder channel is a plunger. It works like a pump which pumps the molten glass downwards. The mass of glass for each pumping process is adjusted to the mass needed for the containers (bottles, etc.).

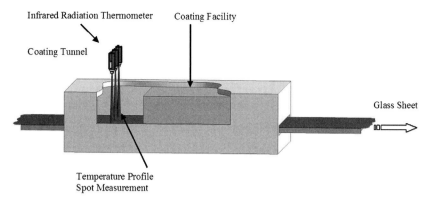

Figure 13   Temperature measurement in the coating station for flat glass production.

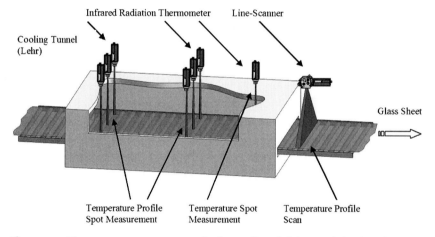

Figure 14   Temperature measurement in the cooling (lehr) tunnel for flat glass production.

Underneath the plunger is a shear that cuts the glass into gobs (pieces) of the specified size. This device is called the cutter. While the gob of glass is falling down into the forming machine, the temperature of the gob is measured (Figure 15) [23]. Typical forming machines have several stages for producing glass containers in parallel. The temperature of the mould is measured with contact thermometers and also with infrared radiation thermometers. This temperature measurement is difficult to perform because the moulds are made of metal with shiny polished surfaces. During the production process, the temperatures of forehearth and the mould are especially critical for the quality of the glass container because the glass viscosity must be carefully controlled. Typically, a 1°C change in temperature results in a 1% change in viscosity.

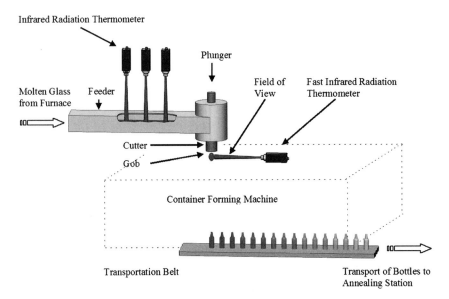

Figure 15 Measurement arrangement at the feeder channel and of the gobs for container glass production.

Once formed, the glass containers are transported to an annealing tunnel for a controlled cool down. Typically, the temperature profiles of the containers are measured by several infrared radiation thermometers or even by line-scanners to monitor and control the temperature inside the tunnel (Figure 16).

4.1.3. Automotive glass

Automotive glass is typically not flat glass. The glass-parts, for example, wind-shields, are cut into the right shape before they are heated in a furnace and bent to the required shape at a bending-station. The bending-station gives the final shape to the glass and it is very important that the temperature of these glass-parts is as uniform as possible during the bending process. On exit from the bending-station, there is typically an inspection-station where line-scanners scan the automotive glass-sheet from above and underneath to ascertain whether the temperature uniformity of the glass and hence desired glass properties have been achieved (Figure 17).

The lamination of wind-shields, that is, the production of a multi-layer unit consisting of a special plastic layer surrounded by two sheets of glass, requires a critical temperature of about 65°C at the interface between the glass and the plastic layer. An ideal way to measure this temperature is to use a radiation thermometer having a wavelength that is suitable for measuring the temperature of the glass interior.

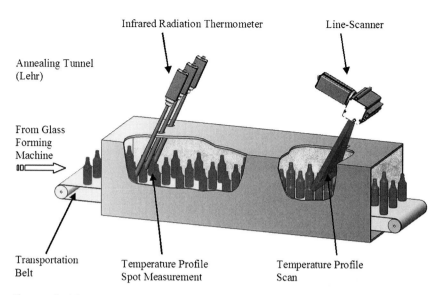

Figure 16 Temperature measurement in the annealing tunnel for container glass production.

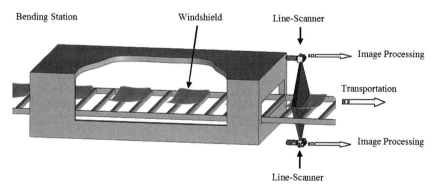

Figure 17 Measurement of the temperature distribution of wind-shields at the exit of the bending-station for automotive glass production.

4.1.4. Glass wool

Glass wool is produced from a thick layer of glass fibres that are cured in an oven at about 200°C to bind the fibres together and hence reduce the thickness of the layer. The molten glass from the furnace is fed by a distribution station into the different forehearths. In the forehearths, the temperatures into the feeders are monitored and typically controlled by silicon detector-based radiation thermometers coupled to the thermal radiation through a fibre optic cable. From the forehearths, the molten glass flows into the rotating fibre spinner, which forms, by centrifugal force,

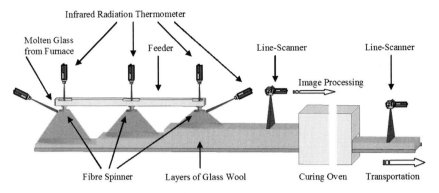

Figure 18   Temperature measurement in the feeder, at the spinners and the entrance and exit of the curing oven for glass wool production.

small pieces of glass fibre. The temperature of the spinner is controlled by either a single or dual waveband radiation thermometer. For quality control purposes at the entrance and exit of the curing oven, the mat of fibres is imaged by infrared line-scanners with a very high measurement speed to detect undesired hot glass gobs inside the insulation material (Figure 18). These gobs could cause problems during cutting, packing and handling the glass wool and also inhibit the thermal insulation properties of the final product.

Figure 19 shows the temperature distribution of glass wool after the cooling process with hot spot detection. By thermal imaging, a gob is detected which could not be seen from the outside of the material. Usually alarm functions are involved in this operation. A cutter can remove this piece of material to avoid problems later in the production process.

4.2. Steel industry

Radiation thermometers are used at all steps of the steel production process — from coke making and blast furnaces to continuous casting, hot and cold milling, forging, annealing and hardening. For control of these processes, a wide temperature range from 100°C to 1,700°C has to be covered. In addition, the emissivity values of the material range from below 0.1 to more than 0.9, over the wavelength range of commercial radiation thermometers, throughout the various steps of the production process. Besides this the process environment is quite hostile. It is, for example, often dusty, and the relative humidity very high. Because of these considerations, ratio- and/or fibre-optic radiation thermometers are frequently used. Detectors and interference/waveband selection filters must be hermetically sealed to protect them from the effects of dust and moisture. Vibration, electrical noise and high ambient temperatures are also common

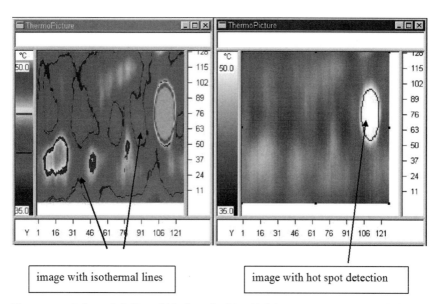

image with isothermal lines | image with hot spot detection

Figure 19 Coloured (left) and black and white (right) temperature image of glass wool after the cooling station with a gob identified by hot spot detection.

in steel processing. Silicon detector thermometers designed for high ambient temperature operation with air or water-cooling are widely used for high-temperature measurements. A detailed description of radiation thermometry used in steel production is given in Chapter 4.

4.3. Aluminium industry

Radiation thermometers are used to control continuous casting, milling, extrusion and forging in aluminium processing. For process control, a temperature range from 5°C to 700°C has to be covered. A specific problem when using infrared radiation thermometers on aluminium or aluminium alloys is their low emissivity — which can be as low as 0.05 — and the resulting high reflectivity. The low emissivity in combination with the relatively low aluminium processing temperatures leads to low radiance levels from the object being measured, whereas the high reflectivity makes it possible for radiation from surrounding objects to be reflected onto the radiation thermometer. Furthermore, due to the rapid and continuous oxidation of the aluminium surface as well as changes in the surface structure (roughness), emissivity variations are a common problem in several production steps.

For metallic materials with ideal surfaces (i.e. optically flat and free of contamination), the Drude free-electron theory [24,25] and also a quantum mechanical treatment [26] may be applied to predict the optical properties from electrical properties, particularly the electrical resistivity. Application of simple classical models within the free-electron theory results in the Hagen—Rubens relation [27] for normal spectral emissivity at long wavelengths ($\lambda \gg 5 \, \mu m$) and high temperatures. In general, the Hagen—Rubens relation will have poor absolute accuracy, but it may be useful to provide the general spectral and temperature behaviour of the emissivity and the form of a relation that can be fitted to the measured emissivity data. Using the Hagen—Rubens relation for normal spectral emissivity, various improved models have been developed to estimate total and angular emissivity of metals from it [28—31]. While the emissivity of pure aluminium evaporated onto optically flat surfaces agrees with the results from the free-electron model at wavelengths longer than $10 \, \mu m$, at shorter wavelengths the measured emissivity is higher and peaks near $0.8 \, \mu m$ [32]. However, it is the emissivity of processed aluminium alloys that is of practical interest for radiation thermometry. The emissivity of processed aluminium strongly changes with surface roughness, oxidation and alloy composition. The effect of surface roughness is to increase the emissivity, as it is the case for magnesium-containing alloys. While the above effects can be qualitatively understood, quantitative models are still needed.

To partly overcome those problems, special approaches and systems solutions are necessary. Manufacturers of radiation thermometers dedicated for use within the aluminium or generally speaking low emissivity primary or secondary metals industry are taking different routes to overcome the problem. The so-called "gold-cup" approach [33] assembles a technique to locally enhance the emissivity of the metal's surface by means of a special reflector in the close vicinity of the detector. The "gold-cup" needs to be placed very close to the surface of the object ($15-20 \, mm$) and is applicable to temperature ranges between $300°C$ and $600°C$. It is worth noting that this particular system is designed to measure the temperature of cylindrical surfaces. It is mainly used to capture the billet temperature prior to pressing (aluminium extrusion process). Other techniques make use of "natural cavities" which are formed by the process itself enhancing the emissivity of the object. As those "cavities" may vary during the process (i.e. coiler), special motorized fixtures are necessary to keep track of the process to ensure that the instruments are positioned in a way which maintains the viewing at the optimum angle.

Modern aluminium mills require the knowledge of the product temperature with a typical uncertainty of $5 \, K$. Such accuracy cannot be met with conventional band and ratio radiation thermometers while the emissivity of the product changes with processing conditions under the hostile environment of a mill. In Ref. [34], it was shown that with *a priori* knowledge of the emissivity behaviour, dual- and multi-wavelength radiation thermometry is able to provide sufficient accuracy for specific aluminium alloys.

Where dual-wavelength thermometers claim to obtain accuracy levels of 10 K immediately after an "optimization" cycle, multi-wavelength thermometers obtain accuracy levels of 3 K for known and pre-determined "alloy sets". Without a knowledge base on the emissivity behaviour, neither dual- nor multi-wavelength radiation thermometry is sufficiently accurate. It may also be noted that at least dual-wavelength instruments need to be shielded against ambient radiation which may be reflected off the surface of the aluminium strip. Generally speaking, both dual- and multi-wavelength instruments make use of highly specialized mathematical algorithms, which describe the spectral and/or temperature-dependent emissivity behaviour of the product (i.e. for the dual-wavelength radiation thermometer, the emissivity is not assumed constant as for the ratio radiation thermometer), to determine the object's temperature based on the very low levels of the radiated energy. In order to "feed" the algorithm with sufficiently accurate data, often the instrument has to repetitively go through teaching cycles. This is especially true if other than standard aluminium products are processed (i.e. different alloys, changing surface structure, etc.).

A desirable method which would provide the means to overcome the need to continuously adjust the radiation thermometer would be to have actively, continuously and accurately measurement of the emissivity of the object, regardless of its optical and mechanical properties. Such method was described and presented at the "European Aluminium Technology Platform" (EATP) conference held in Brussels on 27−28 November 2006 (Figure 20) [35] and is described below.

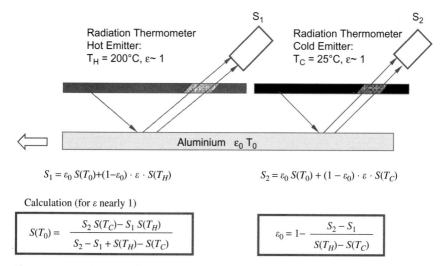

$$S_1 = \varepsilon_0\, S(T_0) + (1-\varepsilon_0) \cdot \varepsilon \cdot S(T_H)$$

$$S_2 = \varepsilon_0\, S(T_0) + (1-\varepsilon_0) \cdot \varepsilon \cdot S(T_C)$$

Calculation (for ε nearly 1)

$$S(T_0) = \frac{S_2\, S(T_C) - S_1\, S(T_H)}{S_2 - S_1 + S(T_H) - S(T_C)}$$

$$\varepsilon_0 = 1 - \frac{S_2 - S_1}{S(T_H) - S(T_C)}$$

Figure 20 Basic principle of active emissivity/temperature measurement for low emissivity secondary metals [35].

Both the hot and the cold emitters in Figure 20 radiate almost hemispherically and with known temperature (e.g. radiance). Their radiation is reflected from the surface of the object whose temperature is to be determined. The radiation measured by either radiation thermometers (S_1 and S_2) assembles the sum of the reflected radiation $(1-\varepsilon_0)\varepsilon S(T_C)$ of the cold emitter and $(1-\varepsilon_0)\varepsilon S(T_H)$ of the hot emitter, respectively, and the radiation of the object $\varepsilon_0 S(T_0)$. Either emitter has a known, very high emissivity of ε and known temperatures of T_C and T_H, respectively. Given that the optical and mechanical properties of the object will not change dramatically during one measurement cycle, the object's emissivity ε_0 and the temperature T_0 can be calculated according to the set of equations depicted in Figure 20. Although promising in its results and described in detail in the literature, it is not reported that this method has yet been commercialized.

Primary and secondary aluminium processing includes hot rolling mills, cold rolling mills, extrusion press and others. Temperature measurement at certain spots along the processing lines helps to run production efficiently while maintaining low energy consumption levels, thus improving quality and increasing productivity. As the processed aluminium material is usually non-stationary, contact temperature measurement is not considered to be the first choice.

4.3.1. Hot rolling
In the hot rolling mill (Figure 21), the so-called billet is heated up to 400°C or to 600°C at the pre-heater depending on the type of alloy. After passing through the reversing hot rolling or "break down" mill, the material is fed through a number of non-reversing tandem mills (400−550°C) where the thickness of the strip is set, before it is wound at the coiler (240−380°C). The material undergoes a reduction in its thickness from typically 600 mm down to 6 mm. The temperature at the coiler is considered the most critical

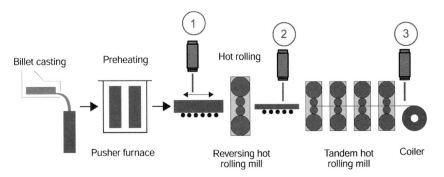

Figure 21 Hot rolling mill process.

one as the coil assembles the common transportation package; thus, oxidation should be limited to a minimum. The requested maximum temperature measurement uncertainty for the processes within the hot mill is $\pm 5\,K$.

4.3.2. Cold rolling

Whenever appropriate, cold rolling (Figure 22) follows the hot rolling process step. Temperature ranges typically from $60°C$ to $180°C$ and the desired reduction in thickness of the material is from 6 mm to 0.06 mm. The process runs continuously at an average speed of 40 m/s. Kerosene is usually utilized as a lubricant; therefore, any electrically powered instrument will have to be supplied in an explosion-proof housing as a minimum. The desired temperature measurement uncertainty for the processes within the cold mill is again $\pm 5\,K$.

4.3.3. Extrusion

The aluminium extrusion process (Figure 23) is used to manufacture profiles for a variety of applications ranging from architectural to consumer

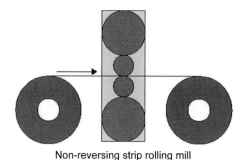

Non-reversing strip rolling mill

Figure 22   Simplified cold rolling mill.

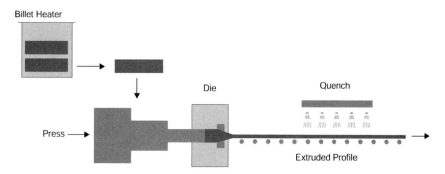

Figure 23   Aluminium extrusion (profiles) process.

products. A hot cylindrical billet is pushed through a shaped die which eventually defines the shape of the final product. Profile widths of 10−500 mm are common, while the speed of the process can reach 40 m/ min. One of the more general challenges despite the known issues with emissivity is that multiple profiles are extruded simultaneously and the created profiles can have "bizarre" shapes.

The desired temperature at the billet heater is in the range from 400°C to 500°C. Thus, maintaining the mechanical properties of the aluminium at an optimum with regard to the flow stress the billet is exposed to during processing is considered to be very important. The oxidized surface of the billet has a rather high but strongly varying emissivity (0.4−0.8).

At the die exit, the temperature ranges from 500°C to 600°C. It is obvious that the mechanical stress in the press leads to an increased temperature of the extruded profile. The temperature gradient along the axis of the extruded profile can be as high as 1 K/cm and can induce unwanted material properties if not controlled tightly. As the material is moving fast, contact probe temperature measurements are not convenient or practical. The desired temperature uncertainty at the die exit is in the range of ±5 K.

Following the die exit, the profile runs through a quench where it is cooled down to roughly 400°C. The surface of the material exhibits non-oxidized properties resulting in emissivity levels of 0.05−0.1. Temperature measurement along the quench is important to control and optimize the unique relationship between the die exit temperature, press speed and quality of the final product. Once the profile has passed the quench section, it cools down from 400°C to ambient temperature. Maintaining a cooling rate of a few degrees per second, especially above 250°C, is important and typical for this final processing step.

4.3.4. Other processes

Other aluminium manufacturing processes which require accurate radiometric temperature measurements include aluminium melting, forging and pouring streams. For an accurate temperature measurement of an aluminium melt directly down in the melting pot, where the temperature exceeds 600°C, the liquid aluminium has to be stirred to avoid oxidation. Forging aluminium between 500°C and 550°C requires forming with a sledge-hammer until the surface is free of oxide films. Basically the forging can only be measured with pyrometers. Due to high flow rates and the material condition in a pouring stream of aluminium, the temperature can only be measured using a radiation thermometer with a short response time. Fortunately, in pouring streams, high-temperature accuracies can be obtained because oxide films do not exist and the temperature measurements are not influenced by emissivity variations.

4.4. Plastics industry

Temperature is one of the most critical parameters in plastics processing (through, as in glass production, the control of viscosity) and radiation thermometers are very important in all parts of the production process. Most processes are carried out at relatively low temperatures (typically below 250°C), as the materials decompose at temperatures above a few hundred degrees Celsius. Infrared radiation thermometers and line-scanners are used for measuring thin-film temperatures, the surface temperatures of bulk materials and the average temperatures over a certain depth of the bulk material, by choosing suitable effective wavelengths for the thermometers.

Thin plastic films, which play an important role in plastics processing, are often semi-transparent; hence, their emissivities vary strongly with the film thickness and the wavelength over the infrared spectral range. For example, it can be seen from Figure 24 that in some narrow spectral ranges, polyethylene and polyester films become almost opaque, whereas the same films are semi-transparent over most of the infrared spectrum. Again similar to glass, in the case of thin plastic films, the spectral responsivity range of the infrared radiation thermometer strongly determines the measurement result. It must

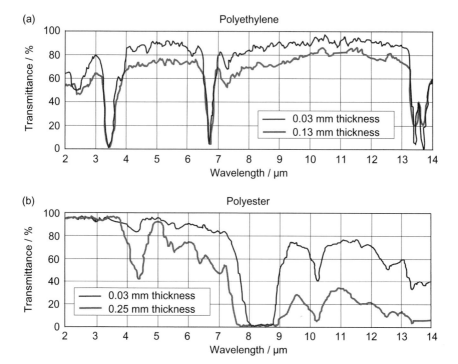

Figure 24 Spectral transmittance of (a) polyethylene and (b) polyester films of different thicknesses.

be noted that for these measurements a general-purpose broadband radiation thermometer will measure thermal radiation from the film and from behind the film, and will not give true film temperatures.

The strong absorption band around 3.43 µm which can be seen in Figure 24a is caused by the saturated hydrogen to carbon molecular bonds (C−H bond) of polyethylene which can also be found in many other plastic materials. Other plastic films, such as polyester, polyurethane and acrylic plastics, do not show this spectral feature, but show a strong absorption band around 7.92 µm, resulting from the carbon to oxygen molecular bonds (C−O bond) in these materials (Figure 24b). As the reflectance is typically 4% for most plastic materials throughout the infrared spectral range, emissivities of up to 0.96 can be achieved for thin-film measurements when using a sufficiently narrow band radiation thermometer with an effective wavelength at 3.43 µm or 7.92 µm. Consequently, such narrow band radiation thermometers are widely used in the processing of thin-film plastics. Manufacturers of radiation thermometers often specify minimum thickness figures for certain plastic materials, in order to achieve a specified emissivity within the C−H or C−O molecular bond absorption band.

General-purpose broadband radiation thermometers can be used in the plastics industry for temperature measurement under certain specific circumstances. Examples of these are determining the temperature of films having a thickness of more than 2.5 mm as these are typically opaque for all wavelengths between 2 µm and 16 µm. Besides thickness, plastics often have other materials added to them to give colour or modify their mechanical properties. These are often opaque over large parts of the infrared spectrum and further enhance the emissivity of plastic films.

Plastic products are sometimes thermally processed by means of high-intensity tungsten filament lamps. The spectral responsivity range of any radiation thermometer used in this process should be where the quartz envelope of the lamps is opaque, in order to avoid interference of the lamp radiation with the emitted thermal radiation from the plastic.

In the film and sheet extrusion processes, infrared radiation thermometers and line-scanners are used to control the product thickness and surface finish as well as the detection of surface non-uniformities and cracks (Figures 25−28). In film printing, stamping, lamination coating and thermoforming, infrared sensors are used to control the adjustment of the product heating and to monitor the cooling process in the cooling tunnels (Figures 25−27). Typical temperature measurement uncertainties for the processes described briefly below are in the range of 6 K.

4.4.1. Blown film extrusion
In the blown film extrusion process (Figure 25), the film is extruded as a continuous tube, air cooled, collapsed (compressed) and wound onto rolls

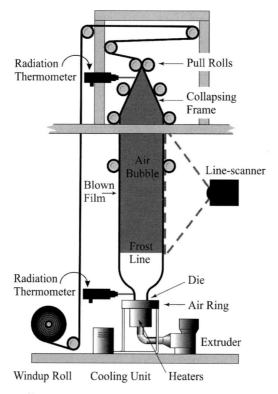

Figure 25 Blown film extrusion process.

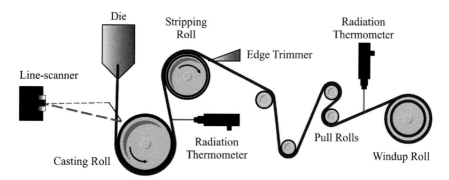

Figure 26 Cast film extrusion process.

as bags or slit into single layer widths. Accurate temperature monitoring in line with the ability to adjust heating and cooling helps maintain the tensile integrity and thickness of the plastic film. Infrared radiation thermometers take spot measurements at the die and the collapsing frame while infrared

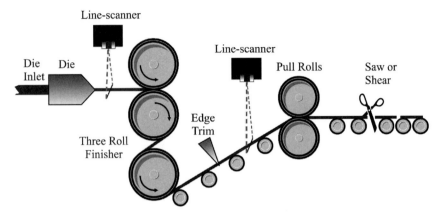

Figure 27 Sheet extrusion process.

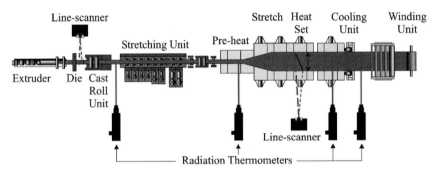

Figure 28 Bi-axially oriented film extrusion process.

line-scanners take temperature profiles between the frost line, the point where the material which forms the bubble starts to become opaque in the visible spectral range and solidification takes place, and the collapsing frame. Temperature measurement is performed to facilitate early detection of any unforeseen temperature drift that might be caused by problems with the die. Also, real time monitoring of the air ring, which is blowing cold air onto the outer surface of the film, is essential to maintain efficiency of the process. In conclusion, non-contact temperature measurement also helps to eliminate gauge bands, achieve consistent flatness and width of the film and most importantly reduce the wastage at the plant.

4.4.2. Cast film extrusion

In the cast film extrusion process (Figure 26), the melt is extruded through a wide die as a thin web and is cooled on a polished metal casting roll. An infrared line-scanner mounted after the die provides early detection of die bolt heater problems or congested die nozzles. Radiation thermometers help control temperatures throughout the process so that proper thickness and finish uniformity are maintained. Other benefits of radiation temperature measurement are improved film thickness uniformity, enhanced surface finish, less product breaks and hence reduced manufacturing down time.

4.4.3. Sheet extrusion

Figure 27 is an example of a typical sheet extrusion process. Note that the material thickness determines the type of thermometer deployed and the type of spatial resolution needed for optimum non-contact temperature measurement. Installing an infrared line-scanner before the three-roll finisher allows the operator to monitor the sheet temperature and adjust the die heater and/or the roll cooling so that product quality is consistent. A line-scanner mounted adjacent to the pull-rolls helps safeguard against tearing the extruded sheet and product irregularities such as possible sheet thickness non-uniformity.

4.4.4. Bi-axially oriented film extrusion

In a bi-axially oriented film extrusion process (Figure 28), where the film material (usually polyethylene) is drawn first in the direction of movement (first stretching unit) and subsequently after the pre-heater orthogonally to the direction of movement (second stretching unit), infrared line-scanners can be mounted at the die to monitor the die bolt heater as well as to take the melt temperature profile. They are also deployed at the heat set, where the now bi-axially oriented film is exposed to a certain temperature (typically $200°C$) in order to prevent the film from "shrinking" back into its original unstretched shape. Radiation thermometers can be mounted at the cast roll unit for chill roll control and at the pre-heater and cooling units, for heating and cooling control. A radiation thermometer mounted after the cooling unit determines if the product is cool enough for finishing.

4.4.5. Extrusion coating

This process applies a thin coating of plastic to paper, film or foil in an extrusion coating process (Figure 29). The distance between the die and the pressure and chill rolls is usually $75-125$ mm. The resin temperature at this location must be very high (i.e. up to $320°C$) for the melt to adhere to the substrate. A radiation thermometer is the ideal temperature sensor for this narrow and often difficult to access measurement area. The operator can monitor and adjust the die heater and the chill roll temperatures either manually or it can be performed automatically. For adhesion to smooth

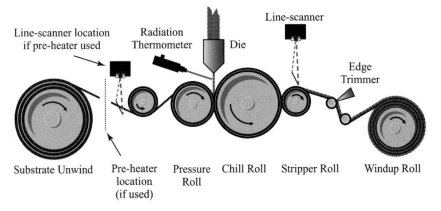

Figure 29 Extrusion coating process.

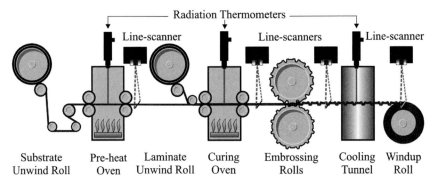

Figure 30 Laminating and embossing process.

surfaces, such as aluminium foil, other processes such as pre-heating the substrate and increasing of melt temperatures can all be controlled through use of radiation thermometers or line-scanners.

4.4.6. Laminating and embossing

Figure 30 illustrates where radiation thermometers and line-scanners are used to perform temperature measurement during the laminating and embossing process. Line-scanners are mounted between the pre-heat and curing ovens. At these points, the line-scanner can monitor cross-web temperatures and facilitate optimum heating of the substrate. At and before the cooling tunnel, a radiation thermometer or a line-scanner monitors the cooling efficiency. At the windup roll, another line-scanner identifies tear or breaks in the final product.

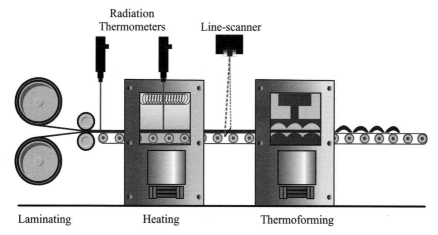

Figure 31 Thermoforming process.

4.4.7. Thermoforming

Thermoforming is a secondary thermoplastics manufacturing process. Raw materials like thermoplastic sheet, film or epoxy composites are converted or "formed" into finished products. Figure 31 shows an example of radiation thermometer and line-scanner locations in a schematic of the thermoforming process. If laminating is part of the process, a radiation thermometer can ensure that the correct temperatures are attained for proper laminating of a multi-layer product and for subsequent forming. A line-scanner positioned between the heater and the forming machine, or mounted as an integral part of the heater, can monitor and inform the control of the product temperature distribution prior to entering the thermoforming apparatus.

4.5. Laser welding and laser cutting of metals and plastics

In metal and plastics processing (e.g. in automotive engineering), more and more parts are connected with one another or cut to size through laser processing. The temperatures at the weld or cutting site have to be measured in order to control the process.

In the case of *metal-welding facilities*, short-wave radiation thermometers and line-scanners in the spectral range from 1.0 μm to 2.2 μm are generally deployed. The radiation thermometers are equipped with blocking filters at the laser wavelengths. The measuring temperature range extends from 250°C to 2,000°C. The instruments must operate at ambient temperatures of 10−70°C. Typical response times are short, typically 2 ms or higher.

In the case of *plastics welding*, measurements are generally carried out by longer wavelength radiation thermometers, usually starting at 5 μm. The temperature range to be measured reaches up to 600°C. Here, too, because of the intrinsic speed of the process, radiation thermometers having a short response time (≥ 5 ms) have to be deployed.

Due to the small and, in part, mobile measurement object (weld or cutting site in the laser focus), ratio radiation thermometers are used in addition to spectral radiation thermometers, particularly for metal laser processing. Light guides with lens attachments are also sometimes used.

4.6. Semiconductor industry

Virtually every process step in semiconductor wafer fabrication depends on the measurement and control of the wafer temperature. Innovations in processing mean that there is a continuous increase in the size of the wafers, and a continuous decrease in the dimensions of the integrated circuits. This means that the cost of each wafer also increases, making in-process temperature monitoring increasingly important. Temperatures from typically 50°C up to 1,250°C have to be measured. Processes such as physical vapour deposition (PVD: 200−600°C) and chemical vapour deposition (CVD: 250−700°C) can be controlled and improved further if the wafer temperature is measured accurately. For rapid thermal processing (RTP) and rapid thermal chemical vapour deposition (RTCVD), temperature measurement is especially important and demanding due to the high temperatures (400−1,250°C), the speed of the process and the necessity of tightly controlling the thermal budget. During RTP, the temperature of the wafer may change with a rate of up to 100°C/s. The International Technology Roadmap for Semiconductors (2004) [36] has established requirements of uncertainties of ± 1.5°C at a temperature of 1,000°C, with temperatures traceable to ITS-90. For RTP of silicon nanoelectronics applications, the requirements on the uncertainty of the measurement of temperatures are even higher. For such applications, the temperature would need to be controlled better than 0.7°C accuracy and reproducibility at six standard deviations [37]. Spectral and band radiation thermometers, operating at short wavelengths from 0.65 μm to 1.6 μm, are used to control wafer to wafer uniformity in both temperature and film thickness. Fibre optics is frequently used and response times of ≥ 1 ms are achieved.

A silicon wafer at room temperature is opaque to wavelengths short of about 0.95 μm. Beyond this wavelength, its transmission is relatively high, unless opaque layers are grown on the wafer. The emissivity of the wafer can vary widely for any given radiation thermometer wavelength due to slight changes in this layer thickness. To minimize the problem of a strongly changing wafer emissivity and transmission during the various process steps, measurements are usually performed at the shortest possible wavelength (typically around 1 μm). Highly sensitive radiation thermometers are used to achieve a good temperature resolution and a fast response time at short operating wavelengths. Furthermore, in some cases the emissivity of all wafers is measured in a separate chamber before they enter the process chamber. A detailed description of radiation thermometry for semiconductor processing is given in Chapter 3.

4.7. Measurement in the tunnel furnace

In industrial production processes, many materials and moulds are heated, annealed, tempered and sintered in continuous tunnel furnaces. In these processes, the parts, as a rule, enter into the tunnel furnace after pre-heating. They are then heated or tempered in the tunnel furnace (or a succession of tunnel furnaces) for a period between varying from minutes to hours. The heating in the tunnel furnace is carried out by means of infrared radiators or by flame-heating or, at lower temperatures, by pre-heated air. In this industrial process, it is usual that the furnace wall temperature is higher than the workpiece temperature.

Figure 32 shows a typical measurement set-up. It is shown that thermal radiation from the furnace wall is measured additionally to the radiation

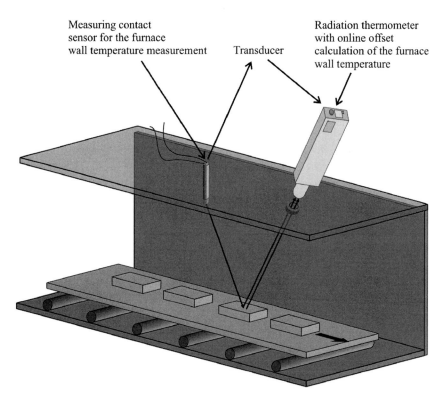

Figure 32   Typical measurement set-up in the tunnel furnace.

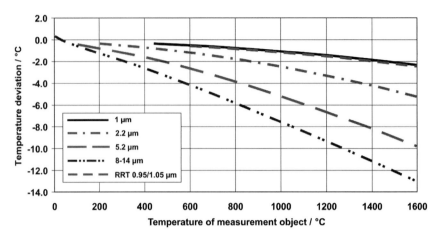

Figure 33 Deviation of the temperature measurement for radiation thermometers of different operating wavelengths when measuring a hot object with an ambient temperature of 30°C. The measurement error is assumed to be 1% of the output signal of the radiation thermometer (0.1% for the ratio radiation thermometer) and the emissivity of the object is 0.90.

temperature of the workpiece itself by the radiation thermometer. The latter measures both the emitted and the reflected radiation of the furnace wall from the workpiece. To correct for the reflected radiation, a continuous contact thermometer measurement of the furnace wall temperature is performed and a correction for the reflected wall radiation made via a mean emissivity of the workpiece in the spectral range of the radiation thermometer. An on-line determination of the temperature of the workpiece from the radiance measured by the radiation thermometer is performed. Equation (2) is applied, with typically $T_a \gg T_o$ (T_o: temperature of the workpiece and T_a: temperature of the surrounding furnace), though this condition is not necessary to ensure the correction is valid.

The influence of different effective wavelengths of the radiation thermometer on the temperature measurement for two different measurement scenarios ($T_a < T_o$ and $T_a > T_o$) is given in Figures 33 and 34.

Figure 33 illustrates the general case where the background temperature (and hence thermal radiation) is low compared to that of the workpiece; then, to obtain an accurate measurement of the workpiece temperature, as short a wavelength radiation thermometer as possible should be used. However, Figure 34 demonstrates that there are significant exceptions to that general rule. In particular, in the situation where $T_a \gg T_o$, using a short-wavelength radiation thermometer actually leads to a larger measurement error (as regards the emissivity uncertainties). In this case, the reflected thermal radiation from the furnace wall dominates the signal. Rather than a short-wavelength device, a better choice would be a measuring instrument

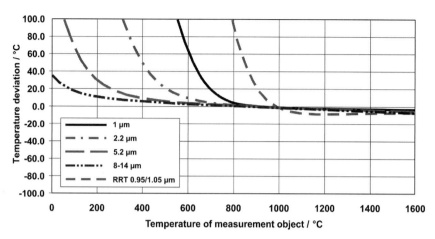

Figure 34 Deviation of the temperature measurement for radiation thermometers of different operating wavelengths when measuring a hot object with an ambient temperature of 1,000°C. The measurement error is assumed to be 1% of the output signal of the radiation thermometer (0.1% for the ratio radiation thermometer) and the emissivity of the object is 0.90.

in the 8 μm range. It should be carefully noted that because of this constraint, measurement with a ratio radiation thermometer into a furnace is practically impossible because ratio radiation thermometers always indicate the higher temperature, in this case, the furnace wall temperature. However, even in this situation, if the furnace wall temperature is taken into account, and the emissivity of the workpiece is sufficiently well known, it is still possible to carry out temperature measurement with short-wavelength thermometers.

4.8. Other applications

Temperature measurement in building thermography and firefighting applications generally does not require stringent temperature uncertainties but utilizes accurate qualitative comparisons (see Chapter 5, this volume). Remote sensing is used to measure the temperature of both land and water temperature on the earth's surface (see Chapter 6, this volume). Measurements of the human body temperature, both surface and sub-surface, are now widely performed on a routine basis (see Chapter 7, this volume). Many other applications which require accurate temperature measurements but are not covered in this series (Vols. 42 and 43) include combustion, detection of coal fires, detection of criminal activities, electrical power generation and distribution, and the manufacturing of asphalt, cement, ceramics, fibre optics, paper, cardboard, tobacco, textiles and various coatings.

5. SUMMARY

In order to control, optimize, repeat and compare industrial production processes, temperature must be measured sufficiently accurately and uniformly worldwide. This is accomplished with the aid of the specifications and regulations of the International Temperature Scale. A non-contact measurement, based on the ITS-90, of surface temperatures using radiation thermometers is possible, over a temperature range from $-100°C$ to over $3,000°C$. In an industrial environment, radiometric temperature measurement offers a range of advantages over contact methods. Radiation thermometers react very quickly, and the measurement is not affected by heat supply or dissipation. Objects that are moving rapidly, or are electrically energized (e.g. microwave heating, electrical induction) or are subject to rapid temperature changes, can be measured this way. As a result, radiation thermometry is being used increasingly for the monitoring and control of thermal processes, for maintenance and in building services engineering.

This chapter provided a survey of the use of radiation thermometers in industrial production processes. The typical set-up and current types of industrial radiation thermometers as well as their possible fields of application were described. The selection criteria and the essential specifications necessary for the practical use of radiation thermometers were introduced. A survey was given of industrial production areas in which non-contact temperature measurement is highly developed.

ACKNOWLEDGEMENT

The authors thank Elzbieta Kosubek and Beate Prußeit for their valuable support in producing the manuscript.

REFERENCES

[1] H. Preston-Thomas, "The International Temperature Scale of 1990 (ITS-90)," Metrologia **27**, 3–10 & 127 (1990).
[2] Z. M. Zhang, B. K. Tsai, and G. Machin (eds.), *Radiometric Temperature Measurements: I. Fundamentals*, Academic Press/Elsevier, San Diego (2009).
[3] M. Henini and M. Razeghi (eds.), *Handbook of Infrared Detection Technologies*, Elsevier Science Ltd., Oxford, UK (2002).
[4] F. Niklaus, C. Vieider, and H. Jakobsen, "MEMS-Based Uncooled Infrared Bolometer Arrays – A Review," Proc. SPIE **6836**, 1–15 (2007).
[5] T. J. Love, "Environmental Effects on Radiation Thermometry," in *Theory and Practice of Radiation Thermometry*, edited by D. P. DeWitt and G. D. Nutter, Wiley, New York, Chapter 3 (1988).

[6]   D. P. DeWitt and J. C. Richmond, "Thermal Radiation Properties of Materials," in Ref. [5], Chapter 2.

[7]   R. Siegel and J. Howell, *Thermal Radiation Heat Transfer*, Taylor & Francis, London, 4th ed. (2002).

[8]   A. Ono, "Methods for reducing Emissivity Effects," in Ref. [5], Chapter 10.

[9]   G. D. Nutter, "Radiation Thermometers: Design Principles and Operating Characteristics," in Ref. [5], Chapter 4.

[10] P. Saunders, "General Interpolation Equations for the Calibration of Radiation Thermometers," Metrologia **34**, 201−210 (1997).

[11] VDI/VDE Guideline, VDI/VDE 3511 Part 4, *"Temperature Measurement in Industry − Radiation Thermometry,"* Verein Deutscher Ingenieure, Düsseldorf (1995).

[12] P. Saunders and H. Edgar, "On the Characterization and Correction of the Size-of-Source Effect in Radiation Thermometers," Metrologia **46**, 62−74 (2009).

[13] J. Hollandt, R. Friedrich, B. Gutschwager, D. R. Taubert, and J. Hartmann, "High-Accuracy Radiation Thermometry at the National Metrology Institute of Germany, the PTB," High Temp.-High Press. **35/36**, 379−415 (2003/2004).

[14] P. Saunders, "Reflection Errors and Uncertainties for Dual and Multiwavelength Pyrometers," High Temp.-High Press. **32**, 239−249 (2000).

[15] P. B. Coates, "The Least-Squares Approach to Multi-Wavelength Pyrometry," High Temp.-High Press. **20**, 433−441 (1988).

[16] G. Neuer, L. Fiessler, M. Groll, and E. Schreiber, "Critical Analysis of the Different Methods of Multiwavelength Pyrometry," in *Temperature: Its Measurement and Control in Science and Industry*, edited by J. F. Schooley, American Institute of Physics, New York, Vol. 4, pp. 787−789 (1992).

[17] J. Hollandt, J. Hartmann, B. Gutschwager, and O. Struß, "Radiation Thermometry − Non-Contact Temperature Measurements," Automatisierungstechnische Praxis (atp) **48**, 70−81 (2006).

[18] IEC/TS 62492-1 Edition 1.0 2008-04, *"Technical Specification − Industrial Process Control Devices − Radiation Thermometers − Part 1: Technical Data for Radiation Thermometers,"* IEC, Geneva, Switzerland (2008).

[19] P. Saunders, "Reflection Errors in Industrial Radiation Thermometry," in *Tempmeko99, The 7th International Symposium on Temperature and Thermal Measurements in Industry and Science*, Delft, The Netherlands, edited by J. Dubbeldam and M. J. de Groot, IMEKO/NMi-VSL, Delft, The Netherlands, pp. 631−636 (1999).

[20] D. P. DeWitt and G. D. Nutter (eds.), *Theory and Practice of Radiation Thermometry, Part IV*, Wiley, New York, Chapters 16−21 (1988).

[21] P. Saunders, *Radiation Thermometry − Fundamentals and Applications in the Petrochemical Industry, SPIE-TT78, Tutorial Texts in Optical Engineering*, SPIE Press, Bellingham, WA (2007).

[22] R. Gardon, "The Emissivity of Transparent Materials," J. Am. Ceram. Soc. **39**, 278−287 (1956).

[23] C. P. Harris and J. O. Isard, "The Emission of Thermal Radiation from Hot Glass. Part 2. Emissivity of Gobs," Glass Technol. **31**, 21−24 (1990).

[24] P. Drude, "Optische Eigenschaften und Elektronentheorie," Ann. d. Physik **319**, 677−725 & 936−961 (1904).

[25] M. P. Givens, "Optical Properties of Metals," Solid State Phys. **6**, 313−352 (1958).

[26] C. Kittel, *Introduction to Solid State Physics*, Wiley, New York, 3rd ed. (1967).

[27] E. Hagen and H. Rubens, "Über Beziehungen des Refelxions- und Emissionsvermögens der Metalle zu ihrem elektrischen Leitvermögen," Ann. d. Physik **11**, 873−901 (1903).

[28] E. Aschkinass, "Die Wärmestrahlung der Metalle," Ann. d. Physik **17**, 960−976 (1905).

[29]   P. D. Foote, "The Emissivity of Metals and Oxides," Bull. Natl. Bur. Stand. **11**, 607 (1914–1915).

[30]   D. Davisson and J. R. Weeks, "The Relation between the Total Thermal Emissive Power of a Metal and its Electrical Resistivity," J. Opt. Soc. Am. **8**, 581–605 (1929).

[31]   E. Schmidt and E. Eckert, "Über die Richtungsverteilung der Wärmestrahlung von Oberflächen," Forsch. Geb. Ing. Wes. **6**, 175–183 (1935).

[32]   H. E. Bennett, M. Silver, and E. J. Ashley, "Infrared Reflectance of Aluminum evaporated in Ultra-High Vacuum," J. Opt. Soc. Am. **53**, 1089–1095 (1963).

[33]   M. D. Drury, K. P. Perry, and T. Land, "Pyrometers for Surface-Temperature Measurement," J. Iron Steel Inst. **169**, 245–250 (1951).

[34]   M. A. Pellerin, B. K. Tsai, D. P. DeWitt, and G. J. Dail, "Emissivity Compensation Methods for Aluminum Alloy Temperature Determination," in *Temperature: Its Measurement and Control in Science and Industry*, edited by J. F. Schooley, American Institute of Physics, New York, Vol. 4, pp. 871–876 (1992).

[35]   R. Gärtner, P. Klatt, H. Loose, N. Lutz, K.-P. Möllmann, F. Pinno, F. Muilwijk, S. Kalz, and H. Stoppiglia, "New Aluminium Radiation Thermometry," ALUMINIUM Int. J. Ind. Res. Appl. **80**, 642–647 (2004).

[36]   Semiconductor Industry Association, "*The International Technology Roadmap for Semiconductors – 2004 Edition*," Washington, DC (2004).

[37]   A. T. Fiory, "Rapid Thermal Processing for Silicon Nanoelectronics Applications," J. Miner. Met. Mater. Soc. **57**, 21–26 (2005).

CHAPTER 2

Experimental Characterization of Blackbody Radiation Sources

Sergey N. Mekhontsev, Alexander V. Prokhorov *and* Leonard M. Hanssen

Contents

1. Introduction   57
2. Reflectometric Determination of the Effective Emissivity of
   Blackbody Radiators   60
   2.1. Laser reflectometry   60
   2.2. Broadband source reflectometry   70
3. Radiometric Measurement of Blackbody Sources   74
   3.1. Characterization of medium temperature blackbodies   74
   3.2. Characterization of high temperature blackbodies   87
   3.3. Characterization of blackbodies in a reduced-background
        environment   93
4. Measurement of Input Parameters for the Calculation of Blackbody
   Radiator Characteristics   98
   4.1. Input parameters for the calculation of blackbody
        characteristics   98
   4.2. Measurement of cavity temperature distributions   99
   4.3. Reflectance measurement of cavity materials   103
   4.4. BRDF measurements of cavity materials   113
   4.5. Spectral emittance measurement of cavity materials   119
5. Conclusions   129
References   130

1. Introduction

Blackbody radiators are used as reference sources for the calibration of radiation thermometers and radiometers because it is possible to calculate

National Institute of Standards and Technology, 100 Bureau Drive, Gaithersburg, MD 20899-8442, USA

Experimental Methods in the Physical Sciences, Volume 43

ISSN 1079-4042, DOI 10.1016/S1079-4042(09)04302-1

their radiation characteristics on the basis of fundamental physical laws. However, blackbody radiators themselves must be accurately examined, preferably by experiment, to determine how their radiation differs from that of a perfect blackbody.

The available literature concerning the general question of the experimental characterization of blackbody radiators is voluminous. However, only a few surveys exist on particular topics, such as Section 12.9 in Ref. [1], devoted to an experimental verification of the results of effective emissivity calculations, and a review [2], in which a considerable part is dedicated to recent methods for the experimental investigation of high-temperature blackbodies.

For a wavelength λ in the medium, the spectral radiance $L_\lambda(\lambda)$, spectral effective emissivity $\varepsilon_e(\lambda, T_0)$, and radiance temperature $T_S(\lambda)$ of a blackbody radiator are related by the following equations:

$$L_\lambda(\lambda) = \varepsilon_e(\lambda, T_0) c_1 n^{-2} \pi^{-1} \lambda^{-5} \left[\exp\left(\frac{c_2}{n\lambda T_0}\right) - 1 \right]^{-1} \tag{1}$$

and

$$L_\lambda(\lambda) = c_1 n^{-2} \pi^{-1} \lambda^{-5} \left[\exp\left(\frac{c_2}{n\lambda T_S(\lambda)}\right) - 1 \right]^{-1} \tag{2}$$

Equation (2) can be solved for $T_S(\lambda)$, that is,

$$T_S(\lambda) = c_2 n^{-1} \lambda^{-1} \left[\ln\left(\frac{c_1}{n^2 \pi \lambda^5 L_\lambda(\lambda)} + 1\right) \right]^{-1} \tag{3}$$

Here c_1 and c_2 are the 1st and 2nd radiation constants, respectively [3] (see also Appendix A of this book), n is the refractive index of an environment medium, T_0 the temperature of an isothermal blackbody radiator, or the reference temperature of a nonisothermal one (see Section 2 of Chapter 5 in the companion volume, *Radiometric Temperature Measurement: I. Fundamentals*, Vol. 42 of this series).

The primary measured quantities of an artificial blackbody are the spectral radiance and the radiance temperature, which are related by Equation (3). If the temperature T_0 of a blackbody can be measured independently of the spectral radiance and radiance temperature (e.g., using one of the contact methods) or assigned using some reproducible procedure, then Equation (1) can be used to calculate the spectral effective emissivity. For an isothermal cavity, Kirchhoff's law [4] provides a way to determine the effective emissivity ε_e by measuring the reflectance ρ_e, since

$$\varepsilon_e = 1 - \rho_e \tag{4}$$

The methods of reflectometric determination of the effective emissivities for blackbody radiators are surveyed in Section 2. The following conditions must be fulfilled in order to use Equation (4): a cavity under investigation must be opaque and isothermal and, for the reflectance measurement, the cavity must be irradiated by radiation with the same polarization state, beam geometry, and in the same medium (air, vacuum, etc.) as for the desired emissivity measurement. Application of the Helmholtz reciprocity principle [5] makes possible two approaches to reflectometric measurements of the directional emissivity. The first, reviewed in Section 2.1, is the irradiation of a cavity by a collimated beam and collection of the radiation reflected by the cavity into the hemispherical solid angle. Therefore, the directional-hemispherical reflectance is measured in this case. The second, reviewed in Section 2.2, is the use of uniform hemispherical irradiation of the cavity and collection of the reflected radiation along a given direction. In this case, the hemispherical-directional reflectance factor will be measured. According to Helmholtz reciprocity, these two quantities are equal.

Generally, the reflectometric methods applicable for cavities are the same as for flat samples. However, reflectometric measurements of cavities have specific features, which define the design of the appropriate measuring devices. First, the level of the radiation flux reflected by a cavity is extremely small; it is usually < 0.01 of the incident flux. Second, the radiation reflected by a cavity can significantly differ in angular distribution from the Lambertian case even for cavities with Lambertian walls. Third, the entire cavity internal surface participates in multiple reflections. Hence the cavity opening can be considered as an extended source of reflected radiation. Finally, to obtain sufficiently precise values for the effective emissivity, ε_e, of a cavity, a relatively large uncertainty $\Delta\rho_e$ is admissible for the measurement of the effective reflectance ρ_e, since $\Delta\varepsilon_e = \Delta\rho_e = \rho_e(\Delta\rho_e/\rho_e)$. For example, for a measured reflectance of 0.001 with an uncertainty $\Delta\rho_e/\rho_e$ of 10%, the equivalent relative uncertainty of the effective emissivity determination $\Delta\varepsilon_e/\varepsilon_e$ is 0.01%. The methods and hardware, which use laser and thermal radiation sources, are considered separately. The majority of these methods require the use of a reflectance standard.

A direct radiometric measurement is the only way to obtain the performance parameters of a blackbody with a minimum of assumptions. Section 3 is devoted to the measurement of spectral radiance, spectral effective emissivity, and radiance temperature of blackbodies. In the first two subsections, the application of these methods to high-, medium-, and low temperature blackbodies is considered. The third subsection is devoted to the radiometric characterization of blackbody radiators in cryo-vacuum chambers, in medium- and low-background environments. These conditions are typical for remote sensing and defense applications (Earth climate monitoring, retrieval of Earth surface and atmosphere properties, radiation budget, missile guidance, detection, and tracking, etc.).

To date, computational methods remain an important tool when the experimental characterization of a blackbody is difficult or even impossible using the present state-of-the-art measurement techniques. In addition, such calculations are necessary in the blackbody design stage. A reliable calculation must be based on an adequate mathematical and physical model for the radiation transfer in the blackbody (and often in the radiation collecting system) being analyzed. The model inputs depend on assumptions, which form the basis of the computational method. The simplest analytical formulas for the effective emissivity of a blackbody cavity obtained within the framework of an isothermal diffuse model, only require knowledge of the geometry and the emittance (or reflectance) of the cavity wall. For more sophisticated models, it is necessary to know the temperature distribution over the radiating surface as well as the spectral and angular characteristics of the radiation emitted and reflected by the radiating surface. These issues are considered in Section 4.1. Section 4.2 is devoted to measurements of temperature distributions. The measurement of the spectral directional-hemispherical reflectance and bidirectional reflectance distribution function (BRDF) of materials suitable for blackbody manufacturing are discussed in Sections 4.3 and 4.4, respectively. In Section 4.5, the spectral emittance measurements of such materials are considered. Section 5 follows with conclusions.

2. REFLECTOMETRIC DETERMINATION OF THE EFFECTIVE EMISSIVITY OF BLACKBODY RADIATORS

2.1. Laser reflectometry

Lasers provide monochromatic and collimated radiation of high intensity. These factors are of great importance for precise measurement of the cavity directional-hemispherical effective reflectance. Modern lasers offer an extensive selection of wavelengths of the emitted optical radiation, from the ultraviolet to the far infrared.

Heinisch et al. [6] and Heinisch and Schmidt [7] described a reflectometer with a hemiellipsoidal mirror and CO_2 laser as a radiation source. The main components of the experimental facility (see Figure 1) are the following: (A) a 2 W CO_2 laser at 10.6 μm with a long-term stability better than 0.1%, (B) a chopper, (C) the cavity under investigation, of various shapes including off-axis cones, coated with 3-M Black Velvet paint, and (D) a hemiellipsoidal mirror to collect the radiation reflected out of the cavity and focus onto a large-area thermopile detector (E). The mirror could be rotated to map the directional emissivity of the off-axis cavities. The absolute values of the incident and reflected power were measured, an absolute calibration curve of the detector was plotted. The

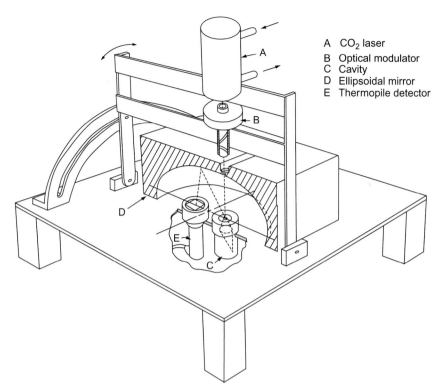

A CO$_2$ laser
B Optical modulator
C Cavity
D Ellipsoidal mirror
E Thermopile detector

Figure 1 Isometric view of an experimental apparatus (adapted with permission from Ref. [6]).

authors claimed that the technique described is capable of providing highly precise emissivity values with a resolution of five significant figures for high-emissivity cavities.

Jones and Forno [8] described a laser reflectometer with an integrating sphere for measurement of the directional–hemispherical reflectance of cavities. The system for the measurement is shown in Figure 2. The cavity was irradiated by the f/11 cone of light obtained from a HeNe laser operating at 632.8 nm. A silicon detector was placed behind a hole cut in the sphere. The baffle between the cell and the cavity prevented the cell irradiation by the radiation reflected directly by a cavity sample. The sphere was coated with 1.5 mm layer of smoked magnesium oxide. The sphere wall served as a reference.

An important source of error, the loss of reflected radiation through the sphere input port was investigated using goniophotometric measurements (see Figure 3). Light from the 633 nm laser was focused onto a pinhole using a microscope objective. This, in turn, was imaged at the plane of the cavity aperture by means of a lens. The incident cone was restricted to f/11 by a stop. The light reflected from the cavity opening was directed by the

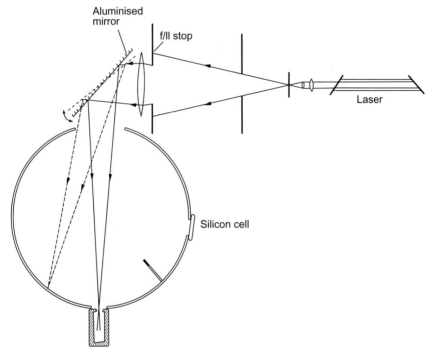

Figure 2 Apparatus for measurement of a cavity reflectance (adapted with permission from Ref. [8]).

beam splitter onto an f/11 collection system. The cavity reflectance was compared with that of a reference smoked MgO disk continuously rotated to avoid speckle effects. Limiting features in the experiment were the stray light and imperfections in the beam splitter. However, this goniophotometer produced unrepeatable results for the very small tungsten cavities, so photographic recording was employed. A correction for stray light was made, based on an additional exposure with no tube. The irradiance distribution across the photographic plates was determined from densitometer scans. The uncertainty achieved in these measurements was about 5%, which was acceptable in view of the low reflectances and small areas involved (the coverage factor was not stated).

Ballico [9] used an integrating sphere coated with $BaSO_4$ and a 5-mW HeNe laser to investigate the reflectance of small fixed-point blackbody cavities (see Figure 4). The laser beam was focused onto a 20 μm pinhole by a 40 × microscope objective and then collimated by a 30 mm focal length lens, focused at the exit port of the integrating sphere ($\sim$ 1 m from the lens). A circular aperture (3 mm in diameter, about 15 cm from the lens) placed with its edge in the minimum of the first minima of the Airy diffraction pattern of the pinhole, was used to strip the higher order modes

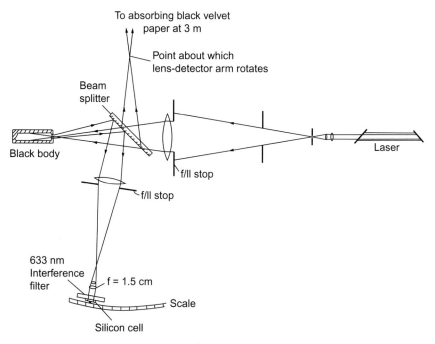

Figure 3 Apparatus for measurement of an angular distribution of radiation reflected by a cavity (adapted with permission from Ref. [8]).

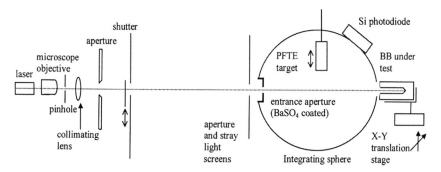

Figure 4 A schematic diagram of the optical components used to measure black-body emissivity (adapted with permission from Ref. [9]).

from the beam. Additional screens were used to prevent reflected stray light passing beyond the shutter. The blackbody under test was held in a cradle mounted on a X–Y translation stage in front of the exit port of the sphere. The signals were normalized using a polytetrafluoroethylene (PTFE) target disk within the sphere, which could be lowered into the beam to measure the total beam power.

Figure 5 shows the measured reflectance of a newly manufactured graphite cylindro-conical cavity of the fixed-point blackbody crucible. Although the base was prepared with a specially sharpened 60° drill bit, the results showed a higher reflectance in a 1−2 mm rounded area at the apex. When finished with a special tool (sharpened wedge of a thin brass sheet), the reflectance became much more uniform. All the graphite cavities were found to have the same effect. The cavities with a 60° base cone exhibited a larger variation across the base than the 120° base cone cavities, as would be expected from rounding of the cone apex.

Sakuma and Ma [10,11] used a similar technique to evaluate the effective emissivity of the cavities of fixed point blackbodies. In distinction to Ballico [9], they placed the reference Spectralon™ panel outside the sphere to switch with the cavity under test. The influence of crucible aging on the effective emissivity was also studied.

Zeng and Hanssen [12] and Hanssen et al. [13] described a reflectance instrument that consisted of several infrared lasers and associated beam shaping optics, polarizing optics, spatial filter, attenuating filters, monitor detector, and chopper. The 200 mm diameter integrating sphere had a 6 mm diameter entrance port for the laser beam, directly opposite the 50.8 mm diameter sample port. The sample cavity was held separate from the integrating sphere on a motorized multiaxis stage, allowing alignment and measurement of the spatial and angular dependence of the reflectance. Several detectors, including Ge, MCT, and pyroelectric, were used. The sphere reflectance measurements were performed relative to specular and diffuse gold standards calibrated using the Fourier transform infrared (FTIR) spectrophotometry facility [14]. Measurements were taken with vertically and horizontally polarized input light, or with circularly polarized input

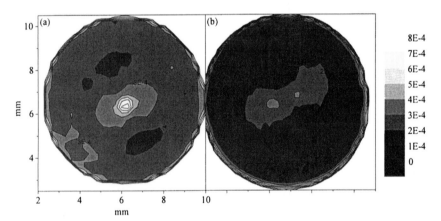

Figure 5 Measured reflectance of a fixed point blackbody cavity: (a) prior to modification and (b) after redrilling with a modified drill (adapted with permission from Ref. [9]).

light. A schematic and photograph of the infrared reflectometric instrument with a cavity of a water bath blackbody in the measurement position are shown in Figure 6. The cavity reflectance measurement sequence included measurements of (a) the previously calibrated gold standard, (b) the port cover with the same coating as the sphere wall, (c) the cavity, and (d) the background incident light level, made with the cavity removed and the beam incident on a distant black surface. For high-emissivity cavities, the background may dominate the signal, but the cavity-reflected component can be obtained accurately by maintaining a high degree of stability, not moving the aperture between measurements (c) and (d), and taking into account the minor change in sphere throughput. Also, for (b) the cavity was retracted slightly and the port cover placed onto the port.

Several approaches were used to perform the reference measurements, including a substitution mode. In that case, the sphere throughput change was separately characterized and corrected for between measurements. Three graphite cavities of fixed point blackbodies with 6 mm apertures were investigated: the first with a $60°$ cone bottom and an inner diameter of 6 mm, the second with a $37°$ cone bottom and an inner diameter of 12 mm, and the third with a V-grooved bottom and an inner diameter of 12 mm. Three larger cylindro-conical black painter copper cavities with inner diameters of 45 and 108 mm were also studied. Multiple copies of the graphite crucibles were manufactured at the same time. One of each type was sacrificed for the reflectance studies. The three graphite fixed point cavities were measured at both 1.32 and 10.6 μm. In order to properly measure the emissivity of the cavity, the entire cavity was measured in place, including the section in front of the aperture.

In operation, the section in front of the aperture is close in temperature to that of the cavity and contributes to the emissivity, while the reflectance

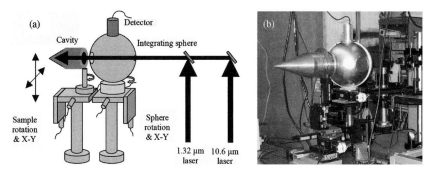

Figure 6   (a) A schematic and (b) photo of the infrared total integrated scatter instrument with a water bath blackbody cavity in the measurement position (adapted with permission from Ref. [13]).

is reduced accordingly. At the same time, for the "aperture only" measurement (d), the front section remained in place for proper subtraction. In order to enable both (c) and (d) measurements, each cavity was physically separated behind the aperture. The variation in the cavity emissivity across the aperture for the three designs at 1.32 μm is shown in the left-hand column of Figure 7. The plots indicate that the emissivity for the small

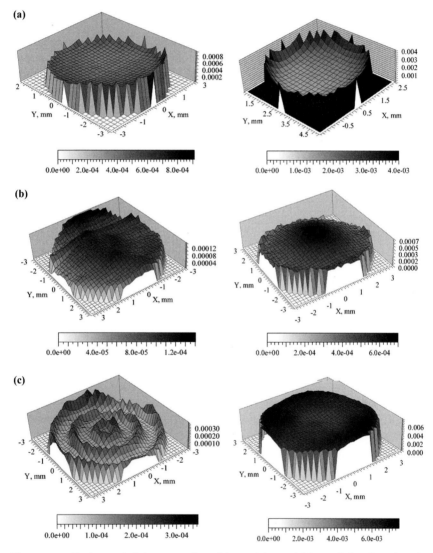

Figure 7 Effective emissivity scan of graphite cavities at 1.32 μm (left column) and at 10.6 μm (right column): (a) 6 mm cylindro-conical cavity, (b) 12 mm cavity, and (c) cavity with V-grooved bottom (adapted with permission from Ref. [13]).

cavity in (a) has a reflectance value, approximately four times greater than that of the large cavity with the conical bottom in (b). Also, a ringed structure corresponding to the V-grooves in (c) was clearly evident. The ridges and valleys in the reflectance map correspond to the valleys and the ridges, respectively, of the V-grooves. This was due to a small shallow or flat region at the bottom of the valleys, where the surface is normal to the incident light. Similarly, the variation of the cavity emissivity across the aperture for the three designs at a wavelength of 10.6 μm is shown in the right hand column of Figure 7.

The reflectance levels of all the cavities were significantly greater than that at 1.32 μm due to the higher graphite reflectance levels at 10.6 μm. Both the small and large cavities with the conical bottom were found to have reflectance levels approximately four times greater than those measured at 1.32 μm. In contrast, the reflectance of the V-groove cavity increased by an order of magnitude. Structures in the map associated with the groove locations exist, but are very shallow. The angular dependence of the cavity emissivity was also measured to be able to predict the performance for pyrometers with differing fields of view or f-number. The angular variation is related to the spatial variation, since the input spot moves across the cavity bottom with change of incidence angle, but the change in angle also means that specularly reflected components of the light will be directed to different locations within the cavity, resulting in additional angular variability.

Thus, the application of laser reflectometry to nondestructive testing of blackbody cavities has been demonstrated by the research described in Refs. [9,12,13]. This issue is of key importance for the onboard calibration of blackbodies in remote-sensing space missions. The accuracy of Earth radiometric observations, including the investigation of climate variability and global warming, directly depends on the stability of satellite onboard calibration blackbodies [15,16]. The stability of their radiation is determined not only by the constancy of the cavity temperature but also by the invariability of the cavity effective emissivity, which, in turn, depends on the ageing processes on the coating of cavity internal surface or contamination of cavity wall.

One possible solution to monitoring the cavity emissivity *in situ* using lasers was proposed by Gero et al. [17]. A method to characterize the emissivity of a spaceborne blackbody using a quantum cascade laser (QCL) based reflectometer has been proposed for the Climate Absolute Radiance and Refractivity Earth Observatory (CLARREO) satellite mission. The main objective of the CLARREO mission is to initiate a high-accuracy record of the Earth's spectrally resolved outgoing infrared radiation, which is needed to detect long-term climate change trends, and to test and improve climate predictions. The prelaunch determination of satellite radiometer uncertainty cannot be valid over the operational lifetime of the

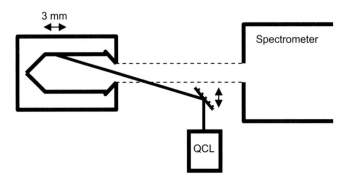

Figure 8 Experimental layout of the reflectometer (adapted with permission from Ref. [17]).

instrument. The harsh conditions of spacecraft launch and the low–Earth orbit environment can lead to drift in the instrument's physical properties due to degradation affecting the blackbodies used for the calibration of a FTIR spectrometer. The proposed scheme of monitoring the blackbody effective emissivity on-orbit was tested in Ref. [17] and is presented in Figure 8. The radiation beam of a QCL with a wavelength of 7.91 μm was directed to the cavity internal wall. The diameter of the laser beam was 4 mm at the point of incidence on the cavity. The cavity had a temperature of 40°C. The laser radiation reflected by the cavity together with its thermal radiation and reflected ambient radiation was registered by an FTIR spectrometer. The incidence point of the beam was varied over a length of 3 mm, measured along the central symmetry axis of the cavity. The entire apparatus was operated inside a purge box under a constant flux of nitrogen gas, in order to reduce absorption by the water vapor, and to stabilize the ambient temperature.

The output power P of a QCL can be measured with high accuracy. In a first approximation, the reflected laser power $P_{\text{reflected}}$ is given by:

$$P_{\text{reflected}} = L_{\text{laser}} A\Omega \tag{5}$$

where L_{laser} is the reflected laser radiance, A the area of the blackbody aperture, and Ω the solid angle into which the blackbody radiates. The cavity effective emissivity for these conditions can be evaluated as:

$$\varepsilon_{\text{e}} = 1 - \frac{P_{\text{reflected}}}{P} \tag{6}$$

The measured spectra consisted of the Planckian curve and a narrow peak corresponding to the wavelength of the QCL radiation (see Figure 9). The shaded area between the peak and the baseline fit represents the reflected laser radiance L_{laser}. The results obtained indicated sufficient sensitivity to be able to monitor changes in a satellite instrument's onboard

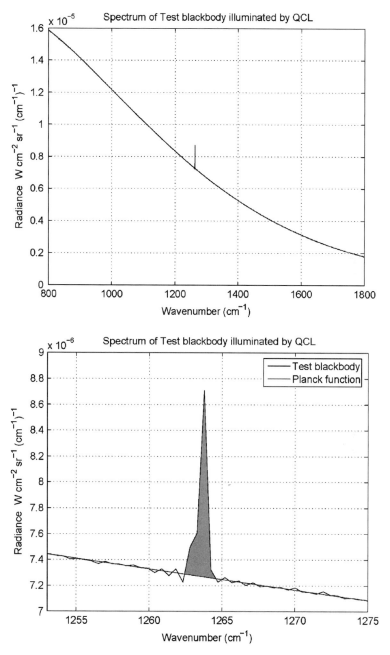

Figure 9 Top plot: Spectrum of the test blackbody illuminated by the QCL, as observed by the Fourier transform infrared (FTIR) spectrometer, showing the sharp laser peak superimposed on the baseline blackbody radiation at 40°C. Bottom plot: Expanded view of the laser peak with the fitted baseline Planck function (adapted with permission from Ref. [17]).

blackbody emissivity on a scale finer than the current levels of radiometric accuracy [17].

2.2. Broadband source reflectometry

It is difficult to obtain a collimated beam of thermal radiation with sufficient flux to measure the directional-hemispherical effective reflectivity
of blackbody cavities. Therefore, for the characterization of cavities,
thermal radiation sources are used primarily for the measurement of the
hemispherical-directional reflectance factor. Cavity reflectometry using
thermal radiation sources has a long history. In 1961, Bauer [18] considered
the use of a hemiellipsoidal reflector for measurement by the direct
scheme and an integrating sphere for measurement by the inverse scheme.
Bauer measured the effective (i.e., after multiple reflections) reflectance of
cylindrical cavities and compared these results with calculations performed
with the help of methods available at the time.

Kelly and Moore [19] also compared their measurement data with the
calculated effective emissivity for a diffuse cylindrical cavity. The room-
temperature reflectance measurements were made in the wavelength range
from 0.4 to 0.75 μm by placing a paper-lined brass cavity of variable depth
into the port of a magnesium oxide-coated integrating sphere attached to
a double-beam recording spectrophotometer. A reference standard was
calibrated against freshly smoked magnesium oxide. The spectral reflectance
was measured under conditions of $6°$ incidence and hemispherical viewing,
which is approximately equivalent to a spectral reflectance factor for
hemispherical irradiation and normal viewing. The normal-hemispherical
spectral reflectance, in turn, is the complement of the normal spectral
emissivity. Three different colored papers were used for cavity linings.
Goniophotometric measurements were made for each in order to
determine its deviation from a Lambertian reflector. After the introduction
of several corrections, the authors found a satisfactory agreement with their
measurements for the simple analytical expression that was tested.

Bauer and Bishoff [20] used the inverse scheme that had been described
in Ref. [18] to determine the effective emissivity of a cylindrical cavity. The
measurement scheme with an integrating sphere is depicted in Figure 10.
The reflectance measurements were conducted against a reflection standard
(surface of $BaSO_4$). Hemispherical irradiation was realized by an integrating
sphere irradiated by an incandescent lamp. The sphere had two ports on
opposite sides, one for the cavity under test and the other for viewing the
cavity. Due to the port H_2, the cavity was not irradiated from the direction
of observation. Additional irradiation was provided by a source S_2 and a
semitransparent mirror M. The filter F combined with a ground glass served
to match the spectral radiance of the source S_2 to that of the source S_1. The
f-number of the viewing beam was 250. The spectral range was limited by

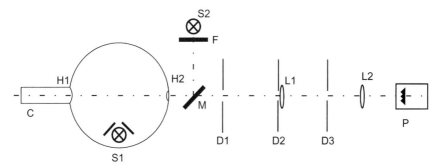

Figure 10   Measurement scheme for cavity reflectance: C, cavity under test; U, integrating sphere with H_1 and H_2 circular ports; B_1 and B_2, baffles; M, semi-transparent mirror; S_1 and S_2, sources; F, filter combined with a ground glass; L_1 and L_2, lenses; D_1, D_2, and D_3, diaphragms; P, photomultiplier (adapted with permission from Ref. [20]).

the emission of the incandescent lamps and the spectral response of the photomultiplier from about 400 to 700 nm. The reliability of the measurement method described was demonstrated by good agreement of the measured values with those calculated using two independent methods.

Ballico [21] described a simple but useful apparatus for evaluating the effective emissivity of blackbodies in the commonly used $7-14\,\mu m$ radiation thermometry wave band. The apparatus described (see Figure 11) was a large black disc with adjustable temperature. The disc was placed in front of the blackbody opening and had a small central hole through which the cavity bottom could be viewed by an infrared thermometer. The measured cavity emissivity could be automatically averaged over the operational spectral band of the radiation thermometer. The radiating surface was a 0.5 mm thick copper disc 250 mm in diameter, painted with a diffuse black paint. A water-cooling/heating coil was placed over disc's surface and enabled the temperature to be controlled between $20°C$ and $80°C$. The temperature was measured using thermocouples distributed over the surface. A water-cooled "cold disc" with a slightly smaller hole separated from the hot disc by two copper baffles was used to reduce the errors due to the pyrometer's size-of-source effect.

The blackbody effective emissivity was obtained according to the following analysis. If we assume that the spectral radiance of the hot-plate surface, $L_{\lambda,HP}$, arises only from its own thermal emission and the reflection of ambient thermal radiation at T_{amp} (i.e., ignoring reflections between the blackbody and the hot plate), its surface spectral radiance will be given by:

$$L_{\lambda,HP}(\lambda) = \varepsilon_{HP}(\lambda)P(\lambda, T_{HP}) + [1 - \varepsilon_{HP}(\lambda)]P(\lambda, T_{amb}) \qquad (7)$$

where $\varepsilon_{HP}(\lambda)$ is the hot-plate spectral emissivity and $P(\lambda, T)$ the spectral radiance of a perfect blackbody at temperature T and wavelength λ.

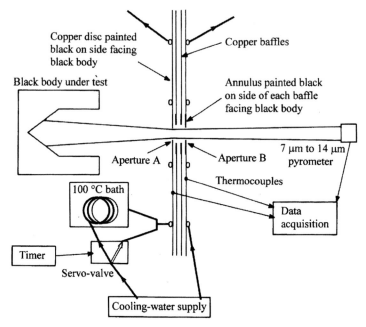

Figure 11 A schematic diagram of the shielded hot-plate apparatus used for the measurement of the effective 7–14 μm reflectance of black-body cavities (adapted with permission from Ref. [21]).

The assumptions for this investigation were as following: (i) the hot plate was sufficiently large to provide hemispherical irradiation to the blackbody aperture, (ii) the hot-plate emissivity was independent of observation angle, and (iii) the blackbody aperture viewed a uniform radiance in all directions. Then the spectral radiance of a blackbody under test could be expressed as:

$$L_\lambda(\lambda) = [1 - r_{BB}(\lambda)]P(\lambda, T_{BB}) + r_{BB}(\lambda)L_{\lambda,HP}(\lambda) \qquad (8)$$

where $r_{BB}(\lambda)$ and T_{BB} are the spectral reflectance, and the temperature of the blackbody, respectively.

From Equation (8) and multiple hot plate measurements as well as integrating over the pyrometer's spectral band, Ballico [21] was able to determine the blackbody effective emissivity and spectral radiance' dependence on the ambient background. The uncertainty analysis performed showed that the method and the apparatus developed enables measurement of the effective emissivity of blackbodies with an expanded uncertainty of 0.0002 ($k = 2$).

Hanssen et al. [22] later implemented Ballico's method [21] in a spectrally resolved configuration for calibration of customer blackbody sources. Figure 12 depicts a schematic diagram and photo of this setup.

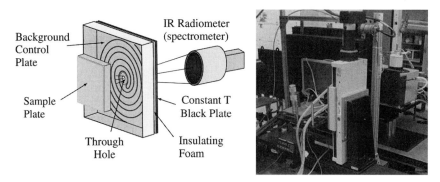

Figure 12 A schematic and photograph of the emissivity measurements setup (adapted with permission from Ref. [22]).

The diagram shows a sample plate (or flat plate blackbody) facing the IR spectroradiometer, with an intervening controlled background plate. The spectroradiometer was mounted on a large translation stage in order to view additional adjacent reference blackbodies. The controlled background plate, held in an aluminum frame, was essentially a sandwich of two black plates, the temperatures of which were independently controlled.

Two measurement methods were used. The first method (referred to as the absolute radiance method, ARM) involved radiometric measurements of the spectral radiance of the sample and the controlled background plate and also requires knowledge of the sample surface temperature. The simplified measurement equation used was:

$$L_{\lambda,S}(\lambda) = \varepsilon_S(\lambda)P(\lambda, T_S) + [1 - \varepsilon_S(\lambda)]L_{\lambda,\mathrm{bg}}(\lambda) \tag{9}$$

where λ is the wavelength, T_S the sample (or blackbody under test) temperature obtained from contact measurements, $P(\lambda, T_S)$ the spectral radiance of a perfect blackbody at the temperature T_S (as defined by the Planck equation), and $L_{\lambda,S}$ and $L_{\lambda,\mathrm{bg}}$ are experimentally measured spectral radiances of the sample and the controlled background plate, respectively. From Equation (9), the emissivity is given by:

$$\varepsilon_S(\lambda) = \frac{L_{\lambda,S}(\lambda) - L_{\lambda,\mathrm{bg}}(\lambda)}{P(\lambda, T_S) - L_{\lambda,\mathrm{bg}}} \tag{10}$$

A second approach was used to measure the spectral radiance of the sample and the enclosure at two different temperatures of the controlled background plate, maintaining the sample (or blackbody cavity) at the constant temperature. For the two measurements,

$$L_{\lambda,S1}(\lambda) = \varepsilon_S(\lambda)P(\lambda, T_S) + [1 - \varepsilon_S(\lambda)]L_{\lambda,\mathrm{bg}}(\lambda, T_{\mathrm{bg},1})$$
$$L_{\lambda,S2}(\lambda) = \varepsilon_S(\lambda)P(\lambda, T_S) + [1 - \varepsilon_S(\lambda)]L_{\lambda,\mathrm{bg}}(\lambda, T_{\mathrm{bg},2}) \tag{11}$$

where the indices 1 and 2 denote the first and second background temperatures, respectively.

The system of Equations (11) were solved for the sample emissivity:

$$\varepsilon_S(\lambda) = 1 - \frac{L_{\lambda,S1}(\lambda) - L_{\lambda,S2}(\lambda)}{L_{\lambda,bg}(\lambda, T_{bg,1}) - L_{\lambda,bg}(\lambda, T_{bg,2})} \tag{12}$$

Knowledge of the sample's absolute temperature was not required, although it was important to maintain its surface temperature constant or measure its change to introduce necessary corrections. These methods were applied to the flat-plate blackbodies. The results were compared with those obtained using radiometric methods and demonstrated good agreement.

The simplicity of the realization makes the method [21,22] useful for onboard monitoring of a calibration blackbody effective emissivity. Dykema and Anderson [16] suggested use of thermal emission sources for onboard monitoring of the effective emissivity of spaceborne blackbody sources for the CLARREO mission, which has stringent requirements for achieving on-orbit SI traceability.

3. RADIOMETRIC MEASUREMENT OF BLACKBODY SOURCES

3.1. Characterization of medium temperature blackbodies

An ideal blackbody is a spatially uniform and temporally stable source with radiance following Planck's law. There are many ways to describe the performance of real-life sources, and before reporting the measurements we need to define our approach. Spectral radiance and wavelength dependent radiance temperature are the only directly measurable quantities characterizing blackbody output for a particular surface element of the exit port and direction (solid angle). Knowledge of the blackbody spatial uniformity enables us to reduce the number of variables through averaging. Spectral radiance or wavelength dependent radiance temperature should be measured for a range of set temperatures. The commonly used term, spectral emissivity, is a derived unit that makes sense only when calculated and quoted with a clearly defined reference temperature, and obtained from the basic measured result of radiance or radiance temperature.

A complete characterization of a blackbody source includes: (1) a radiance uniformity scan (horizontal/vertical linear scan or full matrix); (2) a short-term temporal stability measurement; (3) a spectral radiance measurement at each temperature set point; and (4) an effective emissivity calculation from the measured spectral radiance, using either the set point temperature or a pyrometer radiation temperature as the reference, as well as knowledge of the background radiation temperature. An example of such a comprehensive calibration [23] of an industrial blackbody is shown in Figure 13.

In this section, we will review the methods, hardware, and some results of radiometric characterization for blackbodies with operational temperatures in the range from $0°C$ to $1,000°C$ in a normal laboratory environment. This temperature range is very important for process control in the chemical and

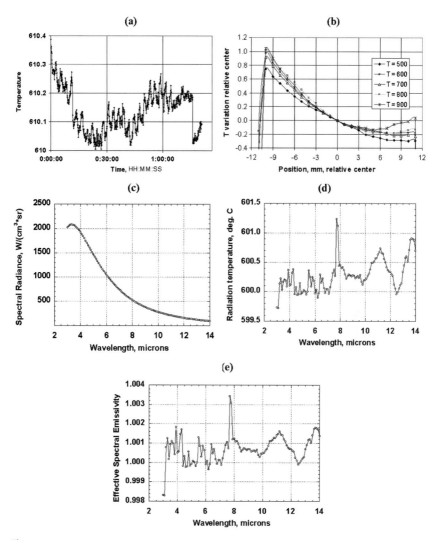

Figure 13 Example of complete calibration of a blackbody: (a) temperature stability, (b) spatial uniformity, (c) spectral radiance, (d) radiance temperature as a function of wavelength, and (e) effective spectral emissivity (adapted with permission from Ref. [23]).

semiconductor industries, for medical applications (ear thermometry, thermal imagery diagnostics and screening) and for military applications (night vision systems, spectral signatures of targets, etc.). One of the difficulties of the radiometric characterization of low temperature (near-ambient) blackbodies is the reflection of background radiation (i.e., thermal radiation of the objects around a blackbody, from the measurement facility components, environment, etc.). Since every blackbody has some level of reflectance, the radiometer or radiation thermometer sighted on the blackbody aperture will measure a combination of the background radiation reflected by the blackbody as well as its own thermal radiation. A number of measurement procedures have been proposed to take into account this effect.

At NIST, the advanced infrared radiometry and imaging (AIRI) facility [24] establishes and disseminates the US primary radiance temperature and spectral radiance scales under ambient background conditions for Planckian sources with thermodynamic temperatures below the silver freezing point of 961°C. A realization of the IR spectral radiance and radiance temperature scales was performed using a redundant set of standard blackbody sources with overlapping temperature ranges, differing principles of operation and cavity geometries. These primary blackbody sources have been characterized using the multicomponent approach, with a number of independent sources internally compared and found to be consistent. The effective emissivity of each cavity was determined by a combined method including reflectometric measurements [12,13] and Monte-Carlo numerical modeling [25]. Temperature measurements are based on contact thermometry and radiation thermometry of fixed-point blackbodies.

The fixed-point blackbodies, all of which were designed and characterized at NIST, employ the phase transition of pure metals Ga, In, Sn, Zn, Al, Ag, and Au. Their design and characterization are presented in Refs. [26,27]. The AIRI variable temperature blackbodies include an ammonia heat pipe blackbody (from −50 to 50°C), two water bath blackbodies (12−75°C) [28,29], a water heat pipe blackbody (55−250°C) [30], and cesium (300−650°C) and sodium (500−1,000°C) heat pipe blackbodies. All variable temperature blackbodies use platinum resistance thermometers (PRTs) for their temperature determination.

For comparison of the blackbodies, AIRI uses a NIST-designed tunable filter comparator (TFC), covering a spectral range from 2.5−13.5 μm, and a short-wave IR pyrometer RT1550 (1.55 μm peak, useful above 120°C) [31]. An internal view and optical system layout of the TFC is shown in Figure 14. It incorporates a temperature-controlled front plate and aperture stop, an elliptical primary mirror, a reflective chopper with an actively stabilized internal reference source, a reflective field stop, a circular variable filter (CVF) mounted on a rotation stage, an elliptic relay mirror, and a LN_2-cooled InSb/MCT sandwich detector. The TFC operates as a comparator, performing radiance interpolation between two reference

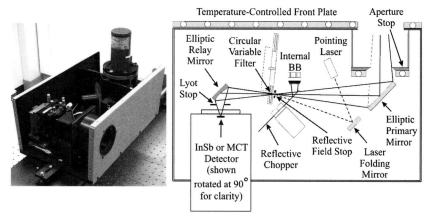

Figure 14 Advanced infrared radiometry and imaging (AIRI) tunable filter comparator (TFC) (adapted with permission from Ref. [24]).

blackbodies ("A" and "C") to calculate the spectral radiance and radiance temperature of a unit-under-test (UUT). The TFC measurement sequence follows a pattern: A−UUT−C−UUT−A, which may be repeated. The radiance of interest is calculated after each measurement cycle to reduce drift effects. This technique only requires short-term stability of the TFC during one cycle of the measurements.

The measurement equation, Equation (13) below, shows how the spectral radiance L_{UUT} is calculated from the measured spectra and known radiances of the reference blackbodies:

$$L_{UUT}(\lambda) = \left[\frac{V_{UUT}(\lambda) - V_A(\lambda)}{V_C(\lambda) - V_A(\lambda)} \right] [L_C(\lambda) - L_A(\lambda)] + L_A(\lambda) \qquad (13)$$

where V_A, V_C, and V_{UUT} are the measured spectra, L_A and L_C are the known spectral radiances of the reference blackbodies A and C, and L_{UUT} is the spectral radiance of the customer blackbody. The actual algorithm, that is used, is slightly more complicated because it takes into account the spectral shape of the CVF instead of assuming a single wavelength for each data point. The spectral transmittance of the CVF was separately measured at multiple wavelength settings. The sensitivity of the temperature interpolation process to differences between the unknown blackbody temperature and closest (in temperature) reference source has been studied in Ref. [32]. To improve the accuracy of interpolation, the temperature difference between the unknown blackbody and one of the reference blackbodies was kept as small as possible, typically within 2°C.

Many of the thermal IR applications rely on flat-plate blackbodies, the design of which makes them especially sensitive to imperfect surface emissivity, radiance temperature nonuniformity, and varying ambient conditions. At the same time, claims of their manufacturers cannot be verified without appropriate standards. Characterization of such sources requires special emphasis on a controlled background radiation environment and the capability to handle large amounts of scattered light originating from a radiating area with typical sizes up to 300×300 mm. AIRI was designed specifically to meet these needs. Figure 15 shows a photograph of the variable-temperature cluster of the AIRI facility. Knowledge of the radiation background temperature enables one to deduce the effective emissivity of a customer's blackbody from experimentally defined apparent spectral radiance of the UUT.

An example of experimentally obtained data for a high-temperature flat-plate blackbody is shown in Figure 16, which includes actual radiance temperature along with one that is anticipated from the manufacturer data. The open symbols denote the anticipated radiance temperatures calculated using the set point temperature, a nominal emissivity of 0.95 and a background temperature of 24°C. The filled symbols denote measured values. As can be seen from this plot, use of this blackbody relying only on the manufacturer data specification of emissivity in the range from 0.9 to 0.95 could easily result in an error up to 20°C, especially in the $3-5\,\mu$m range and near $9\,\mu$m. At the same time, analysis of the effective spectral directional emissivity (as shown in Figure 17) indicates a consistent temperature calibration of the surface temperature, because over a wide

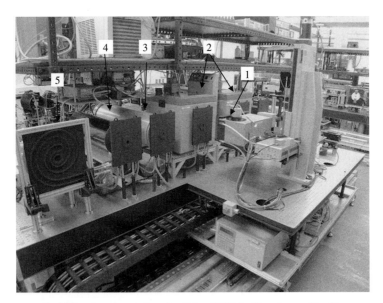

Figure 15 Variable temperature bench of the AIRI facility: 1, tunable filter comparator; 2, water-bath blackbodies; 3, water heat-pipe blackbody; 4, ammonia heat-pipe blackbody; 5, controlled background plate (adapted with permission from Ref. [24]).

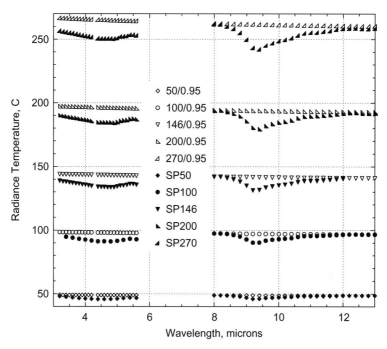

Figure 16 Expected and actual radiance temperature of a high-temperature flat plate blackbody. Light symbols denote the anticipated radiance temperature calculated using the set point temperature, for a nominal emissivity 0.95 and a background temperature of 24°C. Filled symbols denote actual measured values (adapted with permission from Ref. [24]).

temperature range only a negligible shift in the calculated emissivity data is observable.

The implemented capabilities of the AIRI facility for customer IR source calibration include: characterization of the absolute spectral radiance and radiance temperature across the spectral range of $3-13\,\mu m$ and temperature range of -50 to $1,000°C$ with a full calibration uncertainty of $20-70\,mK$ $(k = 2)$ (for the temperatures above $10°C$) and equivalent values for spectral radiance; characterization of the spatial uniformity and stability; and background radiation correction and emissivity (reflectance) evaluation for flat-plate calibrators with emissivities of 0.8 or greater. The NIST developed spectral comparator TFC provide comparisons in spectrally resolved radiance throughout the temperature range of -50 to $1,000°C$ and over a spectral range of $3-13\,\mu m$ at the level of $10-150\,mK$ $(k = 2)$, depending on the temperature, with a relative spectral resolution of $2-3\%$.

Clausen [33] described measurements of the spectral effective emissivity of blackbody calibration sources with flat or grooved surfaces coated by Pyromark® 2500 paint. An FTIR spectrometer is used for the

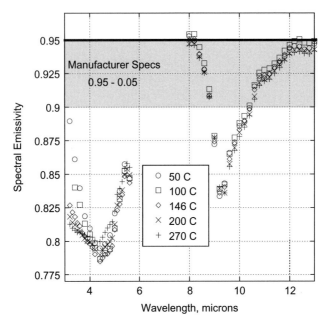

Figure 17 Effective spectral directional emissivity of a high-temperature flat plate blackbody. The manufacturer's specification was 0.95–0.05 (adapted with permission from Ref. [24]).

measurement of spectral radiance in the range from 2 to 20 μm by comparison of blackbody surface calibrators with a high-quality reference blackbody of known temperature and emissivity. The method included elimination of thermal radiation from the spectrometer and takes into account the reflected ambient radiation. To calculate the spectral emissivity of the surface, three measurements were performed: a spectrum S_r of the surface at the set-point temperature, a spectrum S_{am} of the surface at ambient temperature, and a spectrum S_{bb} of the reference blackbody. The three spectra can be described by three equations:

$$S_r(\lambda) = R(\lambda)\varepsilon(\lambda, T)P(\lambda, T) + B(\lambda)$$
$$S_{am}(\lambda) = R(\lambda)\varepsilon(\lambda, T_{am})P(\lambda, T) + B(\lambda) \qquad (14)$$
$$S_{bb}(\lambda) = R(\lambda)\varepsilon(\lambda, T_{bb})P(\lambda, T_{bb}) + B(\lambda)$$

where P is the Planckian function, $\varepsilon(\lambda, T)$ the surface emissivity of the calibrator surface at the temperature T that is the reading of a calibrated contact sensor, $R(\lambda)$ the response function of the entire system, and $B(\lambda)$ the spectral radiance of thermal radiation from the background. The surface at a given temperature, T, was compared with a reference blackbody of known temperature, T_{bb} and known emissivity, $\varepsilon(\lambda, T_{bb})$. Clausen introduced the "apparent emissivity," $\varepsilon_r(\lambda, T)$, that is, a fictitious emissivity computed taking into account the background radiation reflected by a surface, and derived the following equation:

$$\varepsilon_r(\lambda, T) = \frac{S_r(\lambda) - S_{am}(\lambda)}{S_{bb}(\lambda) - S_{am}(\lambda)} = \frac{\varepsilon(\lambda, T)[P(\lambda, T) - P(\lambda, T_{am})]}{P(\lambda, T_{bb}) - P(\lambda, T_{am})} \tag{15}$$

The actual emittance of a surface could be calculated as:

$$\varepsilon(\lambda, T) = \varepsilon_r(\lambda, T) \frac{P(\lambda, T_{bb}) - P(\lambda, T_{am})}{P(\lambda, T) - P(\lambda, T_{am})} \tag{16}$$

The surface under investigation was compared directly with a reference blackbody at a similar temperature. The set-point temperature of the surface was raised or lowered slightly to obtain blackbody radiation similar to the reference blackbody in the spectral range of interest, for example, from 8 to 14 µm. Several reference blackbodies were used: (i) a water heat-pipe blackbody in the temperature range from 50°C to 250°C, (ii) a cavity immersed into an electrically heated stirred salt bath from 250°C to 550°C, and (iii) a cavity mounted in a thermostated water bath at 5−50°C. Measurements were performed with an FTIR spectrometer by averaging over several hundred scans or approximately 3 min. The path between the FTIR spectrometer and the blackbodies was kept constant to minimize the influence of the absorption of water vapor in the air. All measurements were background corrected using a measurement at ambient temperature on a surface under test with a set point temperature of 23°C, or turned off, to eliminate errors due to thermal radiation from the FTIR spectrometer, reflected radiation from surfaces, etc.

The spectral emissivity of a V-grooved aluminum surface coated with Pyromark® 2500 that was used for temperatures up to 550°C was measured and the effect of coating technique was studied. As stated in Ref. [33], the uncertainty of about 0.25% of the emissivity is achievable for the method described ($k = 2$), but this requires great care during measurements. Additional investigations of the size-of-source effect, linearity of the detector of the FTIR spectrometer, improved instrument stability, etc., would be required to improve the accuracy.

Joly et al. [34] described the multiband near-to-far infrared radiance comparator (MIRCO) developed at LNE (Laboratoire National de Métrologie et d'Essais, France) for measurement facilities in radiation thermometry. MIRCO has a mirror-based optical system, an associated set of interference filter wheels, a modular holder for several infrared detectors, and a lock-in amplifier (see Figure 18).

Two filter holders are available, one for the classical bands (including 3−5 µm and 8−12 µm bands) and another to meet custom calibration requests. All the housing's internal walls are coated with 3 M Nextel 811-21 paint to reduce stray radiation. A black painted baffle assembly is placed close to the field stop to avoid direct rays from the source. MIRCO

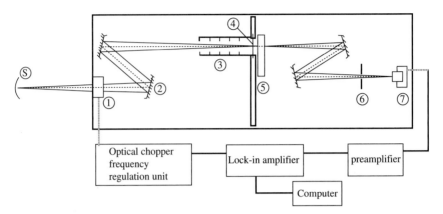

Figure 18 Functional diagram of the near-to-far infrared comparator device; (S) source; (1) optical chopper; (2) mirror equipped with a limiting circular aperture; (3) baffle assembly; (4) field stop; (5) filter wheel; (6) aperture stop (pupil); and (7) detector (adapted with permission from Ref. [34]).

was employed in measurements of radiance temperature of a water heat-pipe blackbody fitted with a standard platinum resistance thermometer (SPRT) close to the cavity back wall.

Tsai and Rice [35] described an application of another filter radiometer, thermal-infrared transfer radiometer (TXR) to compare water-bath and oil-bath blackbodies. The TXR was a two-channel, liquid-nitrogen cooled, transportable filter radiometer (see Figure 19) developed at NIST for NIST-traceable remote sensing-oriented radiometric calibrations in several different thermo-vacuum calibration facilities [36,37]. The TXR components were sealed off in vacuum and cooled to near 77 K. Channel 1 was centered near 5 μm and had a bandwidth of about 0.9 μm, while Channel 2 was centered near 10 μm and had a bandwidth of about 0.95 μm. In Ref. [35], the TXR operated in the ambient mode where the TXR cryostat case, including the ZnSe cryostat window, was near room temperature. During the experiment, the TXR was calibrated against water bath blackbody, then, the radiance temperature of the oil bath blackbody was determined using Channel 2.

Figure 20 shows the dependence $\Delta T(T_c)$, where $\Delta T = T_{X2} - T_c$, T_{X2} is the radiance temperature measured using Channel 2 of the TXR, and T_c is the temperature measured using a SPRT immersed into the oil thermostat. Measurements were conducted for three aperture diameters of both blackbodies: 4, 5, and 10.8 cm (fully open cavities). The evaluated uncertainty in the determination of T_{X2} was 0.06 K ($k = 1$).

Cox et al. [38] described an application of the TXR apparatus to the evaluation of the infrared radiance temperature of a microwave calibration target. The target surface was covered with an array of square-based

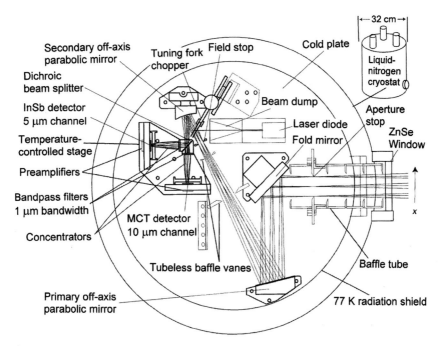

Figure 19 Plane view of the infrared transfer radiometer (TXR) optics from underneath the cryostat (adapted with permission from Ref. [36]).

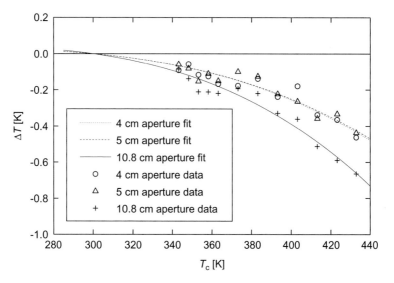

Figure 20 Dependence of ΔT on T_c for an oil bath blackbody for three aperture diameters (adapted with permission from Ref. [35]).

pyramids with an aspect ratio of 4:1. The base material of the target was aluminum, with pyramids formed by electrical discharge machining and coated with ferrous-loaded epoxy to a thickness of about 1 mm. A heating element, adhered to the target base, was used in conjunction with a closed-loop control system to heat the target from 300 to 340 K in 5 K increments allowing time for temperature stabilization between each change. Calibration of the TXR was performed against the water bath blackbody from 17.5°C to 85°C. The measurement for the peak of one pyramid was compared to that for an adjacent valley. TXR data were taken while the TXR field-of-view (around 30 mrad) was centered alternately on the pyramid tip and on the valley. The temperature drop toward the tip (the target was heated at the base) was assessed as approximately 0.03 K. The measurements carried out showed transverse gradients on the order of 1.25 K over distances of about 10 cm.

Nugent and Shaw [39] presented an investigation of the angular variation of large area blackbodies, which were used for calibration of wide angle long wave infrared (LWIR) cameras. The measurement equation used was:

$$\varepsilon(\theta) = \frac{L_{\mathrm{m}}(\theta) - L_{\mathrm{a}}(\theta)}{L_{\mathrm{BB}} - L_{\mathrm{a}}(\theta)} \tag{17}$$

where θ is the angle between the viewing direction and the optical axis which coincides with the normal to the radiating surface, L_{m}, L_{a}, and L_{BB} were measured LWIR (band-limited, approximately $7.5 - 13.5\ \mu\mathrm{m}$) radiances from the blackbody under test, ambient background, and a perfect blackbody at the temperature of the radiator under test, respectively. A 50° field-of-view LWIR camera was used. This camera was calibrated using a grooved-surface blackbody and the manufacturer specified emissivity. Four commercial blackbodies with various surface structures were measured. They were placed inside a chamber on top of a computer controlled rotation stage. The camera remained fixed and observed the blackbody through a port hole in the enclosure while the blackbody was rotated. During the experiments, the blackbody source temperature was set to 60°C to keep the source radiance sufficiently greater than the ambient radiance. During each experiment, the blackbody was rotated in increments of 5° from 0° to 60°, with an image series taken at each increment.

The LWIR average emissivity of four commercial blackbodies was measured versus angle. One of the blackbodies was a flat plate blackbody with a high-emissivity coating, two had a honeycomb surface of approximately 5 mm wide hexagonal cavities with a high emissivity coating, and the fourth was a vertically grooved surface with a groove depth of approximately 1.5 mm and a period of 2.5 mm covered with a high emissivity coating. For all the blackbodies, $\varepsilon(\theta)$ was found to drop at large

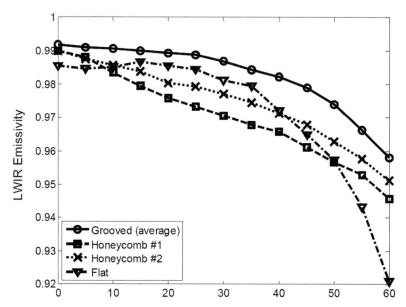

Figure 21 Angular dependences of the long-wave infrared (LWIR) emissivity for four blackbodies (adapted with permission from Ref. [39]).

observation angles (see Figure 21). Because the blackbody with a grooved surface was expected to exhibit different angular emissivity depending on the orientation of the observation relative to the surface, two sets of experiments were conducted with the groove pattern in different orientations to the rotation axis. The grooved surface blackbody demonstrated the highest overall emissivity (averaged for two mutually perpendicular rotation planes) at large angles and was the most uniform with angle. The flat-plate blackbody had the lowest emissivity at large angles, but had slowly varying emissivity for angles $<30°$. The emissivity steadily dropped with angle for the two honeycomb surface blackbodies, whereas the emissivity stayed more constant out to approximately $25°$ for both the grooved surface and flat plate blackbodies.

Ma [40] described a method based on the fact that the effective emissivity of an isothermal cavity must approach unity as its aperture diminishes. Hence, measurements of the spectral radiance of a cavity for different sizes of the aperture can be extrapolated to the zero case. The value obtained in this way is related to the case of a perfect blackbody. The effective emissivity of the cavity for a specific aperture size can be determined by subtracting from unity the difference between the values obtained with and without the aperture. The method is applicable for large aperture high temperature blackbodies where the effect due to changing the aperture size can be more easily measured. Specifically, this method is applicable to fixed point blackbodies since their temperature is defined primarily by the temperature of the phase transition of a material inside a crucible and their temperature field should not change as the aperture size is changed. The method was applied to the copper fixed point blackbody ($T = 1,357.77$ K) at the wavelength $\lambda = 0.65\,\mu$m. The spectral emissivity was measured for a right-circular cylinder with the bottom perpendicular to the cavity axis or inclined at the angle of $30°$ to the cavity axis. The measured effective emissivity progressively decreased with increasing aperture size. In Figure 22, the computed and measured changes in the effective emissivity are shown for a cylindrical cavity with a bottom perpendicular to the cavity axis. The circles and triangles correspond to two independent runs. Dotted curves 1, 2, and 3 show changes calculated under the assumption that the surface of the cavity emits and reflects diffusely and that its wall emissivity is 0.6, 0.7, and 0.8, respectively. The measured values were lower than the computed curves for larger aperture sizes. This was attributed to the deviation from isothermal conditions, due to the increased radiation heat loss through the larger apertures.

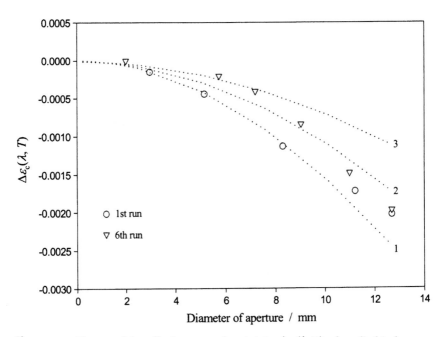

Figure 22 Change of the effective spectral emissivity $\Delta\varepsilon_c(\lambda, T)$ of a cylindrical cavity with the size of the aperture at $\lambda = 650$ nm and $T = 1,357.77$ K (adapted with permission from Ref. [40]).

3.2. Characterization of high temperature blackbodies

We shall consider blackbodies with operational temperature $> 1,000°C$ as high temperature. High temperature blackbodies find use for UV and visible spectroradiometric scale realization, pyrometer calibration in the visible spectral range, high heat flux measurements, and high-temperature process control.

Their characterization has the same purposes and follows the same physical principles, but there are some distinctions, which set it apart from characterization of medium temperature blackbodies. These distinctions include: (1) effects related to the high temperatures involved, including cavity material sublimation and oxidation, as well as other chemical processes of a pyrogenic nature; (2) the feasibility of using visible and even UV radiation for characterization, which enables the application of very mature techniques such as detector-based radiometry; (3) very restricted use of contact techniques, limiting the possibility for cavity wall uniformity evaluation; and (4) very high radiation fluxes, potentially causing large temperature gradients while making difficult their modeling and prediction.

Sperfeld et al. [41,42] demonstrated that under specific conditions absorption bands in the spectrum of graphite high-temperature cavity radiators can be observed, causing deviation of the spectrum from the expected Planckian. One such example is shown in Figure 23. Ref. [42] contains an in-depth analysis of the nature of these features and offers some measures to avoid these absorptions, such as purge gas flow optimization, evacuating, and flashing the blackbody radiator prior to heating, annealing, and the use of high purity purge gas and delivery pipes. It was shown that the use of these measures can eliminate these absorption bands for temperatures below 3,100 K and even at higher temperatures, for some absorptions bands. However, for temperatures above 3,100 K, absorption bands occur that cannot be avoided. The absorptions have been identified as originating from molecular carbon that is increasingly subliming at the highest temperatures. These limit the use of graphite-based high-temperature radiators as ideal Planckian radiators, especially in the ultraviolet and visible spectral ranges.

Friedrich and Fischer [43] described the temperature measurement of the high-temperature blackbody (HTBB) by two methods: using mono-chromatic radiation thermometry relative to the fixed-point blackbody at the freezing point of gold (gold-point blackbody, GPBB) as recommended by the International Temperature Scale, ITS-90 [44] and using absolute radiometry with filter radiometers calibrated against a cryogenic radiometer. For these purposes, the HTBB was operated in radiance and irradiance modes at the same time. In both measurement modes, the same area of the blackbody bottom was viewed by the detector systems. After determination of the HTBB temperature, the value of the unit of the spectral radiance was

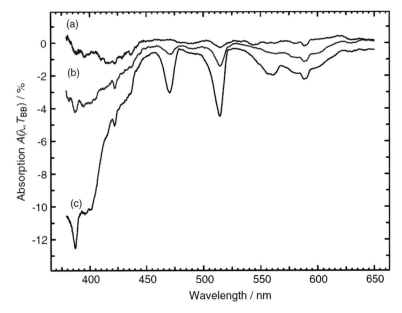

Figure 23 Temperature dependent absorption features in the spectrum of a high temperature blackbody (HTBB) at: (a) 3,090 K, (b) 3,180 K, and (c) 3,300 K (adapted with permission from Ref. [42]).

transferred to the gas-filled tungsten strip lamps, which were used later as transfer standards. A schematic diagram of the measurement setup is presented in Figure 24. The radiating area of the sources was imaged on to the entrance slit of the monochromator. Two filter wheels in front of the monochromator held the filters for order selection. The detectors were mounted on a translation stage and could be exchanged automatically if measurements were to be made across the entire wavelength range.

Following the ITS-90, the temperature, T_{90}, of the HTBB was determined by monochromatic radiation thermometry at 650 nm, directly measuring the spectral radiance ratio, Q, relative to the freezing point of gold. An intermediate step was added and T_{90} was determined by comparison with a secondary standard, a vacuum tungsten strip lamp with a radiance temperature, T_S, calibrated against the GPBB. The equation for the spectral radiances was presented as:

$$\varepsilon_e L_{\lambda,\text{HTBB}}(650 \text{ nm}, T_{90}) = QL_{\lambda,\text{SL}}(650 \text{ nm}, T_S) \tag{18}$$

where ε_e was the spectral effective emissivity (at $\lambda = 650$ nm) of the HTBB, $L_{\lambda,\text{HTBB}}$ was the spectral radiance of the HTBB, $L_{\lambda,\text{SL}}$ was the spectral radiance of the standard lamp, and Q was the ratio of photocurrents of the HTBB and standard lamp. Using Planck's law for $L_{\lambda,\text{HTBB}}$, the HTBB temperature, T_{90}, was determined according to Equation (13). Since the

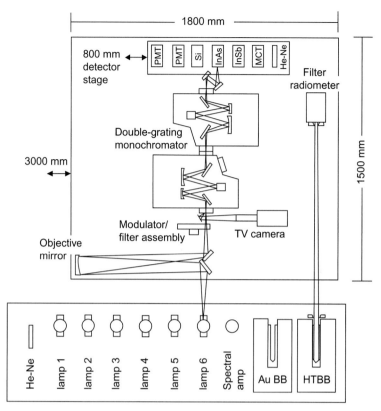

Figure 24 A schematic diagram of the measurement setup (adapted with permission from Ref. [43]).

emission of tungsten strip lamps is polarized, measurements were performed in both s- and p-polarization planes and the appropriate correction was made.

The HTBB temperature determination by absolute radiometry was carried out using filter radiometers calibrated against the cryogenic radiometer. The basic formula for this was given as:

$$i_{FR} = \varepsilon_e G \int_0^\infty L_{\lambda,HTBB}(\lambda, T)s(\lambda)d\lambda \qquad (19)$$

where i_{FR} is the photocurrent of the filter radiometer, G the geometric factor; and $s(\lambda)$ the spectral responsivity of the filter radiometer. The geometric factor depends on the distance between the HTBB and the filter radiometer and on the areas of the apertures in front of the HTBB and in front of the filter radiometer. The temperature, T, of the HTBB was obtained using Planck's law for $L_{\lambda,HTBB}$ in Equation (19). Two filter radiometers were used with the center wavelengths of 676 and 800 nm.

Absolute radiometry has the advantage that there is no need to apply polarization equations.

The evaluation of effective emissivity of the NIST HTBB in a graybody approximation via comparison with the GPBB was described by Yoon et al. [45]. These authors also described the experimental determination of the HTBB with operating temperature near 3,000 K using both a detector scale-based approach and an ITS-based one.

First, the temperature of the HTBB was determined using filter radiometers calibrated against the detector based-scale. The filter radiometer for the temperature determinations consisted of temperature-stabilized, broadband, glass filters, and Si photodiodes. The spectral irradiance responsivity of the filter radiometer was determined using the NIST spectral comparator facility [46], which, in turn, used a scale derived from the NIST high-accuracy cryogenic radiometer [47]. The aperture areas were measured on the NIST high-accuracy aperture measurement facility [48]. The temperatures of the HTBB were found from the measurement equation, which established the following interrelationship between the measured photocurrent, i, and the blackbody temperature, T:

$$i = GD^{-2}\pi r_{BB}^2 \pi r^2 (1 + \delta)\varepsilon_e \int_\lambda S(\lambda)P(\lambda, T)d\lambda \qquad (20)$$

where G is the preamplifier gain, r and r_{BB} are the aperture radii of the filter radiometer and the blackbody, respectively, ε_e the effective emissivity, $S(\lambda)$ the absolute spectral power responsivity, and P the Planck function. The geometric factors D and δ are given by $D^2 = r^2 + r_{BB}^2 + d^2$ and $\delta = r_{BB}^2 r^2 / D^4$, where d is the distance between the filter radiometer aperture and the blackbody aperture. The temperature in the Planck radiance law was calculated using an iterative method.

Using the ITS-90 approach, the spectral radiance of the HTBB was calculated using Planck's law and the knowledge of the detector-based blackbody temperatures determined using the filter radiometer. The radiance temperature of the variable temperature blackbody (VTBB) was assigned at 655.3 nm by the use of radiance ratios to a vacuum tungsten ribbon-filament lamp, held at a radiance temperature of 1,530 K. The radiance temperature of the lamp was, in turn, determined by comparison to a GPBB using a photoelectric pyrometer with a responsivity centered at 655.3 nm. This approach corresponds to the radiance ratio method as prescribed by the ITS-90. The spectral radiance of the HTBB was derived from:

$$L_{\lambda,HTBB}(\lambda) = \frac{g_{HTBB} i_{HTBB}}{g_{VTBB} i_{VTBB}} L_{\lambda,VTBB}(\lambda) \qquad (21)$$

where i, g, and L_λ are the photocurrent, preamplifier gain, and spectral radiance, respectively. Subscripts "HTBB" and "VTBB" correspond to the two blackbodies that were compared. $L_{\lambda,\text{VTBB}}(\lambda)$ was calculated using the VTBB effective emissivity (~ 0.999) and Planck's law for the VTBB temperature transferred from the temperature lamp calibrated against the GPBB.

The agreement between the gold-point based radiance and the detector-based radiance was shown to be within the combined uncertainties of the measurements, but the dominant source of the uncertainty was from the gold-point based measurements. The results of the spectral radiance comparison are shown in Figure 25. The two approaches to determining the HTBB temperature described above, were found to be in agreement within the combined uncertainties of 1.2 K ($k = 2$). The uncertainty budget evaluation performed in Ref. [49] showed that a lower uncertainty can be achieved in the detector-based radiance temperature measurements using broadband filter radiometers as compared to narrow interference filter radiometers.

Sperfeld et al. [50] presented results of an international project investigating the two high-temperature pyrographite blackbodies, BB3200c and BB3200pg, described in detail in Ref. [51]. One of the objectives of the study was the determination of the radiance temperature uniformity of the cavity. When measurements were performed, a high precision aperture of

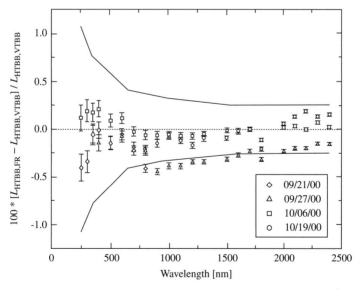

Figure 25 Comparison of the HTBB spectral radiance assigned using the spectral radiance from the variable-temperature blackbody (VTBB) and the spectral radiance from Planck's radiance law using the detector-based radiance temperatures. The solid lines indicate the expanded uncertainties of the spectral radiance determined using the VTBB. The error bars indicate the 0.21 K temperature uncertainty converted to radiance uncertainty (adapted with permission from Ref. [49]).

15 mm diameter in a water-cooled mount was placed immediately in front of the exit aperture of the blackbody. This aperture limited the field-of-view such that only radiation originating from the cavity should be seen by the detectors when the blackbody was viewed along the cavity axis. To scan the uniformity of the radiation passing through the front aperture, the radiance in the blackbody aperture area was imaged on to a filter radio-meter with a 800-nm peak wavelength and about a 20-nm bandwidth. The lens-radiometer system was moved horizontally on a translation stage. The blackbody aperture was imaged onto the radiometer aperture with a diameter of 1 mm. Hence a usable diameter of 13 mm could be scanned without vignetting by the edge of the blackbody aperture. In Figure 26, distributions of the relative spectral radiance for both blackbodies at different temperatures are shown. The values $100 \times [1 - L_\lambda(x)/L_\lambda(0)]$ are plotted in percent as ordinates, where x is the radial coordinate across the

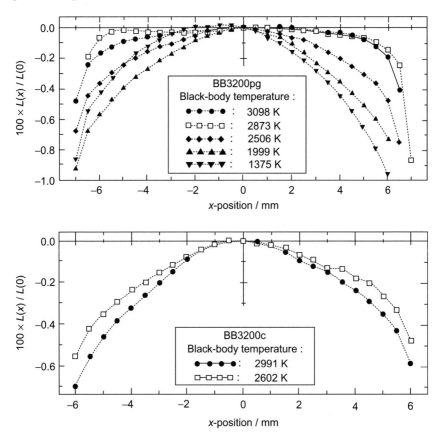

Figure 26 Spectral radiance (at 800 nm) relative distribution across the apertures of the two pyrographite blackbodies at different temperatures (adapted with permission from Ref. [50]).

aperture. The radiance distribution for BB3200pg showed nonuniformity <0.1% for radiometric temperatures higher than 2,800 K. For BB3200c, the nonuniformity was <0.5%.

White et al. [52] described the characterization of a high-temperature graphite blackbody BB22P [51]. The uniformity of the spectral radiance in the horizontal plane orthogonal to the optical axis was measured at wavelengths centered on 407, 468, 647, 800 nm and $V(\lambda)$ using filter radiometers fitted with a 1-mm aperture, positioned so as to align with the optical axis of the system. The resultant scans showed a slight slope across the field (8 mm) with a maximum deviation from the mean of 0.05%. One of the most important criteria in this experiment was that the spatial uniformity remained constant with time. Initial results confirmed that after a burn-in period of 30 h for a new cavity, the spatial uniformity at all wavelengths varied by no more than 0.03% over a period of 200 h over the central 8-mm patch. The measurements also include any nonuniformity caused by temperature gradients within the radiator. The results indicated that the performance of the calibration setup and the cavity under study were sufficient to pursue the detector-based blackbody radiance scale realization.

3.3. Characterization of blackbodies in a reduced-background environment

To imitate conditions experienced in a spaceborne environment, various types of remote sensing and defense equipment are tested in cryo–vacuum chambers, which can provide a reduced radiation background, high vacuum and irradiation by sources with spectra resembling an exoatmospheric solar spectrum.

As mentioned in Sections 3.1 and 3.2, the goals of blackbody calibration and the basic physics behind it remain the same, while operation under vacuum and potentially low-background conditions imposes certain limitations on the measurement techniques. Also, the nature of the final application may call for some extreme conditions, such as the need to measure very faint sources with low temperature, small aperture sizes and at long wavelengths. These may necessitate highly specialized characterization facilities and techniques for modeling and measurement.

At NIST, the low-background infrared (LBIR) facility was established to fulfill the US need for calibrations in the infrared wavelength region in a low-background 20 K environment. Measurements of infrared radiation of 1 nW performed with a standard deviation of random error not exceeding 1%. Since the creation of the facility in 1989, the primary service of the facility has been the broadband calibration of infrared sources [53−55]. The LBIR facility general layout is shown in Figure 27.

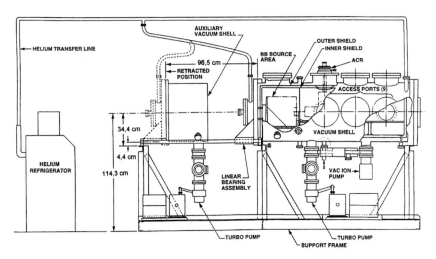

Figure 27 LBIR facility layout (adapted with permission from Ref. [53]).

The central elements of the LBIR facility are the absolute cryogenic radiometers. They have cavity receiver elements and measure the optical power by electrical power substitution. Due to the cavity receiver design, the absolute cryogenic radiometers are spectrally nonselective. The LBIR facility has two low-background test chambers.

The LBIR facility is used to calibrate user-supplied blackbody sources with operating temperatures above 100 K, and capable of functioning in a 20 K environment. The methodology of cryogenic blackbody calibration is considered in detail by Datla et al. [56].

Carter et al. [57] analyzed the LBIR calibrations of several black-bodies that operated with cavity temperatures of $180-800$ K. All of the calibrations were performed entirely within the cryogenic-vacuum chambers at 1×10^{-7} Pa in a $20-30$ K background environment. The radiation temperatures of the blackbodies were calibrated using *in situ* length measurements of the calibration optical geometry, power measurements made by the absolute cryogenic radiometers and the Stefan−Boltzmann law. The correction for the diffraction losses from the defining apertures was made according to Ref. [58].

Figure 28 shows the results of radiation temperature measurements for blackbodies of various designs designated A through F. The temperature error is equal to the difference between the radiometrically measured temperature and the contact temperature. The error bars shown for design D are the expanded $k = 2$ uncertainties and are larger than that for most of the others. The error bars for the others were removed for clarity. Analysis of the results led to the determination of the main sources of errors.

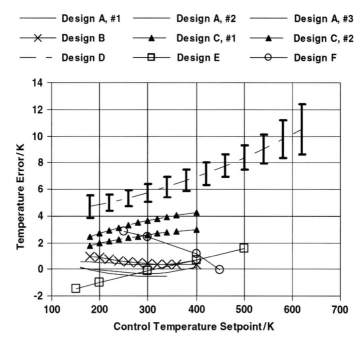

Figure 28 Blackbody cavity radiation temperature calibration data (adapted with permission from Ref. [57]).

These were a nonuniformity of the temperature field along some cavities due to thin walls and strain on the PRT sensors due to mounting methods.

The Strategic Defense Initiative (SDI) and subsequently the Ballistic Missile Defense Organization (BMDO) of the US Department of Defense (DoD) has motivated the work at NIST in support of the defense remote-sensing area [59,60]. In this application, the onboard sensors must detect and measure, on parity with the radiation from the Earth, the faint signals from small targets viewed against the cold background of space. Flux levels at the detectors tend to be orders of magnitudes smaller in such defense remote-sensing applications than in environmental remote sensing applications. To achieve the sensitivity required, onboard sensors are often cooled to 77 K and below. Also, to simulate the cold-space background, shrouds in space-simulation calibration chambers are cooled to a low background such as 20 K using cooled helium gas. NIST's role has been to support prelaunch calibration activities by calibrating blackbodies and measuring other artifacts [61]. They are used as standards in the sector of the aerospace industry performing the prelaunch radiometric calibrations for DoD missile defense programs.

Jung et al. [62] described a portable transfer standard radiometer (BXR) developed at NIST according to the BMDO specification for verifying the scales established at customer calibration facilities. The BXR setup includes a stainless steel chamber, liquid-nitrogen cryoshroud and radiometer. The BXR uses an arsenic-doped silicon blocked-impurity band (BIB) detector behind a rotating filter wheel. The BIB detector operates at temperatures near 12 K and is cooled by a flow of liquid helium. All other elements of the radiometer are cooled to temperatures below 20 K to reduce the background. The BXR is capable of detecting irradiance levels as low as 10^{-15} W/cm^2 over the spectral range from 2 to 30 μm. The standard uncertainty is about 5% for irradiance levels down to 10^{-9} W/cm^2.

A cryogenic FTIR spectrometer developed for the LBIR facility is described in Ref. [63]. This spectrometer was developed for the Missile Defense Agency Transfer Radiometer (MDXR) that is intended for calibration of infrared sources that cannot be transported to NIST for calibration. When used inside the MDXR, the cryogenic FTIR spectrometer provides relative spectral measurements with repeatability better than 1% over the spectral range from 3 to 15 μm and at a spectral resolution of 0.6 cm^{-1}. The interferometer uses a compensated Michelson configuration with flat mirrors and a KBr beam splitter and compensator. The interferometer has an operating temperature range between 10 and 340 K.

Since the last quarter of the 20th century, remote sensing has played a continuously growing role in many areas of human activity. Measurements of the radiant flux from the Earth, Moon, planets, and stars using optical instruments deployed remotely are used to determine a wide range of physical parameters. For the Earth−Sun system, a critical long-term goal is to acquire an accurate and detailed understanding of processes affecting global climate change, in particular the role of human activities. From radiometric data, it is possible to study and quantify the Earth's temperature distribution, energy budget, properties, and dynamics of the atmosphere and oceans, land use, carbon cycle, and other systems that relate to climate [64−66]. Medium-background infrared radiometry deals with radiometric measurements in an approximately 80 K environment. It produces thermal radiation, but this is negligible for the purposes of programs involving blackbody sources with temperatures between 250 and 350 K. One of the initial calibrations was for the TXR radiometer developed at the NIST in support of the Earth Observing System (EOS). The EOS is the key element in the Mission to Planet Earth (MTPE) program of the National Aeronautics and Space Administration (NASA).

One example of a medium-background facility (MBF) is described in Ref. [67] for calibration of blackbody sources and infrared sensors under medium vacuum conditions (10^{-3} Pa) and a medium-background environment (using an LN$_2$-cooled shroud). A schematic of the MBF is shown in Figure 29. The MBF consists of a gallium fixed-point reference blackbody (Ga BB, 29.76°C), a VTBB with operational temperatures from −60°C

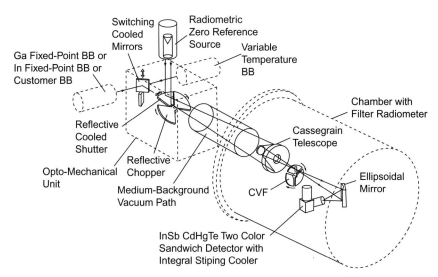

Figure 29 A schematic of the medium-background facility (adapted with permission from Ref. [67]).

to 80°C, a vacuum chamber for the blackbody source under calibration, an opto-mechanical unit, a radiometric zero-reference source (a blackbody with a high effective emissivity at a temperature of about 77 K), a medium-background vacuum path, a vacuum chamber for the filter radiometer or infrared sensor under calibration, a data acquisition and control system and fluid circulators. The radiation from the target source or radiometric zero-reference source is modulated by a chopper operating at 200 Hz. A shutter is held at a temperature of 80 K. The chopper provides background subtraction.

A further development of this approach intended mainly for providing calibration of space-based infrared radiation sources and sensors for remote-sensing experiments in terms of spectral radiance and radiation temperature from −173°C to 430°C is described in Ref. [68]. To exclude the influence of background radiation, the source chamber is connected to the detector chamber via the LN$_2$-cooled beamline. The radiation of the VTBB [69,70] with an operational temperature range from −173°C to 177°C and of the fixed-point blackbodies are alternately directed toward the detector chamber, in which the vacuum infrared standard thermometer (VIRST) and off-axis ellipsoidal mirror are arranged on translation stages. Radiators can be compared with dedicated reference blackbodies in terms of radiation temperature in the spectral band from 8 to 14 µm via the VIRST or in terms of spectral radiance via the FTIR spectrometer. More details on this facility are given in Chapter 6 in the companion volume, *Radiometric Temperature Measurement: I. Fundamentals*, Vol. 42 of this series.

4. MEASUREMENT OF INPUT PARAMETERS FOR THE CALCULATION OF BLACKBODY RADIATOR CHARACTERISTICS

4.1. Input parameters for the calculation of blackbody characteristics

In some cases, it is not feasible or practical to measure the radiation characteristics of a blackbody. Often, it is necessary to know the radiation characteristics *a priori*, for example, in the blackbody design stage. In such cases, calculation is the only method for determination of these characteristics. The effective emissivity is the most universal characteristic of a blackbody radiator. There are many calculation methods for the effective emissivity. They can differ from each other in the calculation approach (analytical or numerical), in the adopted physical assumptions (a "gray surface" model, an isothermal approximation, a "diffuse surface" model, a presumption that radiative fluxes are distributed uniformly over the internal surface of a blackbody cavity after the first or second reflection, etc.), in the radiation registration mode (normal, hemispherical, directional), and so forth. The choice of the computational method determines the set of physical parameters that need to be known or defined in order to perform calculations. A detailed overview of these methods can be found in Chapter 5 in the companion volume, *Radiometric Temperature Measurement: I. Fundamentals*, Vol. 42 of this series. Here, we shall attempt to specify briefly which input parameters are needed for calculations using a particular method for the effective emissivity.

The majority of analytical methods (i.e., methods which allow deriving an explicit formula) are based on a diffuse (Lambertian) model for emission and reflection. In this case, it is sufficient to know the hemispherical or directional emittance (they are equal for diffuse surfaces). And according to Kirchhoff's law for opaque materials, it is sufficient to measure the hemispherical reflectance instead of the emittance. If the hemispherical reflectance depends on incident angle (which is an indication that the diffuse model may be inadequate), its values can be averaged over all possible incidence directions. For the "gray surface" approximation, either the total emittance ε_t must be measured, or the spectral emittance $\varepsilon(\lambda)$ weighted with the Planck function $P(\lambda, T)$ must be averaged as follows:

$$\varepsilon_t = \frac{\int_0^\infty \varepsilon(\lambda)P(\lambda, T)d\lambda}{\int_0^\infty P(\lambda, T)d\lambda} \tag{22}$$

where T is the blackbody temperature.

More accurate methods based on numerical solution of the integral equations of radiation heat transfer theory for diffuse surfaces include accounting for temperature nonuniformities. Therefore, the blackbody cavity temperature distribution must be added to the optical parameters as another input parameter.

Monte-Carlo ray tracing enables inclusion of the angular characteristics of the emitted and reflected radiation and the use of a more sophisticated model of optical properties. Such models include an angle-dependent specular-diffuse model as well as one employing the spectral BRDF of the radiator surface as an input parameter.

In the following subsections of Section 4, we shall briefly consider the methods for temperature distribution determination, as well as the measurement methods and measuring devices used for emission and reflection characteristics applied to the materials of blackbody cavities. General reviews of the methods and techniques for measuring the emission and reflection characteristics of materials can be found in Refs. [71–73]. The primary criterion for the application of these methods to high-emissive surfaces is a requirement to determine the optical characteristics within the temperature range of the operation of the blackbody. For blackbody cavities operating in different temperature ranges, different materials and coatings need to be used. For example, graphite is the most widely used material for high-temperature cavities. For blackbodies that operate at moderate temperature, oxidized metals are frequently employed. Low-temperature cavities are usually internally coated with black paints.

4.2. Measurement of cavity temperature distributions

There are two techniques available for the measurement of temperature distributions over the radiating surface of a blackbody. These are radiation and contact thermometry. Radiation thermometry is applicable only for surfaces that can be viewed through the blackbody opening. Often, the temperature distribution over the remainder of the cavity surface cannot be measured in this way. The exception is when light pipe or fiber radiation thermometers are used; however they cannot be employed in all cases. Even though radiation thermometry gives radiance temperatures, which must be converted to thermodynamic temperatures to be used in the calculation of the effective emissivity, for high-temperature blackbodies, radiation thermometry may be the only way to evaluate temperature distributions.

Hartmann et al. [74] used a pyrometer with a 0.65-µm narrow-band interference filter to measure the temperature distribution along the cavity walls of a pyrolytic graphite HTBB radiator in the temperature range from 1,337 to 3,200 K. Only a section of the cavity wall near the bottom could be investigated. The cavity wall radiance temperature was measured with steps of 5 mm. Figure 30 shows the radiance temperature distributions along the part of the cavity wall near the bottom for several cavity temperatures from 1,337 to 3,200 K. A polynomial fitting was performed and the used in

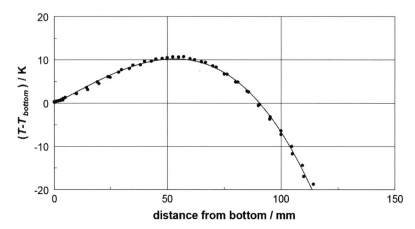

Figure 30 Radiance temperature distribution along the cavity wall at a temperature of 2,300 K. The solid line is a polynomial fitting (adapted with permission from Ref. [75]).

a ray-tracing Monte-Carlo simulation as input for an effective emissivity calculation at various wavelengths. All measured temperature distributions showed a distinct temperature rise with respect to the temperature of the bottom of the cavity, followed by a temperature drop. The width of this temperature bump decreased as the cavity's temperature was increased, whereas its height increased as the cavity's temperature was increased. The measured temperature distributions are axially symmetric. The measured radiance temperatures are another example that blackbodies are never ideal, and show the importance of performing comprehensive characterization of blackbody radiation sources. More details of this work can be found in Chapter 6 in the companion volume, *Radiometric Temperature Measurement: I. Fundamentals*, Vol. 42 of this series.

Hill and Woods [76] measured the temperature distribution along the walls of heat-pipe blackbodies. The temperature distribution along the wall of a water heat pipe was measured with a radiation thermometer at 1.55 μm. The temperature distribution along the wall of a cesium-filled heat pipe at 660°C was measured using the radiation thermometer at 899 nm. The temperature distribution along the wall of the sodium-filled heat pipe at 962°C was measured using a radiation thermometer at 650 nm. The results of the measurements are presented in Figure 31.

Horn and Abdelmessih [77] investigated a graphite cavity heated by passing electric current. The blackbody temperature was variable between 800°C and 2,200°C. The blackbody cavity was a hollow cylinder of 25.7 mm inside diameter. Cavity was 28.9 cm long, with a 5.4 mm thick partition in the middle. The tube wall thickness decreased toward the open ends to compensate for the radiative heat losses. For the temperature

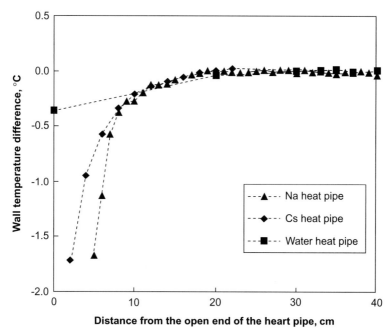

Figure 31 Deviations of the cavity wall radiance temperature from those of the bottom center for the three heat pipe blackbodies (adapted with permission from Ref. [78]).

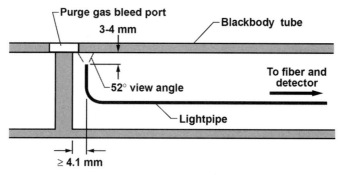

Figure 32 Optical fiber thermometer (OFT) insertion geometry (adapted with permission from Ref. [79]).

distribution measurement, the temperature of 1,100°C was selected because it was the temperature at which the blackbody must be switched from the uninsulated configuration that was used at lower temperatures to the graphite felt and foil insulated configuration that was used at higher temperatures. Axial surface temperature profiles of the blackbody cavity were obtained using an optical fiber thermometer (OFT), with a 90° bend at the tip (see Figure 32). The OFT system included a single crystal sapphire light pipe, a fiber optic cable, and a receiving unit. The OFT system was capable of measuring temperatures from 400°C to 1,900°C with a resolution of 0.4°C. The OFT was mounted on the slide track and crossbar assembly. The axial temperature profiles were obtained by fixing the OFT at several angular locations and then slowly inserting the OFT into the blackbody. The axial temperature distributions for the two configurations are shown in Figure 33. The temperature distribution for the uninsulated configuration was more uniform than that for the insulated configuration. However, this was true only for the target temperature of 1,100°C. For higher temperatures, the temperature distribution for the uninsulated configuration became less uniform while the temperature gradient for the

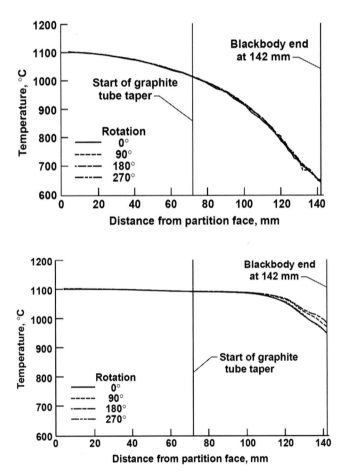

Figure 33 Axial temperature distributions for the uninsulated blackbody configuration (left) and for the insulated blackbody configuration (right) (adapted with permission from Ref. [79]).

insulated configuration remained approximately the same. The measured temperature distributions were used for computation of the effective emissivity of the blackbody.

Contact methods are used mainly for temperature distribution measurements of medium- and low-temperature blackbodies. Two approaches are most common for contact measurements of the temperature distributions of blackbody radiators. The first approach is to scan the radiator's wall using a movable sensor. The second approach is to use multiple sensors placed on (or under) the radiation surface.

Fischer et al. [80] and Hartmann et al. [81] described the double heat pipe large-area blackbody (LABB) that consisted of two coaxial sodium heat pipes. The inner heat pipe forms the radiating cavity. Operational temperatures of this blackbody were up to 1,000°C. The bottom of the inner heat pipe had four bores for SPRTs, which are connected to an automatic thermometry bridge. One of the thermometers was used as the sensor for the fine-control loop. Using a long SPRT, the temperature distribution close to the cylindrical cavity walls was determined for different cavity temperatures by inserting the SPRT into an alumina tube positioned in the cavity and measuring the temperature at different immersion depths. The temperature profiles for four set point temperatures obtained in this way are presented in Figure 34. The immersion depth was measured from the bottom of the cavity. Predicted values were obtained using a simple heat transfer model that takes into account only radiation heat loss of the sensor through the cavity aperture. Initially the cavity was assumed isothermal. The observed monotonic decrease of the cavity temperature toward the cavity aperture could be explained by the simultaneous occurrence of two effects: the nonuniform temperature of the cavity walls and the cooling of the SPRT sensor due to radiation losses through the aperture. To separate the two effects, the axial temperature distribution was measured with opened and closed aperture of the LABB. The significantly smaller temperature drop in the case of the closed aperture identified radiation heat losses through the aperture as the major contribution to the nonisothermal temperature distribution. The radiation heat loss-corrected temperature distributions were used for the Monte-Carlo calculations of the LABB effective emissivity.

4.3. Reflectance measurement of cavity materials

An overview of methods and techniques for reflectance measurements can be found in Ref. [82]. Here, we present some examples of reflectance measurement methods and instrumentation used to characterize black materials and coatings in particular, since that is the information needed for input into modeling codes for blackbody cavities. We will not consider the case of regular (specular) reflection because most of the measurement schemes applicable to diffusely reflecting samples can also be applied to the

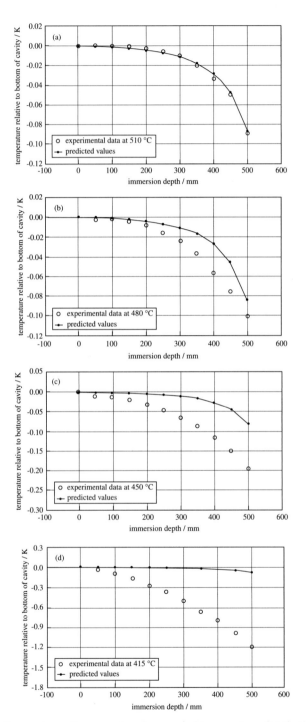

Figure 34 Comparison of measured and computed temperature distributions along the cavity walls (adapted with permission from Ref. [83]).

near-specular samples with moderate or low reflectance. Conventional devices for such measurements are concave mirror concentrators of reflected flux and integrating spheres. Measurements of directional-hemispherical and hemispherical-directional reflectance factor using reflectometers with mirrors shaped by rotation of a conic section around a symmetry axis are considered in Ref. [84].

Betts et al. [85] described the application of a hemispherical mirror to the reflectance measurement of various black coatings for radiometric detectors in the mid- to far-infrared spectral range. These coatings are also suitable for low-temperatures radiation sources. The device described in Ref. [85] is capable of measuring absolute hemispherical reflectance over a wide range of angles. The quantity actually measured is the hemispherical-directional radiance factor, which is numerically equal to the directional-hemispherical reflectance by the Helmholtz reciprocity principle. The directional-hemispherical reflectance can be found as the ratio of the radiance of the sample when it is irradiated diffusely by the radiation reflected from the hemispherical mirror. The scheme of the reflectometer is depicted in Figure 35. The infrared source is a uniformly emitting ceramic tube. Uniform hemispherical irradiation of the sample is achieved by focusing the central portion of this source onto the sample by means of a hemispherical mirror of polished copper. The mirror hemisphere is mounted on gimbals so that the sample can be viewed at any required angle

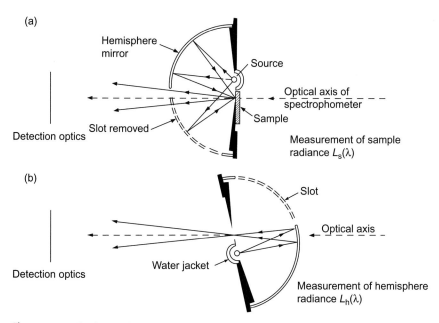

Figure 35 The hemispherical mirror reflectometer (adapted with permission from Ref. [85]).

through a slot in the hemisphere and so that the radiance of most parts of the hemisphere can be measured to test its uniformity. The source and its image on the sample plane are slightly shifted from the center of curvature of the mirror, so that aberrations appear only for large viewing angles. The reading for the incoming radiance of the hemisphere irradiation is obtained by turning the mirror round through nearly 180°. The reflectometer is aligned and clamped in the large sample compartment of a spectro-photometer covering a spectral range from 2.5 to 55 μm (4,000–180 cm^{-1}). The entire apparatus is purged to remove the effects of the H_2O and CO_2 absorption bands with sufficient accuracy in the final radiometric ratios. Two types of spectral reflectance measurements were made on each sample: directional-hemispherical (total) reflectance (including the regular component) and diffuse reflectance (excluding the regular component). The absolute scaling factor for the diffuse component of reflectance was obtained from spectral measurements around 2.5 μm on fresh pressings of $BaSO_4$. The same apparatus was employed for measuring the diffuse and specular reflectances of several black coatings and to study their thermal and solar ageing properties [75]. A Spectralon™ tile served as a reference standard to which the samples were compared.

Persky [78] reviewed measurement data for black surfaces used in spaceborne infrared systems and instrumentation employed to obtain these data. Data presented include the spectral reflectance at wavelengths between 2 and 500 μm for a variety of incident angles from 5° to 80° and angular distributions of the reflected radiation in a plane of incidence at a number of wavelengths between 5 and 300 μm for various incident angles from 0° to 80°.

Persky and Szczesniak [79] studied the spectral directional-hemisphe-rical and spectral hemispherical-directional reflectance of several materials useful for infrared systems and experiments, all using FTIR spectrometers. Three systems were used to obtain the results. One system was an absolute diffuse gold-coated integrating sphere, described by Hanssen and Kaplan in Ref. [14]. The other two devices shown in Figure 36 are commercial instruments, the SOC 400T and SOC 100 reflectometers, manufactured by Surface Optics Corp.

Figure 36(a) shows the setup for measuring the directional-hemi-spherical reflectance. The radiation from a silicon carbide source of an FTIR interferometer after reflections from the mirrors M1 and M2, and the beam splitter, is reflected by a parabolic mirror and a flat mirror M3, and is incident on the sample (or reference) location at 20° ± 10° with a beam width of 5°. The radiation reflected by the sample (or reference) is collected by the compound parabolic concentrators and falls onto the detector. The gold standard for which the value of 0.98 was assumed provides the absolute scale. The chopper reference is periodically inserted into the beam to compensate for a baseline drift. A value for the baseline is obtained by

(a)

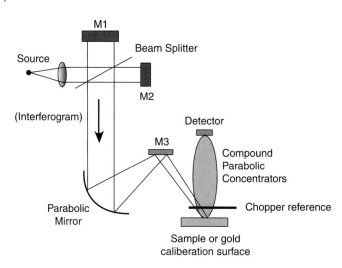

(b)

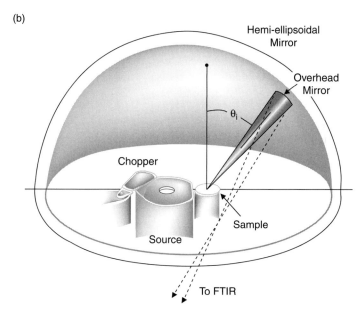

Figure 36 Drawings of the facility with the directional–hemispherical reflectometer (a) and of the hemispherical–directional reflectometer (b) (adapted with permission from Ref. [79]).

allowing the modulated beam to completely exit the instrument through the port, where the sample or reference is otherwise located.

The second device, depicted in Figure 36(b) is intended to measure the hemispherical-directional reflectance factor which is equal to the directional-hemispherical reflectance by reciprocity. The 2π imaging hemiellipsoid diffusely irradiates the sample using a 700°C blackbody. The source is located at one focus and the sample at the other. After reflection from the sample, the beam is directed to an FTIR spectrometer using an overhead mirror at given angles. The beam width is 1°. There are linear polarizers at the entrance to the interferometer and a blocker to provide separation of the reflectance into diffuse and specular components. Calibration is by reference to a 0.98 reflectance gold sample used for both unpolarized and polarized measurements. All the materials were measured in the spectral range of $2-26\,\mu$m, at ambient temperature; in addition, the chrome oxide ceramic was measured at 486 K, and the Pyromark® 2500 paint at four temperatures to 877 K.

Hameury et al. [86] described an infrared reflectometer to measure the spectral directional-hemispherical reflectance of solid materials at ambient temperature. The reflectance can be measured from 0.8 to 14 μm in five directions with an angle of 12°, 24°, 36°, 48°, and 60° with respect to the normal to the surface of the sample. The optical arrangement to collect the reflected flux is based on a hemispherical mirror. In fact, four mirrors cut in a hemisphere are used to collect the flux reflected by the sample. This optical arrangement was chosen to limit the angle of incidence of rays on the detector (38° instead of 90° for the usual hemispherical mirror arrangement). Figure 37 shows a general schematic of the apparatus. Two sources are used for the generation of the incident beam: a lamp for short wavelengths ($\lambda<3\,\mu$m) and a blackbody for longer wavelengths. A monochromator with three gratings or a set of interference filters are used to select the spectral band (between 0.8 and 14 μm). The radiation is modulated by a chopper before the monochromator or the filter. The field stop determines the size of the spot on the specimen. A rotating flat mirror and a set of five flat stationary mirrors are used to focus the incident beam on the surface of the sample with one of the five possible angles of incidence (12°, 24°, 36°, 48°, 60°).

The principle used to collect the reflected flux is based on a set of four mirrors with spherical surfaces instead of a simple hemispherical mirror. The four mirrors are "cut" in a hemisphere, with each of these four mirrors collecting the radiation reflected in a quarter of the half space and focuses it at a point located in the same plane as the surface of the sample. The directional-hemispherical reflectances of several samples including on the black paint Nextel Velvet Coating 811-21 in the spectral range from 4 to 14 μm for the incidence angles of 0° and 12° are presented.

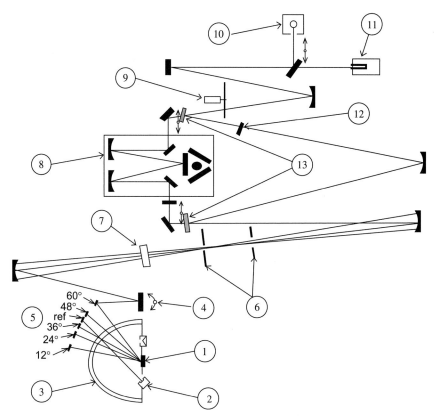

Figure 37 Schematic of the apparatus: (1) sample; (2) detector; (3) set of four mirrors with spherical surface; (4) rotating plane mirror; (5) fixed plane mirrors for the selection of the angle of incidence; (6) field stop and aperture limiting stop; (7) polarizer; (8) grating monochromator; (9) mechanical chopper; (10) lamp source; (11) black-body source; (12) interference filter; (13) flat mobile mirrors for the selection of the monochromator or the interference filters (adapted with permission from Ref. [86]).

There are many publications (see, for instance, Refs. [83,87−89]) on the application of an integrating sphere to reflectance measurements in the visible and near-infrared spectral range. The integrating sphere has become the routine measurement device for diffuse reflectance, since the latter half of the 20th century. The history of the integrating spheres for reflectance measurement in the infrared spectral region beyond 5 μm began with the development of the first rough-gold-coated integrating sphere by Willey [90]. Hanssen and Snail [91] described the design and operating principles of integrating spheres for mid- and near-infrared reflection spectroscopy. In Ref. [14], Hanssen and Kaplan described infrared diffuse reflectance instrumentation and standards. The availability of a well-defined reflectance standard is critical to the establishment of reflectance measurement uniformity across the user community.

Hanssen [92] described the integrating-sphere system with an FTIR spectrometer and a method for the absolute spectral reflectance measurement of specular samples that could be used as reflectance standards for the infrared spectral range from 2 to 18 μm. It was shown that the expanded relative uncertainty ($k = 2$) of the specular reflectance measurements was <0.003.

Hanssen et al. [13], with the help of the integrating sphere system that was described in Ref. [14], measured flat graphite samples with varying degrees of surface roughness. Graphite is a material widely used for blackbody radiators at medium and high temperatures. Four samples of graphite were measured. Three samples were 38 mm in diameter, partially polished, and roughened with sandpaper. The fourth sample was the flat backside of the V-grooved bottom (15 mm diameter) of the fixed-point blackbody graphite crucible. Near normal (8°) directional-hemispherical reflectance spectra of these four samples are shown in Figure 38, with the cavity sample labeled "glossy" in the figure. The measurements were performed at the FTIR spectrometer facility [14]. Error bars indicate the

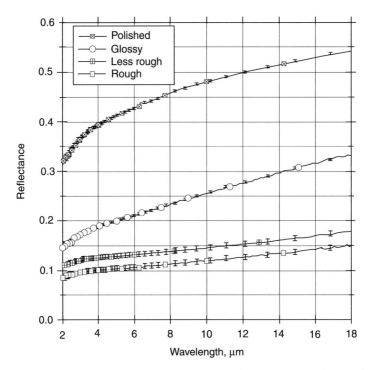

Figure 38 Reflectance spectra of four graphite samples with varying levels of surface roughness (adapted with permission from Ref. [13]).

expanded ($k = 2$) uncertainties. These results show a potential difficulty in predicting the effective emissivity of a graphite cavity: the individual characterization of graphite samples is necessary; moreover, the sample should have the same surface finish as the cavity internal surface. At high temperatures, oxidation may take place, causing the formation of particles that may enhance the emissivity as well as alter its specularity.

Hanssen et al. [27,93] and Cagran et al. [94] described an integrating sphere for the hemispherical-directional reflectance factor measurement at high temperature. The measurements were conducted at near-normal incidence (8°) by comparing with a calibrated reference standard. For accurate relative measurements, the sphere reflectometer employed an isotropic sphere design concept [95], which ensures that sample and reference reflected light are equally treated despite possible differences in their angular distributions of reflected radiation. The cross-sectional view of the integrating sphere reflectometer is shown in Figure 39(a). The sphere had a diameter of 10 in and was coated with a high reflectance sintered PTFE shell (except for the sample/source insert shown in Figure 39(b) which was coated with $BaSO_4$). The assembled integrating sphere is shown in Figure 39(c). A diode laser or a halogen lamp with a gold reflector was coupled to a diffuser located at the sphere wall between the sample and the reference by means of a fiber bundle. The two baffles between the light source and the sample and reference avoided direct illumination of the samples from the source. During a measurement, the directional flux at 8° reflected off of the sample was compared to that of a calibrated reference standard, for which a set of reference samples (with both high and low reflectance, as well as both specular and diffuse) was calibrated in the $0.6 - 2.5\,\mu m$ range. A gold-coated flip mirror in the interface optics is used to switch between the sample and the reference. Two different sample heaters were built: a Cs heat-pipe heater

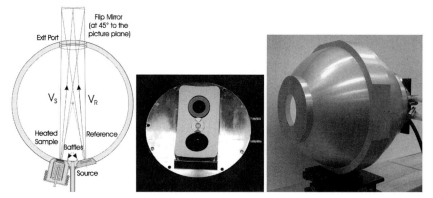

Figure 39 Integrating sphere reflectometer: (a) integrating sphere schematic; (b) interchangeable water cooled sample/source insert that matches the sphere wall when assembled as seen in (c) (adapted with permission from Ref. [27]).

and a Na heat-pipe heater. Each of the small heat pipes was temperature controlled with a center-mount Pt resistance thermometer and has a water-cooled base. They were built to hold opaque, disc-shaped samples up to 19 mm (3/4 in.) in diameter, and the entire unit was designed to fit the sphere reflectometer for temperature dependent reflectance measurements.

"Flat-plate" blackbodies are simply flat surfaces (sometimes extended by grooves, spikes, etc.) with highly emissive coatings. They frequently are too large to be measured with an integrating sphere or other conventional measuring device. In these cases, the "hot plate" method proposed by Ballico [21] and described in Section 2.2 can be a suitable choice. In Ref. [22], the hardware implementation of a somewhat modified Ballico's method ("controlled background plate method") as well as the results of reflectance measurements (and hence emittance indirectly determined) are presented and analyzed. The measurements were under-taken with a 200×280 mm temperature-controlled plate (see also Section 3) and a 50×50 mm sample, both coated with the same diffuse black paint. Figure 40 summarizes the results of the measurements, including the reflectometric measurement performed using the integrating sphere with the FTIR spectrometer (marked as "Reflectometry"), the

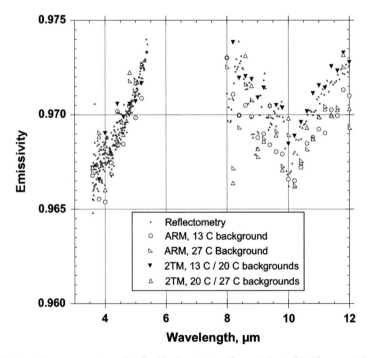

Figure 40 Measurement results for black paint surfaces (adapted with permission from Ref. [22]).

absolute radiance method (marked as "ARM") and the two background temperatures (marked as "2TM") method (see Section 3 for details). The results demonstrated good agreement of all three methods for the black paint surface. The components of uncertainty of the comparison already established include: (i) the expanded ($k = 2$) uncertainty of the reflectometry measurements, which were 0.001 for the black paint; (ii) the variation of the coupon paint emittance values of approximately 0.002; and (iii) the nonuniformity of the target plate emissivity of approximately 0.001. Combined, these factors could easily account for the differences seen in the comparison results. The results obtained validated the independent methods and demonstrated the potential of the controlled background method for measurements of the radiative properties of infrared materials.

4.4. BRDF measurements of cavity materials

The BRDF is the most general characteristic of reflective materials. BRDF was introduced by Nicodemus et al. [96]; it is a function of five variables:

$$f(\lambda, \theta_i, \phi_i, \theta_r, \phi_r) = \frac{dL_{\lambda,r}(\lambda, \theta_r, \phi_r)}{dE_{\lambda,i}(\lambda, \theta_i, \phi_i)} \tag{23}$$

where $L_{\lambda,r}$ is the spectral radiance of the reflected radiation (in $W/(m^3 sr)$), $E_{\lambda,i}$ the spectral irradiance (in W/m^3) created by the incident radiation, λ the wavelength, θ_i, ϕ_i, and θ_r, ϕ_r are the spherical coordinates, which define the directions of incidence and observation, respectively.

BRDF has the dimension of sr^{-1} and describes the angular distribution of reflected radiation for all possible directions of the incident radiation. The usual method of BRDF measurements is to scan the reflected radiation field using a detector collecting the reflected radiation within a very narrow solid angle. The optical energy contained in such a solid angle is a very small portion of the entire energy reflected into the hemisphere. Therefore BRDF measurements, especially of low-reflective (and, consequently, high-emissive) materials, require a strong source of collimated radiation in a given spectral range. Lasers are such sources. In a paper dedicated to the measurements of reflection properties of black coatings for radiometric detectors, Betts et al. [85] described an application of the polar scattering facility at Sira, Ltd. for investigation of the angular reflection properties of coatings at a wavelength of 10.6 μm of the radiation of a stabilized CO_2 laser source. The laser radiation was chopped at a frequency near 10 Hz, coinciding with the minimum noise equivalent power (NEP) of the pyroelectric detector used. After the chopper, the laser radiation was attenuated by several removable plane-parallel discs of calcium fluoride, each 3 mm in thickness and about 40% in transmittance. The collimated beam from the laser was then directed onto the sample and the reflected flux measured as a function of angular position by a 2 mm square pyroelectric detector, which was mounted on an optical bench rotating about an axis through the sample. The detector was well baffled to ensure that only scattered radiation from the sample reached it. The device described allowed measurement of the BRDF only in the plane of beam incidence.

However, reflection from real materials is not constrained to the incident plane and frequently it is necessary to measure reflection throughout the full hemisphere. In addition, the dependence of reflected flux on polarization of the incident and received radiation may be required. The design of a full hemisphere goniometer operating at five wavelengths of laser sources (0.325, 0.6328, 1.06, 3.39, and 10.6 µm), with polarization registration is described in Ref. [97].

Drolen [98] described another facility used for BRDF measurements. Four lasers are used for these measurements: argon ion (0.4545−0.5145 µm, circularly polarized), Nd:YAG (1.06 µm, randomly polarized), HeNe (3.39 µm, vertically polarized), and CO_2 (9.2−11.0 µm, circularly polarized). The beam diameter was about 5 mm to obtain an irradiated spot size < 2.54 cm at near-grazing incidence. A chopper and reference detector were included in the beam path to continuously monitor and correct for drift in the source output power. The solid angles of incidence and reflection were identical 5×10^{-5} sr. The facility allowed investigation of both the in-plane and out-of-plane BRDF.

Ford et al. [99] described a system based on a FTIR spectrometer (see Figure 41) and developed to perform in-plane BRDF measurements simultaneously for multiple infrared wavelengths. The incidence and reflection angles were selected independently by two planar mirrors. The reflectometer optics was symmetric: it consisted of two ellipsoidal mirrors having one joint focal point that coincided with the center of the sample irradiated area. Measurements could be performed for vertical or horizontal polarization states of both incident and reflected radiation.

Hünerhoff et al. [100] described a state-of-the-art robotized gonioreflectometer facility designed at PTB. The gonioreflectometer system consisted of three major parts: a large rotation stage carrying the homogeneous spherical radiator with an internal 250 W quartz tungsten halogen lamp, the five-axis industrial robot as a holder for the sample under test and a monochromator with a mirror-based imaging system for the spectrally selected detection of the reflected radiance of the sample. A spherical radiator was located on a rotation stage with a diameter of 1.5 m and could be rotated 360° around the five-axis robot serving as the sample holder. The main part of the gonioreflectometer was the small five-axis industrial robot. The flexibility of the robot arm provided the ability to measure out-of-plane BRDF with arbitrary irradiating and detection angles. The direction of the detection path was fixed due to the fact that a triple grating half-meter monochromator for spectral selection of the

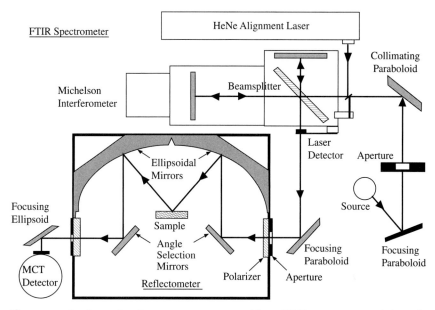

Figure 41 A schematic of measurement setup with an FTIR spectrometer (adapted with permission from Ref. [99]).

reflected radiation was used. The wavelength range for measurements of diffuse reflection was 250–1,700 nm.

Four different detectors behind the monochromator were used for detecting the radiance signals of the incident and reflected beams: a solar blind channel photomultiplier for the measurements between 250 and 350 nm, a yellow enhanced channel photomultiplier between 300 and 450 nm, a silicon photodiode between 400 and 1,100 nm, and a cooled InGaAs photodiode (at −25°C) between 1,000 and 1,700 nm.

The spectral tri-function automated reference reflectometer (STARR) developed at NIST is described in Ref. [101]. This instrument was capable of measuring specular and diffuse reflectance, BRDF of diffuse samples, and both diffuse and nondiffuse transmittance over the ultraviolet, visible, and near-infrared (UV–Vis–NIR) wavelength range (200–2,500 nm).

STARR is the national reference instrument for spectral reflectance measurements of spectrally neutral, nonfluorescent samples at room temperature. Additional information about STARR facility can be found in the NIST Web site.

Usually, BRDF measurements, especially of a complete hemisphere, are time-expensive. The sample spectral irradiance produced by the radiation source must be constant during the entire measurement cycle as well as all other affecting parameters. Otherwise, the uncertainty due to drift can decrease the measurement accuracy significantly. To avoid this uncertainty and reduce the time of measurements, it is possible to use multiple detectors to register the spectral radiance of reflected radiation, or to map the hemispherical distribution of the spectral radiance onto a plane using an optical system and then to register this planar distribution with a sensor array.

The first approach was employed in the prototype design of the Multidetector Hemispherical Polarized Optical Scattering Instrument (MHPOSI) [102]. It employed 28 detectors, each sensitive to only p-polarization, to capture the BRDF over the most of the scattering hemisphere at once. These detectors enabled a partial mapping of the polarimetric properties of samples. An argon–ion laser (35 mW, 488 nm) irradiated the sample at a fixed incident angle and had adjustable polarization. Applications of this instrument included the inspection of optical quality surfaces such as optics, silicon wafers, and magnetic disks, and the analysis of specific types of defects on these surfaces.

An alternative approach was used by Dana and Wang [103] with a bidirectional imaging device (see a simplified scheme in Figure 42) that employed a concave parabolic mirror and a CCD camera for registration of a part of radiation reflected by a sample then by a parabolic mirror. The same mirror was used to direct the incident illumination ray to the sample at the desired angle. Each pixel of a camera image corresponded to a different viewing direction. A beam splitter was used to direct the incident beam parallel to the parabolic mirror axis without occultation of the reflected radiation.

Ghosh et al. [104] proposed a more sophisticated device, which contained a convex parabolic and a concave ellipsoidal mirror. However, the application of imaging bidirectional reflectometers are limited to the visible and near-infrared spectral ranges since there are very sensitive and uniform large-area sensor arrays available for these spectral regions. In the

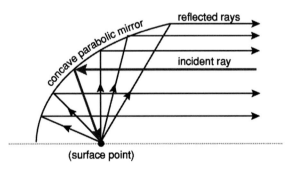

Figure 42 The focusing property of a concave parabolic mirror is exploited to simultaneously measure reflected rays from a large range of angles over the hemisphere.

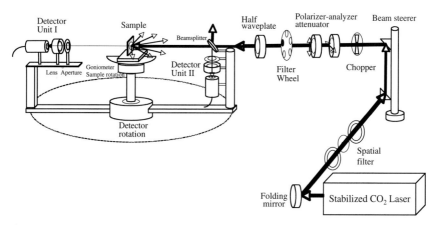

Figure 43 A schematic of the bidirectional reflectance distribution function (BRDF) instrument (adapted with permission from Ref. [103]).

infrared range, mechanical scanning remains the dominant method for BRDF measurement.

Zeng and Hanssen [105] described an infrared optical scattering instrument for BRDF measurement that has been developed at NIST. The instrument described has out-of-plane and retro-reflection capabilities. The BRDF instrument as shown in Figure 43 consisted of a stabilized CO_2 laser (10.6 µm), beam manipulation and control, a motorized sample manipulation system which includes rotation, goniometer, and translation stages, motorized detector positioning stages, and data acquisition and motion control electronics and software. Two detection geometries were employed as shown in Figure 43. The simpler Detector Unit I consisted of a defining aperture, focal lens, and detector. It was limited to measurements outside $\pm 5°$ of the incident beam. The vertical geometry of Detector Unit II enabled performing BRDF measurements over the completed angular range including the retro-reflection region not accessible with Unit I. Unit II was composed of similar components to Unit I, with optional substitution of a beam splitter for the mirror. The edge limitation of the beam splitter was around $\pm 4°$.

Example performance of the instrument is illustrated by BRDF measurements at 10.6 µm for a graphite radiator with a flat back surface and a front surface with five-concentric V-grooves. The measurements were performed for s- and p-states of polarization. In Figure 44, the BRDF of the flat graphite surface (left) presents a sharp specular peak ($\pm 2°$) at its center, with symmetrical sloped shoulders on either side. The BRDF of a V-grooved surface (right) shows a broader peak ($\pm 10°$) at its center with symmetrical flat shoulders on either side.

The BRDF instrument described in Ref. [103] was later improved with the addition of laser sources with wavelengths at 1.32 and 1.55 µm.

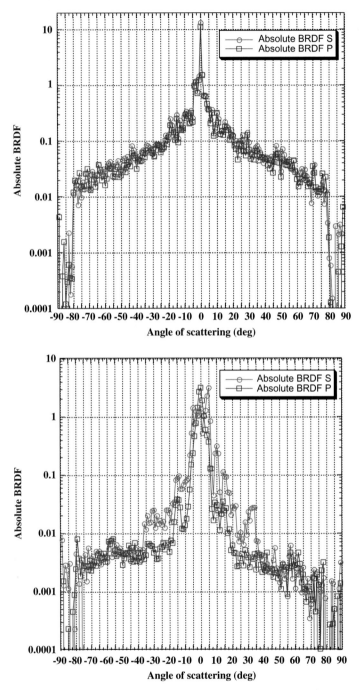

Figure 44 BRDF results of graphite flat (top) and V-grooved surface (bottom) at normal incidence (adapted with permission from Ref. [103]).

Figure 45 presents the in-plane BRDF of graphite sample (flat back of V-groove bottom section of the blackbody cavity) at (a) 1.32 μm and (b) 10.6 μm, for different incidence angles. Each curve was calculated as a mean of two mutually perpendicular polarization states. The scattering angle was defined with respect to the normal to the sample. No data were available within 10° of the incidence angle. The graphite sample was reasonably diffuse at the short wavelength, but became much more specular and had a higher level of reflectance at the longer infrared wavelength. This difference had a dramatic effect on the effective emissivity of a graphite cavity.

Shen et al. [106] described a three-axis automated scatterometer (TAAS) that was intended for measurements of the BRDF of rough surfaces, had extremely high angular resolution ($0.001-0.005°$ for three axes), and allowed polarization-dependent measurements. The schematic diagram of TAAS apparatus is shown in Figure 46. A temperature-stabilized, fiber-coupled diode laser system provided coherent radiation with output wavelengths of 635, 785, and 1,550 nm. The power at the output of the fiber was about 5 mW. The beam divergence was <0.22 mrad (0.0126°). Two highly linear photodiode detectors of the same type were used for scanning the reflected radiation field and to monitor the incident power through a cubic beam splitter next to the polarizer. The very narrow field-of-view of the scatterometer (the solid angle is 1.84×10^{-4} sr, the half-cone angle is 0.45°) reduced the stray light in comparison with the wide-angle systems. A lock-in amplifier sent reference signals to the laser diode controller so that outputs of two detectors were phase-locked to these signals to eliminate the effect of variations of background radiation. The comparison of measurements performed using TAAS with those obtained using the STARR instruments [101] showed the excellent agreement. However, due to its higher angular resolution, the TAAS had the advantage in the investigation of narrow specular lobes. Application of the TAAS apparatus to the BRDF measurement of low-reflectance coatings fabricated from vertically aligned multiwalled carbon nanotubes is described in Ref. [107]. Such high-technology coatings are suitable for thermal detectors and emitters due to their low reflectance and weak spectral selectivity.

4.5. Spectral emittance measurement of cavity materials

In this subsection, we shall consider the direct method of spectral emittance measurements, that is, measurements, which are conducted according to the definition of spectral emittance as the ratio of the sample spectral radiance at a given temperature, wavelength, and direction to the spectral radiance of a perfect blackbody at the same temperature and wavelength. Experimental measurements must take into account that the emittance of a material depends on temperature, wavelength, and viewing angle.

(a)

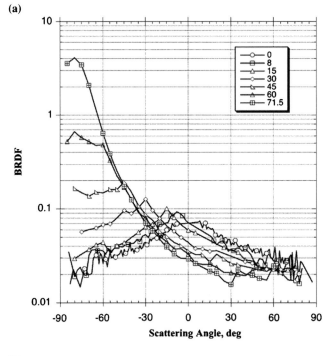

(b)

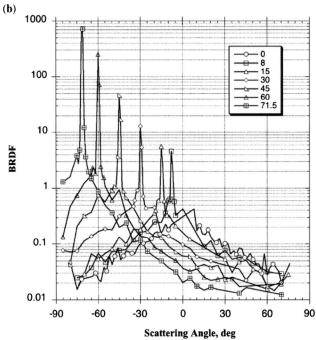

Figure 45 BRDF of a graphite flat surface at $1.55\,\mu m$ (top) and $10.6\,\mu m$ (bottom) for different incidence angles. Legend shows the incidence angle, in degrees, onto the sample for each curve (adapted with permission from Ref. [13]).

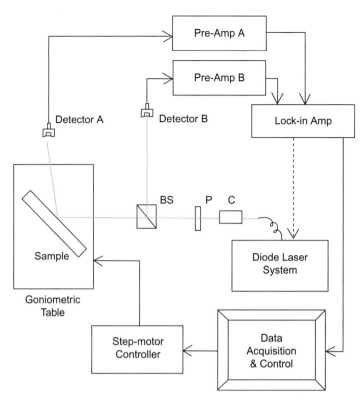

Figure 46 Experimental arrangement of the three–axis automated scatterometer (TAAS). BS, beam splitter; P, polarizer; and C, collimation lens (adapted with permission from Ref. [106]).

Moreover, the emittance also depends on the surface roughness, oxidation state, contamination, etc. Therefore, emittance data published for some material may differ significantly from a specific sample under test.

The direct method of spectral emittance measurements can vary in the manner in which spectral data is obtained, in sample heating or cooling, and in the sample temperature measurements. An FTIR spectrometer is the most suitable modern instrument for spectral emittance measurements in the infrared spectral range. Clausen [108] described the theoretical basis of such measurements, an experimental setup for their implementation, and results obtained for some materials that are used for infrared blackbody sources. The radiance observed by the FTIR was given using a complex formulation:

$$f'(v, T) = R(v)e^{i\phi(v)}[\varepsilon(v, T)P(v, T)E + I(v)\rho(v, T)E + G(v, T_1)E_I e^{i\psi(v)}] \quad (24)$$

where f' is obtained from the measured interferogram as described in Ref. [109], $R(v)$ and $\phi(v)$ are the spectral response function and the phase response of the spectrometer at wave number v, respectively, E the throughput, $\varepsilon(v, T)$ the spectral emittance of the sample at temperature T, $P(v, T)$ the radiance from a blackbody at the same temperature, $I(v)$ the incoming ambient radiation, $\rho(v, T) = 1 - \varepsilon(v, T)$ is the spectral reflectance of the sample, $\psi(v)$ the phase response for internal instrument radiation, and $G(v, T_I)$ is the offset from instrument emission, referred to the input. The offset term $G(v, T_I)E_I e^{i\psi(v)}$ is eliminated from Equation (1) by subtraction of a measurement of the sample at ambient before the Fourier transformation and phase correction are performed. Formally, Equation (24) is valid if incident radiation is independent of angle or for a perfectly diffuse sample. The ambient radiation is divided into two terms: one with radiation from a blackbody at ambient temperature T_a and one describing the deviation of the incoming radiation from blackbody radiation:

$$I(v) = P(v, T_a) + O(v) \tag{25}$$

By subtraction of a measurement of the sample at ambient temperature from Equation (24) and insertion into Equation (25) one can obtain:

$$\begin{aligned} f'(v, T) - f'(v, T_a) = R(v)e^{i\phi(v)}E\{\varepsilon(v, T)[P(v, T) - P(v, T_a)] \\ + O(v)[\varepsilon(v, T) - \varepsilon(v, T_a)]\} \end{aligned} \tag{26}$$

The last term vanishes when the incoming radiation from the source can be described adequately as blackbody radiation or when the emissivity of the source is almost identical at the two temperatures.

$$f'(v, T) - f'(v, T_a) = ER^*(v, T)e^{\phi(v)}[P(v, T) - P(v, T_a)] \tag{27}$$

where $R^*(v, T) = R(v)\varepsilon(v, T)$.

Equation (27) is the usual basic equation for calibration of an FTIR spectrometer in order to determine its response function as well as for measurement of the emittance of a sample. The transmittance of the atmosphere may be included for defining the apparent response function R^*. The phase term $e^{\phi(v)}$ in Equation (27) disappears after phase correction of the measurement. The spectral emittance of a sample can be found from $\varepsilon(v, T) = R^*(v, T)/R(v)$.

The method described above was implemented in the measurement setup that was based on a commercial FTIR spectrometer with a pyroelectric detector. The optical components that collimated radiation from the sample were arranged on a small platform mounted on the spectrometer in front of the emission port. Radiation was collimated with two off-axis paraboloidal concave mirrors. The first mirror collimated radiation into the spectrometer. An iris placed in the focus of the first mirror limited the 22 mrad half-angle field-of-view of the spectrometer. The image of the sample was focused by the second mirror on the iris. The field of view was selected within a circle of 3−15 mm diameter at the sample surface. The interferometer and top compartment of the spectrometer was purged but the path between the spectrometer and sample over a distance of about 820 mm was not purged.

The infrared spectral emittance characterization facility developed at NIST has been described by Hanssen et al. [110]. Hanssen and Cagran [93] and Cagran et al. [94] used the near-infrared integrating sphere and filter radiometers, described in Section 4.3, to measure the sample temperature for the direct measurement of spectral emittance of samples in the system shown in Figure 47. The spectral emittance measurements involved only the sample, blackbody reference sources, and the FTIR spectrometer. After determination of the sample and blackbody temperatures, the relative spectral radiance was calculated from the results of three measurements: the sample, and two reference blackbodies at different temperatures. The temperature of one blackbody was chosen to match the sample temperature as closely as possible whereas the second one was close to room temperature. This second blackbody measurement was used to subtract out the FTIR spectrometer' self-emitted radiance component. The spectral radiance of the sample was

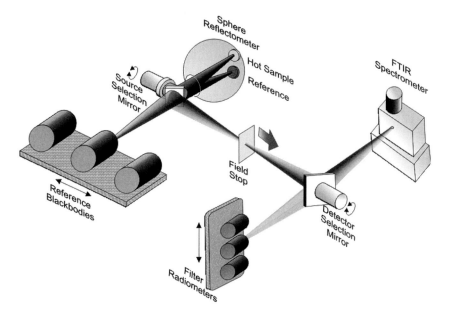

Figure 47 A schematic of the infrared spectral emittance characterization facility. Radiation from blackbody reference sources (to the left) is compared with that from a sample (located opposite the blackbodies) via rotation of a flat source selection mirror. The radiation proceeds through a set of interface optics and is measured either with a filter radiometer or with a FTIR spectrometer, selected by means of the elliptical detector selection mirror. For the temperature-dependent emittance measurement of opaque samples, a sphere reflectometer with a calibrated reference sample is shown (adapted with permission from Ref. [110]).

given by the real part $\Re$ of the complex FT spectra by:

$$L_C(v) = \Re\left\{\frac{(V_C(v) - V_B(v))}{(V_A(v) - V_B(v))}\right\} \cdot (L_A(v) - L_B(v)) + L_B(v) \qquad (28)$$

where v is the wave number, $L(v)$ the spectral radiance, $V(v)$ the FTIR spectrometer measured complex spectrum, A represents the higher temperature blackbody, B represents the ambient temperature blackbody, and C denotes the sample. The spectral radiance of the blackbodies was given by Planck's radiation law and the emissivity as obtained from calculation or comparison with a fixed-point cavity. After obtaining the sample spectral radiance, the unknown spectral emittance of the sample was obtained from L_C, ratioing to Planck and correcting for the ambient radiation reflected off the sample. The measurements were performed in repeated cycles of an A−B−C−B−A sequence to reduce the effects of drift. Depending on the level of emitted radiation from the sample at different temperatures, the number of measurement cycles was adjusted to optimize the signal-to-noise ratio. In Ref. [94], results of infrared spectral emittance were shown for candidate emittance standard materials: SiC and Pt−10%Rh over a temperature range of 300−900°C. A commercial large-area flat-plate source (with a built-in thermometer) with a nominal emittance of 0.95 was measured at the 500°C setting. The measured apparent emissivity (i.e., not corrected for reflected background radiation) as shown in Figure 48 is seen to vary substantially from the nominal value depending on wavelength.

Hanssen et al. [110] presented a Hohlraum device as a complement to the system described above. The device, the design of which is shown in Figure 49, was a blackened temperature-controlled sphere designed to provide a well-defined uniform radiance to the sample. The sample mounted at the center of a sphere could be rotated for angle-dependent measurements, and translated in and out for viewing the sphere wall directly. A second rotation of the sample mount could switch between sample and a calibrated reference. The Hohlraum device could be used for spectral emittance measurements using both direct and indirect methods. The sphere would provide radiation that uniformly irradiates the sample from all directions, except the viewing port. The sphere wall would be grooved to enhance its effective emissivity and coated with high-emittance paint. Motorized stages could change the viewing angle of the sample, view the enclosure wall, and view a reference sample. A spectrometer would view radiation leaving the sample that is a combination of reflected and emitted light. The sample emittance could be obtained in two ways with this system: through a pair of measurements with two different sample temperatures, or with two different enclosure temperatures. Then the difference between the two measurements would be calculated leaving

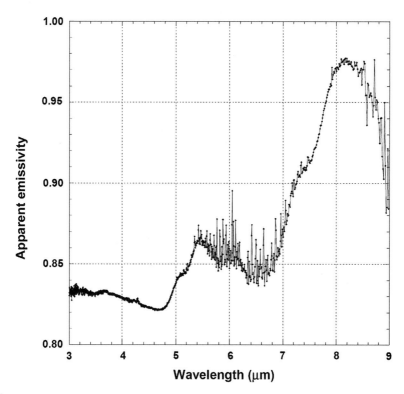

Figure 48 Apparent emissivity of a commercial flat-plate source with a nominal emissivity of 0.95, for a temperature setting of 500°C (adapted with permission from Ref. [27]).

only the sample-emitted radiation or the sample-reflected radiation, respectively. From either result, the sample emittance could be determined.

Del Campo et al. [111] described a computer controlled, fully automated vacuum setup which allows measurement of the spectral directional emittance of opaque samples as a function of temperature and environmental conditions by the direct radiometric method, using an FTIR spectrometer. It had the advantage of having all the optical paths for the sample and blackbody radiation in vacuum or a controlled atmosphere in order to prevent H_2O and CO_2 absorption as well as to assure better thermal stability. In addition, heating the sample in a controlled atmosphere enabled *in situ* emissivity studies during oxidation processes or, in general, to analyze its dependence on environmental conditions. Angular measurements could be obtained for a continuous change of the viewing angle. In addition, there was no need to move the sample and blackbody in order to measure their signals. The directional spectral emissivity could be measured for angles of viewing from 0° to 80°. The measurements could be carried out in a controlled atmosphere for pressures from 30 to 101.3 kPa.

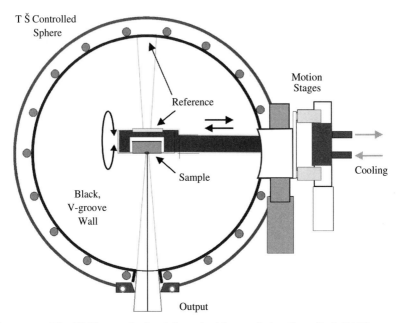

Figure 49 The Hohlraum device (adapted with permission from Ref. [110]).

The experimental device is shown in Figure 50 and has four modules: a sample chamber, a blackbody, an FTIR spectrometer, and an optical entrance box that guides the radiation coming from the sample and the blackbody toward the spectrometer.

The spectral emittance of a sample under test was calculated with the use of the interferograms I_S, I_{BB}, and I_{sh}, which were obtained using the FTIR spectrometer when the radiation from a sample, a blackbody, and a shutter, respectively, were measured. Additionally, the temperatures of the above-listed components T_S, T_{BB}, and T_{sh} as well as the spectral emittance ε_{sur} and the temperature T_{sur} of surrounding were determined. The spectral emittance of the surrounding was considered to be independent of wavelength and temperature and close to unity, so that the interreflections were neglected. Since the shutter and the chamber inner surface, which were considered as part of surroundings, were coated with the same black coating, for the sample spectral emittance the following expression was obtained:

$$\varepsilon_S(\lambda, T_S) = \frac{\mathrm{FT}^{-1}(I_S - I_{sh})}{\mathrm{FT}^{-1}(I_{BB} - I_{sh})} \frac{P(\lambda, T_{BB}) - P(\lambda, T_{sh})}{P(\lambda, T_S) - \varepsilon_{sur}P(\lambda, T_{sur})} + \frac{\varepsilon_{sur}P(\lambda, T_{sh}) - \varepsilon_{sur}P(\lambda, T_{sur})}{P(\lambda, T_S) - \varepsilon_{sur}P(\lambda, T_{sur})}$$

(29)

where P is the Planck function and FT^{-1} represents the inverse Fourier transform.

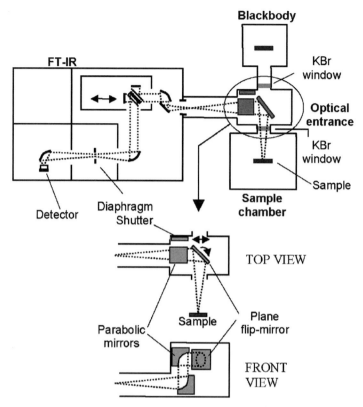

Figure 50 A schematic view of the emissivity measurement setup (adapted with permission from Ref. [111]).

As an example, the directional emissivity for the specimen of Armco iron at 740 K as a function of the emission angle up to 80° is shown in Figure 51 for three wavelengths. The emissivity remained nearly constant up to an angle of about 60°, and then started to increase.

Monte et al. [68] described a reduced-background facility with a capability for spectral emittance measurements. Figure 52 presents a transparent view of this facility to show the beam path for directional spectral emissivity measurements. By using an off-axis ellipsoidal mirror that could be moved to the detector position on the optical axes of the facility, the radiation from various sources could be also be imaged onto a vacuum FTIR to allow for spectrally resolved measurements of both blackbodies and emissivity samples. A dedicated sample holder for emissivity measurements under vacuum could be controlled in the range from 0°C to 430°C. The heated sample was located inside a spherical enclosure of 250-mm diameter made of copper that could be temperature controlled from about

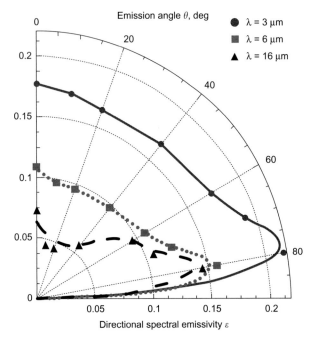

Figure 51 Directional spectral emissivity for the Armco iron specimen at 740 K, at three wavelengths (adapted with permission from Ref. [111]).

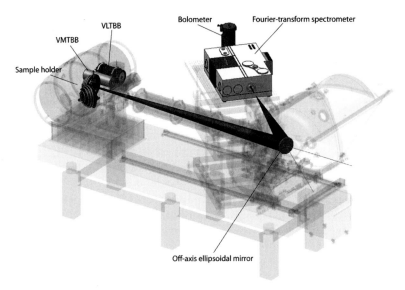

Figure 52 Transparent view of the reduced-background facility to illustrate the position of the blackbodies and the sample holder, as well as the beam path, for spectrally resolved measurements. Radiation from either of the two blackbodies or the sample holder is imaged via the off-axis ellipsoidal mirror into the FTIR spectrometer (adapted with permission from Ref. [68]).

liquid nitrogen to room temperature and as such provides a known thermal environment. The inner surface of the sphere was covered by circular grooves (60°) and coated with black Nextel 811-21. The sample could be rotated within the sample holder by a DC motor-driven rotation. The useful angular range for emissivity measurements was about ±75°.

5. CONCLUSIONS

In this chapter, the methods and measuring techniques for the experimental characterization of blackbody radiation sources were reviewed, with emphasis on recent experimental work. A review of the most recent 20 years has revealed the following major trends: (1) extensive use of both laser-based and broadband reflectometers to evaluate and monitor cavity emissivity, including built-in ones for spaceborne applications; (2) use of multiple and complementary techniques for blackbody characterization, including spectral comparison of different blackbodies; (3) proliferation of detector-based techniques, which are presently dominating in the visible spectrum and are being extended to the near- and long-wave infrared regions; (4) construction of several dedicated facilities for blackbody characterization; (5) the ubiquitous use of Monte-Carlo modeling techniques in nonisothermal and nondiffuse approximations, along with understanding of the need for validation of these results; and, finally, (6) an emerging trend to acquire variable angle reflectance and BRDF parameters of materials and coatings and use them for the prediction of blackbody performance.

The material reviewed here reflects a multitude of methods and experimental approaches, which are being used for the characterization of the nonideality of real-life blackbody sources, as well as the increasing attention being paid to this problem by the international community as the accuracy requirements continue to increase and the number of commercially available products continues to grow. At the same time, it is difficult not to notice the absence of a definitive and internationally accepted set of recommendations for commercial blackbody specification and evaluation procedures. This has led to the presence of commercial blackbodies with unclear and contradictory manufacturer specifications.

It is hoped that this review will facilitate further development of this area of optical radiation metrology and offer some practical help in the navigation of the multitude of available publications on this subject matter.

REFERENCES

[1]   R. E. Bedford, "Calculation of Effective Emissivities of Cavity Sources of Thermal Radiation," in *Theory and Practice of Radiation Thermometry*, edited by D. P. DeWitt and G. D. Nutter, Wiley, New York, pp. 653−772 (1988).

[2]   J. Hartmann, "High-Temperature Measurement Techniques for the Application in Photometry, Radiometry and Thermometry," Phys. Rep. **469**, 205−269 (2009).

[3]   P. J. Mohr, B. N. Taylor, and D. B. Newell, "CODATA Recommended Values of the Fundamental Physical Constants: 2006," Rev. Mod. Phys. **80**, 633−731 (2008).

[4]   F. J. Kelly, "On Kirchhoff's Law and its Generalized Application to Absorption and Emission by Cavities," J. Res. Natl. Bur. Stand. − B. Math. Math. Phys. **69B**, 165−171 (1965).

[5]   F. Clarke and D. Parry, "Helmholtz Reciprocity: Its Validity and Application to Reflectometry," Light Res. Technol. **17**, 1−11 (1985).

[6]   R. P. Heinisch, J. K. Andersson, and R. N. Schmidt, *Emittance Measurement Study*, Report No. NASA CR-1583, Washington, DC (1970).

[7]   R. P. Heinisch and R. N. Schmidt, "Development and Application of an Instrument for the Measurement of Directional Emittance of Blackbody Cavities," Appl. Opt. **9**, 1920−1925 (1970).

[8]   O. C. Jones and C. Forno, "Reflectance Measurements on Cavity Radiators," Appl. Opt. **10**, 2644−2650 (1971).

[9]   Ballico, M. "High Precision Measurement of the Emissivity of Small Fixed Point Blackbodies," in *Proceedings of TEMPMEKO 2001, The 8th International Symposium on Temperature and Thermal Measurements in Industry and Science*, edited by B. Fellmuth, J. Seidel, and G. Scholz, VDI Verlag, Berlin, pp. 233−237 (2002).

[10] F. Sakuma and L. Ma, "Cavity Emissivity of Fixed-Point Blackbody," in *SICE Annual Conference in Fukui*, Fukui University, Japan, pp. 2187−2190 (2003).

[11] F. Sakuma and L. Ma, "Evaluation of the Fixed-Point Cavity Emissivity at NMIJ," in *TEMPMEKO 2004-Proceedings of 9th International Symposium on Temperature and Thermal Measurements in Industry and Science*, D. Zvizdić, Editor-in-Chief. Laboratory for Process Measurement, Faculty of Mechanical Engineering and Naval Architecture, Dubrovnik, Croatia, Zagreb, pp. 563−568 (2005).

[12] J. Zeng and L. Hanssen, "An Infrared Laser-Based Reflectometer for Low Reflectance Measurements of Samples and Cavity Structures," Proc. SPIE **7065**, 70650F (2008).

[13] L. M. Hanssen, S. N. Mekhontsev, J. Zeng, and A. V. Prokhorov, "Evaluation of Blackbody Cavity Emissivity in the Infrared Using Total Integrated Scatter Measurements," Int. J. Thermophys. **29**, 352−369 (2008).

[14] L. M. Hanssen and S. G. Kaplan, "Infrared Diffuse Reflectance Instrumentation and Standards at NIST," Anal. Chim. Acta **380**, 289−302 (1999).

[15] J. A. Dykema and J. G. Anderson, "A Methodology for Obtaining On-Orbit SI-Traceable Spectral Radiance Measurements in the Thermal Infrared," Metrologia **43**, 287−293 (2006).

[16] J. A. Dykema and J. G. Anderson, "Infrared Standards in Space," Proc. SPIE **6678**, 66781B1−66781B10 (2007).

[17] P. J. Gero, J. A. Dykema, J. G. Anderson, and S. S. Leroy, "On-Orbit Characterization of Blackbody Emissivity and Spectrometer Instrument Line-Shape Using Quantum Cascade Laser Based Reflectometry," Proc. SPIE **7081**, 7081Q1−7081Q11 (2008).

[18] G. Bauer, "Reflexionsmessungen an Offenen Hohlraumen," Optik **18**, 603−622 (1961).

[19] F. J. Kelly and D. G. Moore, "A Test of Analytical Expressions for the Thermal Emissivity of Shallow Cylindrical Cavities," Appl. Opt. **4**, 31−40 (1965).

[20] G. Bauer and K. Bischoff, "Evaluation of the Emissivity of a Cavity Source by Reflection Measurements," Appl. Opt. **10**, 2639−2643 (1971).

[21] M. Ballico, "A Simple Technique for Measuring the Infrared Emissivity of Black-Body Radiators," Metrologia **37**, 295−300 (2000).

[22] L.M. Hanssen, S.N. Mekhontsev, and V.B. Khromchenko, "Validation of the Infrared Emittance Characterization of Materials through Intercomparison of Direct and Indirect Methods," Int. J. Thermophys. Online First, doi:10.1007/s10765-008-0507-9.

[23] S. Mekhontsev, M. Noorma, A. Prokhorov, and L. Hanssen, "IR Spectral Characterization of Customer Blackbody Sources: First Calibration Results," Proc. SPIE **6205**, 650203 (2006).

[24] S. N. Mekhontsev, V. B. Khromchenko, and L. M. Hanssen, "NIST Radiance Temperature and Infrared Spectral Radiance Scales at Near-Ambient Temperatures," Int. J. Thermophys. **29**, 1026−1040 (2008).

[25] A. V. Prokhorov, "Monte Carlo Method in Optical Radiometry," Metrologia **35**, 465−471 (1998).

[26] V. B. Khromchenko, S. N. Mekhontsev, and L. M. Hanssen, "Design and Evaluation of Large-Aperture Gallium Fixed-Point Blackbody," Int. J. Thermophys. **30**, 9−19 (2009).

[27] L. Hanssen, S. Mekhontsev, and V. Khromchenko, "Infrared Spectral Emissivity Characterisation Facility at NIST," Proc. SPIE **5405**, 1−12 (2004).

[28] J. Geist and J. B. Fowler, *A Water Bath Blackbody for the 5 to 60°C Temperature Range: Performance Goal, Design Concept, and Test Results*, NBS Technical Note 1228 (1986).

[29] J. B. Fowler, "A Third Generation Water Bath Based Blackbody Source," J. Res. Natl. Inst. Stand. Technol. **100**, 591−599 (1995).

[30] M. Noorma, S. Mekhontsev, V. Khromchenko, M. Litorja, C. Cagran, J. Zeng, and L. Hanssen, "Water Heat Pipe Blackbody as a Reference Spectral Radiance Source Between 50°C and 250°C," Proc. SPIE **6205**, 620502 (2006).

[31] M. Noorma, S. Mekhontsev, V. Khromchenko, A. Gura, M. Litorja, B. Tsai, and L. Hanssen, "Design and Characterization of Si and InGaAs Pyrometers for Radiance Temperature Scale Realization between 232°C and 962°C," Proc. SPIE **6205**, 620501 (2006).

[32] V. B. Khromchenko, S. N. Mekhontsev, and L. M. Hanssen, "Tunable Filter Comparator for Spectral Calibration of Near-Ambient Temperature Blackbodies," Proc. SPIE **6678**, 66781E (2007).

[33] S. Clausen, "Spectral Emissivity of Surface Blackbody Calibrators," Int. J. Thermophys. **28**, 2145−2154 (2007).

[34] J. Joly, P. Ridoux, J. Hameury, M. Lièvre, and J.-R. Filtz, "Development of a Multiband Near-to-Far Infrared Radiance COmparator (MIRCO)," Int. J. Thermophys. **29**, 1094−1106 (2008).

[35] B. K. Tsai and J. P. Rice, "Comparison of an Oil-Bath Blackbody to a Water-Bath Blackbody using the NIST TXR," in *TEMPMEKO 2004. 9th International Symposium on Temperature and Thermal Measurements in Industry and Science*, edited by D. Zvizdic, Laboratory for Process Measurement, Faculty of Mechanical Engineering and Naval Architecture, Zagreb, Croatia, pp. 859-865 (2004).

[36] J. P. Rice and B. C. Johnson, "A NIST Thermal Infrared Transfer Standard Radiometer for the EOS Program," The Earth Observer **8**, 1−5 (1996).

[37] J. P. Rice and B. C. Johnson, "The NIST EOS Thermal-Infrared Transfer Radiometer," Metrologia **35**, 505−509 (1998).

[38] A. Cox, J. O'Connell, and J. Rice, "Initial Results from the Infrared Calibration and Infrared Imaging of a Microwave Calibration Target," in *IGARSS 2006 Symposium. IEEE International Conference on Geoscience and Remote Sensing*, pp. 3446–3449, Denver, CO, USA (2006).

[39] P. W. Nugent and J. A. Shaw, "Large-Area Blackbody Emissivity Variation With Observation Angle," Proc. SPIE **7300**, 7300Y1–7300Y9 (2009).

[40] C. K. Ma, "Method for the Measurement of the Effective Emissivity of a Cavity," in *TEMPMEKO 2004. 9th International Symposium on Temperature and Thermal Measurements in Industry and Science*, edited by D. Zvizdic, Laboratory for Process Measurement, Faculty of Mechanical Engineering and Naval Architecture, Zagreb, Croatia, pp. 575–580 (2005).

[41] P. Sperfeld, S. Galal Yousef, J. Metzdorf, B. Nawo, and W. Moller, "The use of Self-Consistent Calibrations to Recover Absorption Bands in the Blackbody Spectrum," Metrologia **37**, 373–376 (2000).

[42] P. Sperfeld, S. Pape, B. Khlevnoy, and A. Burdakin, "Performance Limitations of Carbon-Cavity Blackbodies due to Absorption Bands at the Highest Temperatures," Metrologia **46**, S170–S173 (2009).

[43] R. Friedrich and J. Fischer, "New Spectral Radiance Scale from 220 to 2500 nm," Metrologia **37**, 539–542 (2000).

[44] H. Preston-Thomas, "The International Temperature Scale of 1990 (ITS-90)," Metrologia **27**, 3–10, 107 (1990).

[45] H. W. Yoon, C. E. Gibson, and B. C. Johnson, "The Determination of Emissivity of the Variable-Temperature Blackbody Used in the Dissemination of the National Scale of Radiance Temperature," in *Proceedings of TEMPMEKO 2001, The 8th International Symposium on Temperature and Thermal Measurements in Industry and Science*, edited by B. Fellmuth, J. Seidel, and G. Scholz, VDI Verlag, Berlin, pp. 221–226 (2002).

[46] T. C. Larason, S. S. Bruce, and A. C. Parr, *Spectroradiometric Detector Measurements*, US Government Printing Office, Washington, DC, NIST Special Publication 250-41 (1998).

[47] T. R. Gentile, J. M. Houston, and C. L. Cromer, "Realization of a Scale of Absolute Spectral Response using the National Institute of Standards and Technology High-Accuracy Cryogenic Radiometer," Appl. Opt. **35**, 4392–4403 (1996).

[48] J. B. Fowler, R. D. Saunders, and A. C. Parr, "Summary of High-Accuracy Aperture-Area Measurement Capabilities at NIST," Metrologia **37**, 621–623 (2000).

[49] H. W. Yoon, C. E. Gibson, and J. L. Gardner, "Spectral Radiance Comparisons of Two Blackbodies with Temperatures Determined Using Absolute Detectors and ITS-90 Techniques," in *Temperature: Its Measurement and Control in Science and Industry*, edited by D. C. Ripple, American Institute of Physics, New York, Vol. 7, pp. 601–606 (2003).

[50] P. Sperfeld, J. Metzdorf, N. J. Harrison, et al., "Investigation of High-Temperature Black Body BB3200," Metrologia **35**, 419–422 (1998).

[51] V. I. Sapritsky, B. B. Khlevnoy, V. B. Khromchenko, et al., "High-Temperature Blackbody Sources for Precision Radiometry," Proc. SPIE **2815**, 2–10 (1997).

[52] M. White, N. P. Fox, V. E. Ralph, and N. J. Harrison, "The Characterization of a High-Temperature Black Body as the Basis for the NPL Spectral-Irradiance Scale," Metrologia **32**, 431–434 (1995/96).

[53] R. U. Datla, M. C. Croarkin, and A. C. Parr, "Cryogenic Blackbody Calibrations at the National Institute of Standards and Technology Low Background Infrared Calibration Facility," J. Res. Natl. Inst. Stand. Technol. **99**, 77–87 (1994).

[54] S. R. Lorentz, S. C. Ebner, J. H. Walker, and R. U. Datla, "NIST Low-Background Infrared Spectral Calibration Facility," Metrologia **32**, 621−624 (1995/96).

[55] A. C. Carter, T. M. Jung, A. Smith, S. R. Lorentz, and R. Datla, "Improved Broadband Blackbody Calibrations at NIST for Low-Background Infrared Applications," Metrologia **40**, S1−S4 (2003).

[56] R. U. Datla, E. L. Shirley, and A. C. Parr, "Example: Calibration of a Cryogenic Blackbody," in *Optical Radiometry*, edited by A. C. Parr, R. U. Datla, and J. L. Gardner, Academic Press, Amsterdam, pp. 535−546 (2005).

[57] A. C. Carter, R. U. Datla, T. M. Jung, A. W. Smith, and J. A. Fedchak, "Low-Background Temperature Calibration of Infrared Blackbodies," Metrologia **43**, S46−S50 (2006).

[58] E. L. Shirley, "Diffraction Effects in Radiometry," in *Optical Radiometry*, edited by A. C. Parr, R. U. Datla, and J. L. Gardner, Academic Press, Amsterdam, pp. 409−451 (2005).

[59] J. P. Rice and B. C. Johnson, "NIST Activities in Support of Space-Based Radiometric Remote Sensing," Proc. SPIE **4450**, 108−126 (2001).

[60] A. C. Parr and R. U. Datla, "NIST Role in Radiometric Calibrations for Remote Sensing Programs at NASA, NOAA, DOE and DOD," Adv. Space Res. **28**, 59−68 (2001).

[61] R. Datla, "Best Practice for Pre-Launch Characterization and Calibration of Instruments for Remote Sensing," Proc. SPIE **7082**, 70820V1−70820V12 (2008).

[62] T. M. Jung, A. C. Carter, S. R. Lorentz, and R. U. Datla, "NIST-BMDO Transfer Radiometer (BXR)," Proc. SPIE **4028**, 404−410 (2000).

[63] P. Lagueux, M. Chamberland, F. Marcotte, A. Villemaire, M. Duval, J. Genest, and A. Carter, "Performance of a Cryogenic Michelson Interferometer," Proc. SPIE **7082**, 7080Q1−7080Q11 (2008).

[64] D. B. Pollock, T. L. Murdock, R. U. Datla, and A. Thompson, "Radiometric Standards in Space: the Next Step," Metrologia **37**, 403−406 (2000).

[65] B. C. Johnson, S. W. Brown, and J. P. Rice, "Metrology for Remote Sensing Radiometry," in *Proceedings of the International Workshop on Radiometric and Geometric Calibration*, 2−5 December 2003, edited by S. A. Morain and A. M. Budge, Leiden, A. A. Balkema Publishers, Gulfport, MS, pp. 7−16 (2004).

[66] G. T. Fraser, S. W. Brown, R. U. Datla, B. C. Johnson, K. R. Lykke, and J. P. Rice, "Measurement Science for Climate Remote Sensing," Proc. SPIE **7081**, 708101−708112 (2008).

[67] S. P. Morozova, P. A. Morozov, V. I. Sapritsky, B. E. Lisiansky, N. L. Dovgilov, L. Y. Xi, L. Yanmei, and Y. Wenlong, "Low-Temperature Blackbodies and Facility for Calibration of Them," Proc SPIE **4927**, 118−124 (2002).

[68] C. Monte, B. Gutschwager, S. P. Morozova, and J. Hollandt, "Radiation Thermometry and Emissivity Measurements under Vacuum at the PTB," Int. J. Thermophys. **30**, 203−219 (2009).

[69] S. P. Morozova, N. A. Parfentiev, B. E. Lisiansky, et al., "Vacuum Variable-Temperature Blackbody VTBB100," Int. J. Thermophys. **29**, 341−351 (2008).

[70] S. A. Ogarev, M. L. Samoylov, N. A. Parfentyev, and V. I. Sapritsky, "Low-Temperature Blackbodies for IR Calibrations in a Medium-Background Environment," Int. J. Thermophys. **30**, 77−97 (2009).

[71] D. P. DeWitt and J. C. Richmond, "Thermal Radiative Properties of Materials (Chapter 2)," in *Theory and Practice of Radiation Thermometry*, edited by D. P. DeWitt and G. D. Nutter, Wiley-Interscience, New York, NY, pp. 102−187 (1988).

[72] M. F. Modest, *Radiative Heat Transfer*, Academic Press, New York, 2nd ed., pp. 107−121 (2003).

[73] J. M. Palmer, "The Measurement of Transmission, Absorption, Emission and Reflection," in *Handbook of Optics II*, edited by M. Bass, McGraw-Hill, New York, 2nd ed., pp. 25.1−25.25 (1995).

[74] J. Hartmann, S. Schiller, R. Friedrich, and J. Fischer, "Non-Isothermal Temperature Distribution and Resulting Emissivity Corrections for the High Temperature Blackbody BB3200," in *Proceedings of TEMPMEKO 2001, The 8th International Symposium on Temperature and Thermal Measurements in Industry and Science*, edited by B. Fellmuth, J. Seidel, and G. Scholz, VDI Verlag, Berlin, pp. 227-232 (2002).

[75] M. R. Dury, T. Theocharous, N. Harrison, N. Fox, and M. Hilton, "Common Black Coatings − Reflectance and Ageing Characteristics in the 0.32−14.3 μm Wavelength Range," Opt. Commun. **270**, 262−272 (2007).

[76] K. D. Hill and D. J. Woods, "Characterizing the NRC Blackbody Sources for Radiation Thermometry from 150°C to 962°C," Int. J. Thermophys. **30**, 105−123 (2009).

[77] T. J. Horn and A. N. Abdelmessih, *Experimental and Numerical Characterization of a Steady-State Cylindrical Blackbody Cavity at 1100 Degrees Celsius*, Report No. NASA/TM-2000-209022 (2000).

[78] M. J. Persky, "Review of Black Surfaces for Space-Borne Infrared Systems," Review of Scientific Instruments **70**, 2193−2217 (1999).

[79] M. J. Persky and M. Szczesniak, "Infrared, Spectral, Directional-Hemispherical Reflectance of Fused Silica, Teflon Polytetrafluoroethylene Polymer, Chrome Oxide Ceramic Particle Surface, Pyromark 2500 Paint, Krylon 1602 Paint, and Duraflect Coating," Appl. Opt. **47**, 1389−1396 (2008).

[80] J. Fischer, J. Seidel, and B. Wende, "The Double-Heatpipe Black Body: A Radiance and Irradiance Standard for Accurate Infrared Calibrations in Remote Sensing," Metrologia **35**, 441−445 (1998).

[81] J. Hartmann, D. Taubert, J. Fischer, "Characterization of the Double-Heatpipe Blackbody LABB for Use at Temperatures Below 500°C," in *Proceedings of TEMPMEKO 1999, The 7th International Symposium on Temperature and Thermal Measurements in Industry and Science*, edited by J. F. Dubbeldam and M. J. de Groot, NMi Van Swinden Laboratorium, Delft, pp. 511-516 (1999).

[82] A. Springsteen, "Reflectance Spectroscopy: An Overview of Classification and Techniques," in *Applied Spectroscopy. A Compact Reference for Practitioners*, edited by J. Workman and A. W. Springsteen, Academic Press, New York, pp. 194−224 (1998).

[83] J. M. Davies and W. Zagieboylo, "An Integrating Sphere System for Measuring Average Reflectance and Transmittance," Appl. Opt. **4**, 167−174 (1965).

[84] K. A. Snail and L. M. Hanssen, "Accurate Measurements of Spectral Reflectance: Directional Hemispherical and Hemispherical Directional Reflectance Measurements Using Conic Mirror Reflectometers," in *Applied Spectroscopy. A Compact Reference for Practitioners*, edited by J. Workman and A. W. Springsteen, Academic Press, New York, pp. 269−298 (1998).

[85] D. B. Betts, F. J. J. Clarke, L. J. Cox, and J. A. Larkin, "Infrared Reflection Properties of Five Types of Black Coatings for Radiometric Detectors," J. Phys. E: Sci. Instrum. **18**, 689−696 (1985).

[86] J. Hameury, B. Hay, and J. R. Filtz, "Measurement of Infrared Spectral Directional Hemispherical Reflectance and Emissivity at BNM-LNE," Int. J. Thermophys. **26**, 1973−1983 (2005).

[87] J. A. Jacquez, W. McKeehan, J. Huss, J. M. Dimitroff, and H. F. Kuppenheim, "An Integrating Sphere for Measuring Diffuse Reflectance in the Near Infrared," J. Opt. Soc. Am. **45**, 781−785 (1955).

[88] J. C. Morris, "Integrating Sphere for the Infrared," Appl. Opt. **5**, 1035−1037 (1966).

[89]   G. A. Zerlaut and A. C. Krupnick, "An Integrating-Sphere Reflectometer for Determination of Absolute Hemispherical Spectral Reflectance," AIAA J. **4**, 1227–1232 (1966).

[90]   R. R. Willey, "Fourier Transform Infrared Spectrophotometer for Transmittance and Diffuse Reflectance Measurements," Appl. Spectrosc. **30**, 593–601 (1976).

[91]   L. M. Hanssen and K. A. Snail, "Integrating Spheres for Mid- and Near Infrared Reflection Spectroscopy," in *Handbook of Vibrational Spectroscopy*, edited by J. M. Chalmers and P. R. Griffiths, Wiley, New York, pp. 1175–1192 (2002).

[92]   L. M. Hanssen, "Integrating-Sphere System and Method for Absolute Transmittance, Reflectance and Absorptance of Specular Samples," Appl. Opt. **40**, 3196–3204 (2001).

[93]   L. M. Hanssen, C. P. Cagran, A. V. Prokhorov, S. N. Mekhontsev, and V. B. Khromchenko, "Use of a High-Temperature Integrating Sphere Reflectometer for Surface-Temperature Measurements," Int. J. Thermophys. **28**, 566–580 (2007).

[94]   C. P. Cagran, L. M. Hanssen, M. Noorma, A. V. Gura, and S. N. Mekhontsev, "Temperature-Resolved Infrared Spectral Emissivity of SiC and Pt–10Rh for Temperatures up to 900°C," Int. J. Thermophys. **28**, 581–597 (2007).

[95]   K. Snail and L. M. Hanssen, "Integrating Sphere Designs with Isotropic Throughput," Appl. Opt. **28**, 1793–1799 (1989).

[96]   F. E. Nicodemus, Richmond J. C, Hsia J. J., Ginsberg I. W., and Limperis T. *Geometrical Considerations and Nomenclature for Reflectance*, NBS Monograph 160, US Department of Commerce, National Bureau of Standards (1977).

[97]   F. M. Cady, J. C. Stover, D. R. Bjork, M. L. Bernt, M. W. Knighton, D. J. Wilson, and D. R. Cheever, "A Design Review of a Muitiwavelength, Three-Dimensional Scatterometer," Proc. SPIE **1331**, 201–208 (1990).

[98]   B. L. Drolen, "Bidirectional Reflectance and Specularity of Twelve Spacecraft Thermal Control Materials," J. Thermophys. Heat Transfer **6**, 672–679 (1992).

[99]   J. N. Ford, K. Tang, and R. O. Buckius, "Fourier Transform Infrared System Measurement of the Bidirectional Reflectivity of Diffuse and Grooved Surfaces," J. Heat Transfer **117**, 955–962 (1995).

[100] D. Hünerhoff, U. Grusemann, and A. Höpe, "New Robot-Based Gonioreflectometer for Measuring Spectral Diffuse Reflection," Metrologia **43**, S11–S16 (2006).

[101] J. E. Proctor and P. Y. Barnes, "NIST High Accuracy Reference Reflectometer-Spectrophotometer," J. Res. Natl. Inst. Stand. Technol. **101**, 619–627 (1996).

[102] T. A. Germer, "Multidetector Hemispherical Polarized Optical Scattering Instrument," Proc. SPIE **3784**, 296–303 (1999).

[103] K. J. Dana and J. Wang, "Device for Convenient Measurement of Spatially Varying Bidirectional Reflectance," J. Opt. Soc. Am. A **21**, 1–12 (2004).

[104] A. Ghosh, S. Achutha, W. Heidrich, and M. O'Toole, "BRDF Acquisition with Basis Illumination," in *Proceedings of IEEE 11th International Conference on Computer Vision (ICCV)*, Rio de Janeiro, Brazil, pp. 1–8 (2007).

[105] J. Zeng and L. Hanssen, "IR Optical Scattering Instrument with Out-of-Plane and Retro-Reflection Capabilities," in *Conference on Lasers & Electro-Optics (CLEO)*, Baltimore, MD, pp. 1882–1884 (2005).

[106] Y. J. Shen, Q. Z. Zhu, and Z. M. Zhang, "A Scatterometer for Measuring the Bidirectional Reflectance and Transmittance of Semiconductor Wafers with Rough Surfaces," Rev. Sci. Instrum. **74**, 4885–4892 (2003).

[107] X. J. Wang, J. D. Flicker, B. J. Lee, W. J. Ready, and Z. M. Zhang, "Visible and Near-Infrared Radiative Properties of Vertically Aligned Multi-Walled Carbon Nanotubes," Nanotechnology **20**, 215704 (2009).

[108] S. Clausen, "Measurement of Spectral Emissivity by FTIR Spectrometer," in *Proceedings of TEMPMEKO 2001, The 8th International Symposium on Temperature and Thermal Measurements in Industry and Science*, edited by B. Fellmuth, J. Seidel, and G. Scholz, VDI Verlag, Berlin, pp. 259–264 (2002).

[109] S. Clausen, A. Morgenstjerne, and O. Rathmann, "Measurement of Surface Temperature and Emissivity by a Multitemperature Method for Fourier-Transform Infrared Spectrometers," Appl. Opt. **35**, 5683–5691 (1996).

[110] L. M. Hanssen, S. N. Mekhontsev, and S. G. Kaplan, "NIST Program for the Infrared Emittance Characterization of Materials," in *Thermal Conductivity 29/Thermal Expansion 17 Joint Conference,* DES*teeh* Publications, pp. 523–534 (2008).

[111] L. Del Campo, R. B. Pérez-Sáez, X. Esquisabel, I. Fernández, and M. J. Tello, "New Experimental Device for Infrared Spectral Directional Emissivity Measurements in a Controlled Environment," Rev. Sci. Instrum. **77**, 113111 (2006).

CHAPTER 3

Radiation Thermometry in the Semiconductor Industry

Bruce E. Adams[1], Charles W. Schietinger[2] *and* Kenneth G. Kreider[3]

Contents

1. Introduction	138
1.1. Temperature measurement	141
1.2. Semiconductor terms	141
1.3. Semiconductor processes	142
2. Basics of Optical Fiber Thermometry (OFT)	143
2.1. Major challenges	144
2.2. A brief patent history	146
2.3. The lightpipe as the controlled collection optics	146
2.4. Transmission fiber	149
2.5. Light to temperature conversion	150
3. Temperature Measurements in the Semiconductor Industry	152
3.1. Limits from tool design	154
3.2. Limits of contact devices	154
3.3. Temperature measurement requirements in the semiconductor industry	155
3.4. Optical properties of silicon	156
4. Applications	159
4.1. High-density plasma chemical vapor deposition (HDPCVD)	161
4.2. Metal organic chemical vapor deposition (MOCVD)	162
4.3. Epitaxial	162
4.4. Solar photovoltaics	163
4.5. Nonoptical techniques	163
4.6. Wafer measurement applications	163
5. Calibration	164
5.1. Lens system radiation thermometers	164

[1] Applied Materials, Santa Clara, CA, USA
[2] Lopez Island, Washington, DC, USA
[3] Process Measurements Division, National Institute of Standards and Technology, Gaithersburg, MD, USA

Experimental Methods in the Physical Sciences, Volume 43
ISSN 1079-4042, DOI 10.1016/S1079-4042(09)04303-3

    5.2. Lightpipe radiation thermometers (cold calibration, hot
          calibration, *in situ* calibration)   165
    5.3. Blackbody simulators   166
    5.4. Process repeatability standards (PRS) in RTP   168
  6. *In situ* Calibration of Radiation Thermometers in RTP Tools   170
    6.1. Proof wafer TC thermometry   170
    6.2. Calibration of TFTCs on silicon wafers   174
    6.3. Calibration of RTs on RTP tools using silicon wafers with TFTCs   178
    6.4. Transient thermal response of TFTCs   185
    6.5. Transient response of sensors in PEB process   185
  7. Emissivity Methods   189
    7.1. Problem of surface finish   189
    7.2. Use of virtual blackbodies   191
    7.3. Pyrometric sensitivity and difficulties with ratio pyrometry   192
    7.4. RTP reflector plate enhancement theory   193
    7.5. RTP *in situ* emissometer application   195
    7.6. Low-temperature measurements in RTP   199
    7.7. Ripple Technique theory   200
    7.8. Ripple Technique application   202
    7.9. Other methods in lamp-heated systems   203
    7.10. Laser methods   206
  8. Problems to Solve   208
  9. Summary   208
  Acknowledgments   209
  References   209

1. INTRODUCTION

Today near-infrared (NIR) pyrometers are used to measure and
control wafer temperatures in high-temperature integrated circuit (IC)
manufacturing, such as rapid thermal processing (RTP) [1]. The chapter
reviews the vast subject of temperature measurement and control within
the semiconductor industry, with an emphasis on the process-specific
techniques used to measure wafer temperatures (Table 1). Semiconductor
wafer processing spans a huge range of conditions from subambient
temperatures to above the melting point of silicon at just over $1,400°C$.
The most common temperature measuring devices include thermocouples,
fluoroptic decay sensors, and a wide range of optical pyrometers. One of
the most difficult optical temperature measurements occurs when the
amount of emitted thermal radiation from the hot wafer is small compared
to the background light. This larger background condition is the case in a
number of semiconductor process environments, including nearly all

Table 1 A brief summary of temperature measurement technologies used in the semiconductor industry.

Measurement technology	Processes installed	Temperature range	Advantages	Disadvantages
No temperature measurement, power control only	Has been tried in most processes	All	Low initial cost, easy to implement	Drift, poor thermal stability, lost product can be costly
TC in the susceptor or wafer holder	HDP-CVD, plasma etch	$-50°C$ to $800°C$	Simple and low cost	Heat transfer issues, doesn't measure real wafer temperature, TC subject to EMI in some processes, not all tools have susceptors. Susceptor/wafer time lag
Fluoroptic, spring loaded against the wafer or susceptor	HDP-CVD, plasma etch	$-50°C$ to $380°C$	Low-temperature measurement, EMI immunity	Some heat transfer issues, contaminating, susceptor to wafer lag time
TC-embedded wafers	Calibration, many used for development and test only	$-50°C$ to $1,250°C$	Very good measurement in most applications	Costly, not product wafers, rotating wafers are difficult, slow for some applications, lamp effects, heat transfer issues, drift over time

Table 1. *(Continued)*

Measurement technology	Processes installed	Temperature range	Advantages	Disadvantages
Lens pyrometer, long wavelength	CMP pad, poly-single wafer	25°C to 1,250°C	Good signal strength, some are very inexpensive	Very emissivity-dependent, drift, responsive to background emission from heaters and chamber, line-of-sight access, often narrow field of view
Lens pyrometer, short wavelength	RTP, MBE, MOCVD, PVD, crystal growth	350°C to 1,250°C	Reduced emissivity dependence, reduced drift and chamber emission effects	Line-of-sight access, often narrow field of view, sometimes large emissivity dependence, deposition on viewports
Optical fiber thermometry	RTP, plasma etch, CVD	350°C to 1,250°C	Wide field of view, versatile installation, allows emissometer applications	Costly, plasma light is a large error term, changing wafer optical properties

plasma processing and lamp-heated systems. We cover one of these processes, RTP, in greater detail, because RTP temperature control has been achieved with high repeatability and accuracy, whereas other applications have yet to be solved to that degree. We review the key features of one of the most sensitive and robust techniques, optical fiber thermometry (OFT). This chapter also describes the development of calibration wafers based on thin-film thermocouples (TFTCs).

The following subsections contain the lists of acronyms, abbreviations, and definitions for the commonly used terms in this chapter.

1.1. Temperature measurement

EMF — Electromotive force is the energy per unit charge provided by a device to a circuit.

FOT — Fluoroptic thermometry uses the decay time of a thermal phosphor to determine temperature.

OFT — Optical fiber thermometry utilizes a pyrometer with fiber optics. The fiber optics could be a lightpipe or a flexible fiber-optic cable.

LPRT — Lightpipe radiation thermometer is one version of OFT where a lightpipe is used for the collection optics. Typically the lightpipe is made from single-crystal sapphire.

NA — Numerical aperture is a dimensionless number that characterizes the range of angles over which the system can accept or emit light.

Ripple Technique — The Ripple Technique is a powerful optical method that uses the variation in light caused by AC powered lamps to measure emissivity and temperature real time.

RT — Radiation thermometer is a noncontact device that measures thermal radiation to determine the temperature of a remote object.

RTD — Resistance thermal devices are sensitive electronics that change based on changes in temperature.

TC — Thermocouple is a device used to measure temperature based on the Seebeck effect.

TFTC — Thin-film thermocouple is a thermocouple produced by deposition of thin layers of metals onto a substrate.

1.2. Semiconductor terms

Device — Semiconductor integrated circuit, examples include memory devices and processors.

Fab — Fabrication facility where integrated circuits are manufactured. These are typically very large with significant clean room area. A small fab today can cost several hundred million dollars to build and a more typical fab a few billion dollars.

IC — Integrated circuit is a miniaturized electronic circuit created on the surface of a thin substrate of a semiconductor material.

1.3. Semiconductor processes

ALD — Atomic layer deposition is a relatively newer family of deposition techniques which allow extremely thin layers of almost any material to be deposited.

CMP — Chemical—mechanical polishing is used many times throughout the fabrication of ICs. The process uses a combination of low pH chemicals and colloidal slurry to mechanically polish the wafer flat between deposition steps.

Crystal growth — Bulk crystal growth is used to produce silicon ingots which are sliced into wafers and is also used to produce other wafer types including GaAs, sapphire, InP, and GaN. Crystal growth techniques include the Czochralski (CZ) method and float-zone method.

CVD — Chemical vapor deposition is a wide range of techniques for growing thin films using gases as the source material for the films. These processes range from near room temperature to 1,250°C. Some of the more common films grown by CVD are poly-silicon, silicon nitride, and silicon carbide.

Epi — Epitaxial growth produces a deposited film that is nearly the same perfect crystal structure as the underlying material.

Implant — Implant is a process where dopants are driven into the top surface of the wafer.

Lithography — Process where the integrated circuit is printed onto a photosensitive resist using short-wavelength light to expose the resist.

MBE — Molecular beam epitaxial systems are ultra high vacuum deposition systems which allow growth of very thin high purity layers of many materials. These systems are often used in compound semiconductors.

MOCVD — Metal organic chemical vapor deposition is used mostly in compound semiconductors for growing thin films from metal organics. These reactors are used to grow many different types of films for a wide range of applications including light emitting diodes (LEDs), high-preference solar cells, and cell phone ICs. MOCVD is also used to grow some high k dielectrics.

PVD — Physical vapor deposition is a broad range of thin-film growth techniques where the source material comes from a solid source. These methods include thermal evaporation, e-beam evaporation, and most importantly sputtering. The sputtering process uses plasma to grow a film by bombarding a solid target. PVD and CVD account for most of the large-scale production deposition in the manufacturing of silicon-based ICs.

RTA — Rapid thermal annealing fixes the dopants into the silicon crystal lattice before they are significantly spread from thermal diffusion. This is accomplished in an RTP chamber.

RTO — Rapid thermal oxidation is the growth of a very thin high-quality silicon oxide film in an RTP chamber.

RTP — Rapid thermal processing is a family of processes using lamps that quickly heat the wafer.

2. BASICS OF OPTICAL FIBER THERMOMETRY (OFT)

Optical fiber thermometry (OFT) is simply radiation pyrometry where one or more of the collection optics are fiber optics [2–4]. The fiber optic could be a rigid or flexible lightpipe used to collect the emitted light from the hot object. A common configuration uses a flexible fiber-optic cable, typically clad quartz, to take the collected light from a rigid sapphire lightpipe to the photodetector and electronics instrument (see Figure 1). OFT also includes lens-based collection assemblies, where the assemblies are connected to the photodetector with a flexible fiber optic. The use of fiber optics often facilitates a solution to a temperature measurement problem, for example, by allowing a much smaller optic element to intrude on a process or penetrate a vacuum system [5].

This advantage of a fiber-optic lightpipe was especially important as semiconductor process chambers have become smaller in an effort to achieve fast and effective purging and also to decrease gas use. With these evolving constraints, the use of a fiber-optic lightpipe allowed access where lens systems could not go. The lightpipe in Figures 1–4 also collects a nearly 180° cone of radiation. This wide collection has two big advantages in measuring semiconductor wafer temperatures: it increases the amount of light collected, thus allowing the measurement of lower temperatures; and

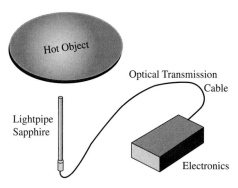

Figure 1 Typical optical fiber thermometry utilizes a sapphire lightpipe for collection and a flexible fiber optic to conduct the light back to the electronics [4].

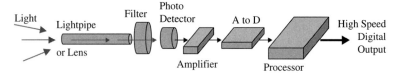

Figure 2 Typical optical fiber thermometry utilizes a sapphire lightpipe and no flexible fiber-optic cable. The lightpipe connects directly to the electronics. Removing the flexible cable increases the amount of light to the detector allowing lower temperatures to be measured.

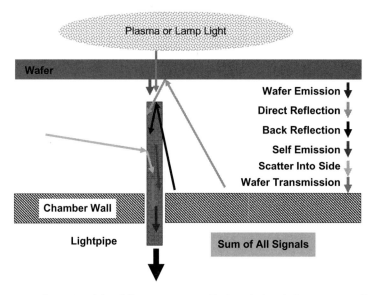

Figure 3 Schematic of the different sources of light which a lightpipe generally collects in most semiconductor applications. The light measurement is the sum of all of these sources of light not simply the desired wafer emission. The most robust measurements quantify and minimize the sources of light other than direct wafer emission.

it reduces the sensitivity to surface irregularities because it averages over a larger area, and a wider range of angles. Typically, the NA of the pyrometer is determined by optical apertures and not the lightpipe as shown in Figure 4.

2.1. Major challenges

Accurate pyrometry in a given application requires that one consider the design of the system as an integral whole. Having the pyrometer included early in the design of any new equipment helps one eliminate many practical difficulties and allow for solutions that reduce stray light and

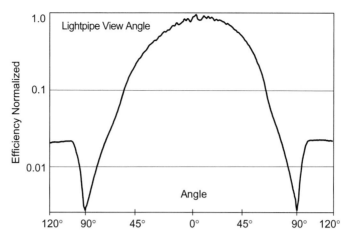

Figure 4 The view of a sapphire lightpipe pyrometer. The graph shows the amount of light collected versus angle. Light collection efficiency (normalized) versus angle of incidence (view half-angle) for a 3-mm-diameter sapphire lightpipe pyrometer [5].

emissivity issues, which are two of the toughest problems. The list of important considerations often includes:

1. optimum line of sight access,
2. optimum positioning of vacuum seal or window,
3. possible light losses due to contamination on optical surfaces (dirt accumulation) or to other sources (electromagnetic interference, or EMI; movement of the flexible optical cable; transmission to photodetector),
4. temperature limitations of the collection optics, and
5. mechanical stability.

The four largest sources of uncertainty in semiconductor pyrometry are:

1. wafer transmission,
2. stray light,
3. unknown or changing emissivity, and
4. low thermal emission at low wafer temperatures.

The hardware issues in OFT include errors introduced by the movement of the flexible quartz clad fiber optics. When the optical cable is moved, it can induce changes in how the light is propagated, and it can affect the amount of light incident upon the photodetector. This is often the largest source of hardware-induced error with OFT. Flexible quartz fiber-optic cables have much lower temperature limits than the 1,900°C of the sapphire lightpipe which demands consideration in the mating of the fiber sensor and the transmission fiber to avoid temperature-induced failures.

2.2. A brief patent history

Pyrometer-based temperature measurement systems have a long history. For example, early pyrometers were invented before 1930 [6−8]. In 1933, a pyrometer was developed and patented which employed an optical lightpipe [9]. In 1955, following the development of synthetically grown sapphire, the first pyrometer was described which used a sapphire lightpipe (bent or straight) for the collection optics [10]. Quartz and sapphire lightpipes with a blackbody sensor tip were used in 1971 [11]. Then in 1978, a modern flexible fiber-optic thermometer was introduced [12]. In the1980s, electronic improvements were made to pyrometers, including an integrating photodetector output circuit [13]. Further improvements were made to electronics, fiber optics, sapphire rods, and blackbody emission temperature measurements [2,14,15].

In the 1990s, many patents and reports were issued that describe the use of pyrometers in semiconductor processing. One paper describes the use of a sapphire lightpipe looking at the backside of a wafer, where the cavity formed by the wafer and a susceptor creates a virtual blackbody [16]. The susceptor is an additional thin disk placed strategically in the wafer processing system to create a more uniform temperature field and thus ultimately reduce the temperature uncertainties when using radiation thermometry. In another example, fiber-optic lightpipes were used for wafer temperature measurements in RTP applications [17]. In 1992, semiconductor wafer temperatures were measured by a fiber-optic thermometer with wavelength-selective mirrors and modulated light [18]. In 1998, an integrating amplifier chip and fiber optic were used to measure wafer temperatures [19], and an infrared (IR) sensing thermometer utilized an integrating amplifier [20]. The benefits of cooling pyrometer detectors were explained in Ref. [21], and another fiber-optic pyrometer for measuring wafer temperatures was described in Ref. [22]. A technique for making wafer temperature measurements in plasma environments is described in Ref. [23].

2.3. The lightpipe as the controlled collection optics

Lightpipes transmit light from the hot semiconductor wafer (or any object being measured) back to the photodetector, where the optical signal is converted into an electrical signal that is interpreted as a temperature reading. Lightpipes are usually placed close to the hot object allowing a large collection angle that improves the measurement in two ways: the proximity increases the solid angle of acceptance and the area of observation thus increasing the amount of light and reducing the sensitivity to surface irregularities, respectively. Suitable lightpipes can collect nearly 180° of incoming light and deliver the light with an exit angle equal to the entrance angle. The collection of $180°$ is generally reduced by the photodetector acceptance angle and other possible optical elements in the path, such as possible limitations posed by a flexible fiber-optic cable.

Solid fiber optics or lightpipes can be made of quartz, YAG (yttrium aluminum garnet), or, preferably, single-crystal sapphire (Al_2O_3). The best lightpipes are manufactured from ultra high purity, low-defect Czochralski grown (CZ) sapphire ingots. These cylindrical ingots are grown in sizes up to about 150 mm in diameter and about 500 mm in length. The ingots are diamond-sawed into long square rods, and then the outer diameter (OD) is ground and polished smooth. The structure of single-crystal sapphire is hexagonal and therefore has a number of orientation planes. The crystal orientation of the sapphire cut from the ingot for the lightpipes is important in producing the best lightpipes. In crystallography various axis are defined to physically orient a crystal. Typically, one chooses the C-axis as the optic axis, but the N-axis is also a possibility. YAG lightpipes can also be manufactured from CZ YAG and also diamond sawed or core drilled and then OD ground and polished.

There are many advantages to using high-purity single-crystal CZ sapphire for lightpipe manufacturing. CZ sapphire:

1. has properties that change very little, if at all, over time (This is critical in most measurements and applications.),
2. will survive high temperatures up to about $1,800°C$ (YAG has a lower temperature limit, about $1,500°C$.),
3. will tolerate large thermal gradients and rapid thermal change (YAG is much more subject to thermal shock.),
4. is inert to most chemicals (Therefore, chemical cleaning is possible and is useful in applications prone to dirt or deposition),
5. is one of the hardest materials other than diamond and chemical vapor deposition (CVD) SiC (Therefore, it can be highly polished and not easily scratched in use or with repeated cleanings. YAG is much softer and can be more easily scratched.), and
6. has a very low number of defects.

This sixth feature is perhaps the most important for any good lightpipe for two reasons. First and most intuitive, absence of defects means that the collected signal will not be absorbed or scattered out of the lightpipe. This is true for any type of defect, including crystal defects, inclusions, voids, impurities, surface flaws, and scratches on the polished outside surface. The cleanliness of the outside surface is particularly critical, because the dirt or surface film absorbs the optical signal as the light travels down the lightpipe. The second problem with defects is often more important in certain high-temperature applications, especially when the lightpipe is itself hot. Defects in the lightpipe will scatter stray light from the heated surroundings into the lightpipe, causing a measurement error. In addition, the defect, if hot, will self-emit adding to the measurement and causing an error dependent on the lightpipe temperature. This source of error can be minimized by a design that limits the heating of the lightpipe. A good example of such a design is the Applied Materials RTP system, in which the lightpipes are embedded in a cool base-plate. In this application the wafer temperature is measured with numerous sapphire lightpipes which connect to a silicon photodetector operating in the NIR spectral range. In another tool manufactured by Mattson Technology, a pyrometer is used which exploits the alternating current (AC)-driven lamp light to measure wafer emissivity and amount of background light. The details are described under the Ripple Technique™.

2.3.1. Edge-defined film feed growth sapphire
Edge-defined film feed growth (EFG) is a process that produces low-cost and low-quality sapphire which typically has too many defects for high-quality lightpipes. In EFG the sapphire is pulled through a high-temperature die in the shape of the finished product. EFG can be manufactured in thin sheets or tubes or nearly any shape that the die is made. EFG sapphire is, however, widely used for the protective sheaths covering CZ grown sapphire lightpipes. The protective sheath adds mechanical strength to reduce the risk of breakage and prevents contamination or deposition from forming on the lightpipe. In some applications the sheaths are purged [3]. There are a couple of techniques for plugging EFG tubes to form closed-end sheaths. One uses a tight-fitting tapered sapphire plug fused to the sheath with a hydrogen torch. In another method the end can be closed as part of the EFG process.

2.3.2. Quartz
Quartz has been used in optical lightpipes for years. However, its use has some severe limitations. Quartz will start to change optically if heated above 650°C, making it poorly suited for high-temperature lightpipes. The strength and hardness of quartz is less than that of sapphire, making it more prone to being easily broken or scratched. Quartz has a smaller field of view than sapphire owing to its smaller index of refraction. There has been some work on using clad quartz as lightpipes, and there are good applications for this design if the lightpipe's temperature can be adequately controlled. The refractive index of the cladding can be designed to give the desired optical characteristics.

2.3.3. Laser pedestal grown sapphire fibers
This material offers most of the advantages of CZ sapphire, along with the significant advantage of being produced in continuous lengths of

high-quality, low-defect, flexible or rigid fiber [24]. However, due to low demand with this complex and difficult technology, the product is too expensive. Consequently, this technique is still in its infancy and is available only as demonstration of concept. The very best CO_2 laser grown fibers are produced from CZ starting material with a ring-focused CO_2 laser melting the sapphire as the fiber is slowly pulled upwards. The pull rate and melt temperature controlled by laser power define the fiber diameter. In order to avoid waviness on the fiber surface (which will affect transmission), the pull rate and melt temperature must be carefully controlled. At diameters less than about 600 μm, the sapphire will start to become flexible and at around 200 μm it becomes very flexible. The huge advantages of laser pedestal grown sapphire for lightpipes are:

1. access into the smallest locations or chambers and minimum thermal impact by the fiber, owing to the small diameter,
2. less breakage because the fiber is flexible,
3. potentially very low defects and impurities,
4. elimination of one optical interface by integration of the collection and transmission fiber, and
5. generally simpler coupling of the small diameter fiber with the photodetector.

To summarize, laser pedestal grown sapphire fibers offer advantages over rigid sapphire rods currently being used in many semiconductor applications including RTP. The cost to develop and manufacture this promising material has yet to be off-set by the market forces.

2.4. Transmission fiber

In OFT a fiber-optic cable is sometimes used to connect a collection lens or a lightpipe to the detector and electronics. This flexible optical transmission element can be either a single optical fiber or a bundle of cable of fibers, as discussed in the following.

2.4.1. Flexible cable

The flexible fiber-optic cable, typically clad quartz and sometimes plastic, was first used on commercial pyrometer systems in the 1980s by Ircon. There have been many advances in the technology since that time. A flexible fiber-optic cable can be used to transmit the light (signal) from the lightpipe or lens collection system. The optical cable can also be used as the collection optics in application where it remains cool, and here the collection area is defined by the numerical aperture (NA) of the fiber optic.

There are three advantages of using a flexible fiber optic with both lightpipes and lens collection systems:

1. The electronics can be made remote from the hot or harsh environment.
2. Less space is needed near the measurement location.
3. The NA of the fiber can be used in the design of the collection optics.

On the other hand, the disadvantages can be significant:

1. The loss of light (signal) is typically 80% or more, due to the difference in NA or view angle between the lightpipe and the optical cable.
2. When the lightpipe is disconnected from the cable, there can be errors introduced from changes in the connection from alignment, spacing or gap between the cable and the lightpipe. These possible errors may be calibrated "out" of the measurement if the entire system is calibrated.
3. Using different cables will possibly change the calibration. Changing cables is a necessated by damage due to bending or other environmental stresses.
4. Bending the cable can change how the light propagates through the cable. One can reduce such errors by using smaller diameter cables at the expense of increasing light loss.

2.4.2. Single fiber or a bundle

There has been an ongoing debate in the pyrometer field as to the advantages and disadvantages of using bundles versus single fibers to connect the collection optics to the photodetector. In practice, there are some applications that favor either one over the other. The bundle has two key advantages over the large single fiber. The bundle can be made from smaller diameter fibers which are therefore more flexible and induce less micro-stress optical losses; and the bundle can be made to have a larger overall area (diameter minus voids) than the single fiber, thus transmitting more light to the photodetector. One disadvantage of the bundles is the possibility that one of the fibers breaks without knowledge of the user causing an unrealized measurement offset. In a closed-loop control application, this offset will be in the wrong direction (the pyrometer will read low) and will cause an over-temperature.

2.5. Light to temperature conversion

An accurate and extremely wide dynamic range ($10-13$ orders of magnitude) conversion from an optical signal to an electrical signal was one of the keys to solving most of the wafer temperature measurement applications within the semiconductor industry. This accomplishment was the result of the team Ray Dils assembled at Accufiber, following Dils' work at the National Bureau of Standards [25]. Prior to this, the pyrometer manufacturers had used a number of methods to convert the collected light signal to an electrical signal, but none of these techniques gave accurate results over a sufficient dynamic range. The use of solid-state photo-detectors with low-noise amplifiers has improved the sensitivity of this technique. Figure 5 shows the improvement in noise threshold and hence sensitivity plotted as minimum diode current versus year. The most sensitive pyrometers use quantum photodetectors, generally silicon or InGaAs detectors, operating in the NIR wavelengths [23,25−27]. These photodetectors have high conversion efficiency, are reliable, and can be well characterized. If the bandgap edge is filtered out (around 1 μm for silicon), then there is much less temperature dependency of the photodetector. The temperature of the photodetector should be carefully measured and its thermal variations in response compensated for, or the detector should be temperature controlled with a TE cooler [26].

The amplifier design is critical for efficient and low-noise light to electrical conversion. Designs using either a transimpedance or an integrating amplifier have been successful. Integrating amplifiers allow the smallest optical signal to be converted into electrical signals while providing a large dynamic range. Generally 16 bits of analog to digital conversion resolution coupled with gain compression technology is required to span the 200−1,200°C required by the industry. Figure 5 illustrates the wide dynamic range in radiance covered by this temperature span.

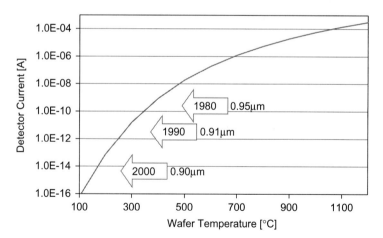

Figure 5 The gray line is photodetector diode current verses wafer temperature for a typical OFT pyrometer. The arrows show the minimum detectable diode current (amps) versus the year of each new generation of pyrometer used with the semiconductor industry. The wavelength used for each of these three generations of pyrometers is showed to the right of the year. The wavelength continued to be shortened to decrease the errors from light transmitted through the wafer.

3. TEMPERATURE MEASUREMENTS IN THE SEMICONDUCTOR INDUSTRY

With the discovery of the transistor in 1947 and the IC in 1960, the manufacturing of semiconductors has grown exponentially to become a global market of over 250 billion dollars by 2008. The heart of the industry is the production of many transistors onto a thin flat silicon crystal slightly larger than a postage stamp. The fabrication of an IC consists of many sequential steps, from the creation of the transistors to the construction of the tiny conductors that link them into a purposeful network. A typical microprocessor with 45-nm features may have nearly a billion transistors on a few hundred square millimeters. From Moore's Law, this density was predicted to double every 24 months since 1968 [28]. It will commonly take some 30 different processes, repeated in a series of 300−600 steps over 4−8 weeks to produce an advanced IC. The production steps to produce an IC include the general technologies of lithography, photoresist, wet and dry (plasma) etch, ashing, cleaning, ion implanting, RTP, CVD, physical vapor deposition (PVD), chemical−mechanical planarization or chemical−mechanical polishing (CMP), and epitaxial (Epi) regrowth. Many of the products that are critical to our society originate from this technology, and precise temperature measurement and control is mandatory for its success.

Wafer temperature is critical to nearly all IC processes. Fab managers have long wanted the complete thermal history of every wafer from incoming inspection through tested finished device. Temperature is hypercritical for some processes including: crystal growth, lithography, rapid thermal implant anneals, poly-silicon deposition, oxidation, Epi film growth, metal organic chemical vapor deposition (MOCVD, typically used in compound semiconductors), silicide formation, some CVD, and plasma etch. The remaining processes are all sensitive to the wafer temperature including PVD, cleans, CMP metal depositions, atomic layer deposition (ALD) as well as nearly all the rest. As processes are changed and refined, better temperature measurement and control is generally necessary.

There are varying degrees of temperature-control difficulty in the various processes in the semiconductor industry. Some processes are easier to control than others, due mostly to their low ramp rates and the extent of their thermal equilibrium with the process chamber. For example, lithography is sensitive to extremely small changes in wafer temperature, but the process occurs near room temperature and all the chamber conditions are well controlled. The thermal conductivity of silicon wafers is high, and this helps reduce wafer thermal nonuniformities if the process speed and temperature uniformity are well controlled. Process repeatability is the key to making consistent ICs, and temperature control is needed to assure that repeatability.

Up until the mid 1990s, dopants were diffused into wafers by large batch furnaces with slow ramp rates (5°C/min) and long soak times. Although temperature uniformity issues indicated that they did not reach equilibrium, they did reach a steady-state condition which allowed for repeatable calibrations. Furnaces were calibrated with long-profile thermocouples and sometimes by wafers instrumented with TCs. The furnaces were controlled by TCs mounted in the furnace wall. The TCs were routinely checked for drift with pure metal freeze point calibration standards to established traceability to the international temperature scale. Large fabs often had a number of pure metal freeze point furnaces.

As processing times became too short for diffusion furnaces, cold-wall, single-wafer, lamp-heated systems evolved. These RTP tools offered many technical challenges for temperature measurement which will be covered extensively in Sections 4.8, 5.4, 6, 7.5, and 7.6. With each generation of semiconductor technology, the thermal budget becomes lower, requiring the control of temperatures beyond the capability of short-wavelength radiation pyrometers. This has led to the development of hybrid systems which use methods exploiting the optical properties of silicon for low temperature control and radiation thermometry at process temperature [5,29–31].

The fluoroptic thermometer which derives temperature by measuring the fluorescent decay time of a thermal phosphor has found use in low-temperature applications where strong EMI fields are present, thereby preventing the use of thermocouples. Generally, the phosphor-tipped sensor is embedded into a wafer holder or electrostatic chuck in environments where the wafer is heated by process plasma, RF induction, microwaves, or a heated susceptor. This technology has the advantage of good resolution at lower temperatures where short-wavelength radiation pyrometers struggle for signal, but it suffers the disadvantages of a contact sensor. Measuring the wafer susceptor to infer wafer temperature assumes equilibrium conditions exist, but there may be a significant heat flow to the wafer which depends on the thermal resistance of the wafer–susceptor gap, gas flows, and the emissivity of both the front and backside of the wafer. To reduce this heat transfer problem, spring-loaded sensors have been designed that push a small high-conductivity "cup" against the backside of the wafer. The inside of the approximately 1-mm diameter cup is coated with the phosphor material, and a small optical fiber both excites the phosphor with LED light and detects the decay to determine wafer temperature. This wafer temperature measurement technique has proved extremely useful when the probe can be calibrated to the process.

3.1. Limits from tool design

The processing of complex thin films presents some of the most challenging temperature measurement conditions in industry. Chambers designs do share some common requirements within the industry even though they vary greatly by manufacturer and process. In recent years, most wafers are processed individually rather than in batches, which requires tools to have rapid cycle time. Also, the existence of even a few parts per million of some contamination can modify the electrical properties of a semiconductor. The transition metals are particularly problematic in that they create electron and hole traps in the middle of the silicon bandgap that results in recombination centers and junction leakage. Gold, copper, and iron with their high diffusivity in silicon are of major concern [28]. With some very limited exceptions, only silicon, carbon, nitrogen, argon, hydrogen, oxygen, and ultra pure de-ionized water are considered benign enough to directly contact the wafer, and so the use of any contact temperature sensor is greatly restricted. The wafers are often rotating or spinning to obtain thermal or depositional uniformity, and processing times are short. The chamber environment may be highly corrosive, of reduced or elevated pressure, and contain plasma, inert gas, or ambient air depending on the process. In some CVD systems the chamber walls and view ports may progressively accumulate deposits which modify the heat transfer from the wafer and make pyrometry difficult. In dry-etch tools, the presence of the plasma interferes with the standard implementation of radiation pyrometry, where the light from the plasma is often significantly brighter than the emission from the wafer. In general, RTP chambers are cold-wall systems to prevent contamination from redepositing onto the wafer, but are highly reflective to create a more efficient uniform irradiance from the heating lamps. As lamp-heated systems obtain peak heating rates up to several hundreds of degree centigrade per second, the irradiance onto the wafer reaches megawatts per meter squared. This intense radiation field makes classical radiation thermometry impossible.

3.2. Limits of contact devices

Wafers are heated by thermal conduction from a hot chuck (susceptor), plasma heated or radiatively from lamps and lasers. Contact devices such as TCs and RTDs have not been successful in the radiatively heated environments. Direct contact with the wafer is unacceptable for contamination constraints, and the wafer is often spinning. The opaque sensors create shadows, and therefore lead to temperature nonuniformities on the target surface. The optical absorption characteristics of contact sensors differ from wafers, and as a result it is difficult for them to establish thermal equilibrium with the wafer. The target temperature for control is the surface of the wafer where the chemical reactions or transformations take place. With nonequilibrium environments present, heat conduction down the electrical leads becomes an important factor.

The National Institute of Standards and Technology (NIST) developed a thin-film-instrumented wafer for calibration and verification of chamber models that sidestepped many of the issues listed above [32−41]. This device traces its calibration to NIST temperature scales. This technology is addressed in detail in Section 6.

For applications where it is appropriate to heat the wafer with a chuck (hot plate), contact sensors can be embedded directly into the heater. The wafers are often held flat with electrostatic or vacuum chucks, which also help reduce thermal contact resistance. When heat flows from the chuck through the wafer and irradiates the chamber walls, the contact resistance, wafer emissivity, and chamber wall absorbance may be dynamic variables that limit the accuracy and repeatability of the embedded sensor measurement.

3.3. Temperature measurement requirements in the semiconductor industry

If each chip is to be nearly identical, each process must be spatially uniform across the wafer and repeatable from wafer to wafer. In general, process variations must be less than 1% (3σ). Each wafer will have a thermal budget depending on the specific device structure and the stage of processing. This budget is the allowable amount of temperature over time that can be tolerated before previous processing steps are degraded. Historically, semiconductor devices have become smaller, resulting in a decrease in thermal budgets. The lower thermal budgets mandate a reduced time at process temperature. One example is the activation of dopants implanted into the silicon surface for an active semiconductor material to form functional transistors. During the 1970s and 1980s, processing times decreased from many hours to several minutes, and in the next two decades RTP reduced the times from minutes down to a second. To create the devices of the future, thermal cycles will evolve from seconds to microseconds to meet the requirements for deposition, dielectrics, annealing, and activation.

Semiconductor processing tools operate around the clock, and to shut a processing line down is extremely expensive and, thus, reliability is a crucial issue. Successful radiation thermometry technologies will combine industrial robustness with long-term repeatability and increased cost effectiveness. Wafer uniformities of 1% or better are a requirement expected to continue into the future, with the additional requirement of increased measurement bandwidth as the processing times shorten. If these high standards cannot be met, then thermally insensitive processes will have to be developed.

3.4. Optical properties of silicon

The thermal optical properties of semiconductor materials are very complex and therefore offer many challenges for radiation thermometry [29−31,42,43]. Single-crystal silicon exhibits a bandgap energy structure. Any photon with energy greater than the semiconductor bandgap can be absorbed by the process of transferring electrons from the valence to the conduction band. Photons with less energy than the bandgap can interact with the intrinsic electrons that have enough thermal energy to be in the conduction band or with the lattice vibration modes. This gives a strong temperature dependency to the emissivity at wavelengths longer than about 1 µm for silicon. This is illustrated by Figure 6, which is a family of curves for various temperatures of emissivity as a function of wavelength and temperature for single-crystal silicon [29,30]. The transmission of silicon wafers is also largely determined by the bandgap, with near zero transmission at wavelengths less than 0.90 µm and with increasing transmission at wavelengths longer than the bandgap (see Figure 7). The transmission is also strongly temperature dependent. A small percentage of transmission variation can cause very large temperature measurement errors when background radiation is large as compared to the emission from silicon.

The dopant, its concentration, and the degree of activation will all influence semiconductor band structure and therefore the emissivity of the substrate. Backside wafer roughness and film stacks which are used to trap defects and to passivate the wafer can greatly influence the effective

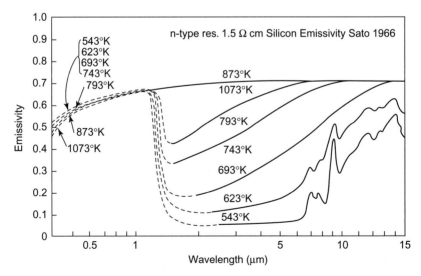

Figure 6   Silicon emissivity. Family of curves for various temperatures as a function of wavelength for single-crystal silicon [30].

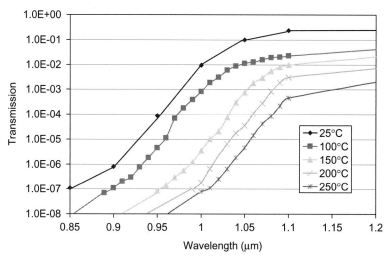

Figure 7   Silicon wafer transmission near the bandgap. This graph shows how transmission decreases as the wafer is heated versus wavelength. These data are for a lightly doped thin silicon wafer.

emittance of the surface. The effects of surface roughness on emittance is discussed in Section 7.1 However, the device structures and film stacks on the process side of the wafer, with physical dimensions near the pyrometer bandpass, have a strong influence on emittance by creating large diffraction effects. Figure 8 is a graph of emissivity modeling results for the growth of single-layer films for silicon nitride, silicon oxide, and polysilicon. The repeated cycles are from thin-film interference at 950 nm.

Figure 9 [31] shows the modeling results for a film stack of varying thickness of polysilicon over varying thickness of silicon oxide on a silicon substrate. These data have been confirmed with measurement to a high degree of accuracy and are used in the development of emittance standards. Even a casual observation of the graphs should suffice to convince the reader of the major influence that film stacks can have on temperature measurement. Wafers patterned with semiconductor device structures also can act as diffraction gratings resulting in angular and wavelength dependence effects. These effects manifest themselves in the beautiful rainbow of colors characteristic of device wafers, but create a major source of complexity in determining surface emittance.

The emittance of the wafer often varies during processing. The wafer emissivity can change simply as a function of wafer temperature [44]. The growth of an oxide layer will also affect the emissivity (Figure 8). Figure 10 shows a real-time measurement of wafer emissivity as the oxide is growing. Note the resolution of the emissivity on Figure 10 as measured with the Ripple Technique (see Section 7.7). Emissivity changes with

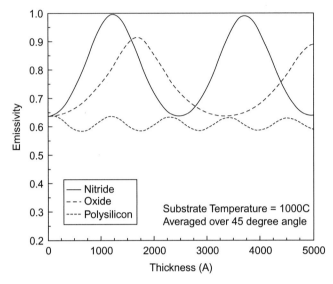

Figure 8 Emissivity modeling results for the growth of single-layer films for silicon nitride, silicon oxide, and polysilicon. The repeated cycles are from thin-film interference at 950 nm.

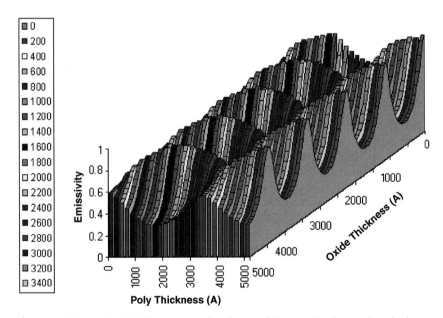

Figure 9 Spectral emissivity contour showing modeling results for varying thickness of silicon oxide over varying thickness of polysilicon in a crystalline silicon substrate at 1,000°C with a 45° averaging angle at 0.9 μm.

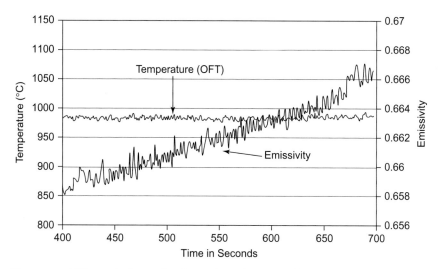

Figure 10   This graph shows the change in emissivity with time during the formation of an oxide layer on a silicon substrate in an RTP chamber at 1,000°C with dry air. The emittance value was determined by a Ripple pyrometer and is used to calculate temperature for stable control during film growth. Note the resolution of the emissivity measurement.

deposition of any film, epitaxial growth from the crystalline substrate, or transformation of NiSi or CoSi phases (Figure 11) will result in a temporal dependency of the optical properties [31]. Silicon and filmed wafers have a large angular emission dependence as can be seen in Figure 12.

Crystalline silicon is by far the most common semiconductor material, but the same optical complexities exist for polycrystalline, amorphous silicon, as well as other semiconductor materials like gallium arsenide. The combination of the wide range of variables for semiconductor materials and the stringent requirements for temperature measurements create one of the most challenging applications in industry today.

4. APPLICATIONS

The processes needed to produce semiconductor devices have evolved into the most tightly controlled industrial processes on the planet. These processes produce huge volumes of materials that are pure to parts per trillion. The semiconductor devices produced have size features that are on a nanometer scale. The yields in these manufacturing plants (called fabs) are designed to maximize profits. The fact that an increase or decrease in a fab yield of just 0.1% can equate to millions of dollars in a very short time drives tighter process control. The most universal process variable within

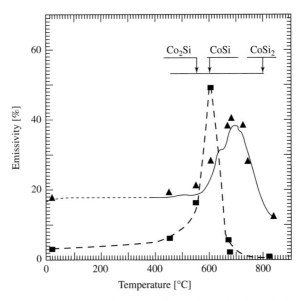

Figure 11 The emissivity during the phase change from Co_2Si to $CoSi$. The triangles and squares are measured at 2.4 and 10 μm wavelength, respectively. Silicides are commonly used in the semiconductor industry to diminish electrical contact resistance [42].

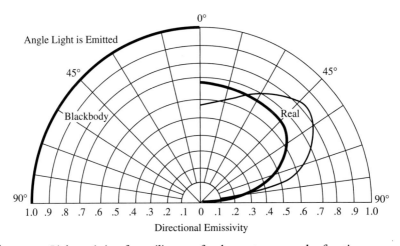

Figure 12 Light emission from silicon wafers has a strong angular function as compared to an ideal blackbody. The two curves on the right are two different wafers. The wide line is a bare silicon wafer with only a thin native oxide. The thin line is a wafer with polysilicon on oxide (SiO_2). Here the emission has a lobe and less emission perpendicular to the surface than at 45°. Both wafers have low emissivity at shallow angles, and therefore shallow angles would be poor for pyrometer collection [44].

semiconductor processing is temperature. Therefore, tighter temperature control has direct monetary value in this industry and has driven the development of improved pyrometers for semiconductor applications. Large efforts have been made for improving pyrometric measurement to meet the needs of the semiconductor industry. Pyrometric temperature measurement improvements in the last 20 years have been driven by the semiconductor industry because of its economic impact. The demand of silicon technology for solar cells has evolved, and 2009 is the first year that more silicon is being used for solar photovoltaics (PV) than ICs.

The pyrometer manufactures are seeing this change due to the growth of solar PV and are focusing some their efforts on the needs of solar cell manufacturing. In one solar cell application, a solder joint is made at about 250°C on each PV cell. The solar PV manufacturer thought there was a $\pm 20°C$ variation in their process, based on the thermocouple measurement. A manufacturer's representative with a pyrometer quickly showed a deviation of more than $\pm 50°C$. The process control for this PV process is now better controlled by pyrometric monitoring by an order of magnitude. This sort of success was achieved earlier in the semiconductor industry.

There are an increasing number of semiconductor processes that have temperature controlled by pyrometric measurements due to their minimally intrusive accurate temperature sensing. Most IC processes require controlling the wafer temperature as tightly as possible, and today's pyrometers meet this requirement best.

The biggest pyrometer challenges today are the same as they were 30 years ago. The technical hurdles are:

1. stray light problems,
2. unknown and changing emissivity,
3. low temperature and hence low emission, and
4. end–user training in proper pyrometric techniques.

4.1. High-density plasma chemical vapor deposition (HDPCVD)

A major success has been using pyrometers in the measurement and closed-loop control of wafer temperature within HDPCVD. CVD dates to the early years of semiconductor devices. The addition of high–density plasma has improved the quality of the films and driven tighter control of wafer temperature during the film growth. Most HDPCVD tools use a sapphire lightpipe embedded in a hole in the susceptor looking at the backside of the wafer. These lightpipes are typically connected to the photodetector with a flexible quartz fiber-optic cable. In some cases the lightpipe goes directly to the photodetector eliminating the optical cable loses. The photodetector is generally a silicon detector with a bandpass filter in the NIR. The bandpass is chosen to optimize signal and to eliminate background signals from the plasma. Typically, HDPCVD pyrometers operate in the wavelength regions of 0.9 or 1.1 µm depending on process temperatures or wafer features.

4.2. Metal organic chemical vapor deposition (MOCVD)

Until recently, MOCVD was used only for compound semiconductor devices, but now the use has spread to other application areas. The use of MOCVD has increased for new high k dielectrics, high-efficiency solar cells, multi-junction PVs, and LEDs. These film growth chambers called reactors typically have multiple wafers rotating in planetary motion that makes temperature measurement a challenge. The reactors grow films which cause changes in wafer emissivity during the film growth process. The wafers often sit on an SiC-coated puck (susceptor) which rotates on air bearings on top of a larger susceptor which rotates in the opposite direction. The rotation and multiple susceptors make backside measurements nearly impossible. Frontside measurements are complicated by the ever-changing emissivity. The solution has involved carefully measuring the RF-heated susceptor temperatures from the backside and at the same time measuring the wafer frontside using an emissivity-corrected pyrometer. The pyrometer used by one vendor measures the wafer reflection real time, with similar collection angles, and uses this reflection measurement to determine the emissivity to use for the pyrometer. This reflection information is also used for film thickness monitoring [43].

4.3. Epitaxial

In Epi growth, a film is deposited which is nearly the same crystal structural as the underlying material. In decades past, blanket silicon epi was conducted at temperatures of $1,000-1,200°C$. Today there is a wide range of selective and lattice stressed epi processes with many materials other than slightly doped silicon. These selective and strained epi processes are much lower in temperature, around $500-600°C$, than blanket epi processes. The lower temperature epi processes are very temperature sensitive and require precise control. Pyrometers are being used, and some at longer wavelengths (3.3 and 3.4 µm) in an attempt to avoid the lamp light peak at ~ 1 µm. These epi applications have most of the conventional pyrometer problems including stray light from the lamps and an unknown and changing wafer emissivity.

4.4. Solar photovoltaics

The rapidly growing market for PV is exciting and demanding for the pyrometer suppliers. The market is not mature, and there are many competing manufacturing processes and cell technologies and, like in IC, manufacturing cost is critical factor. There are processes that deposit silicon, compound semiconductors, metals, and organics onto glass, stainless steel, and organic plastics. When depositing silicon or CdTe onto glass, a longer wavelength, a ~ 5-μm pyrometer, can be used to measure the glass temperature. When depositing onto stainless, a shorter wavelength is likely to be part of the solution. Solar technologies are being developed worldwide and there are opportunities for many aspects of pyrometry development.

4.5. Nonoptical techniques

There are two wafer temperature measurement methods worth noting that are being used which are complementary to optical techniques. These two methods are particularly applicable to new low-temperature processes. The first method involves fine wire thermocouples embedded into the test wafers with an adhesive that has an emissivity which approximates that of the wafer. These test wafers have some severe limitations but are widely used for equipment development. It is worth noting that these are fine-wire thermocouples not TFTCs. The other method uses wafers that are built into complete temperature data loggers with multipoint solid-state measurement, record and remote read circuits measurement capability sandwiched between two thin wafers. These wafer data loggers store all the temperature measurement information from the complete test run. These wafers generally have many RTDs to allow mapping the wafer temperature through the process. These are typically used in low-temperature plasma processes or CMP. The temperature data is read after the test wafer is processed and unloaded. These wafers allow for full rotation and complete process cycle to be monitored and logged. These temperature-monitoring wafers only function for lower temperature processes.

4.6. Wafer measurement applications

Wafer processing requires knowing and controlling wafer temperature throughout the process. Wafer temperature is the most significant process variable in many critical process steps [1,45]. The vast assortment of equipment (tools) used to process wafers are always changing as new and improved processes come on line. This changing process and tool landscape drives an ongoing development effort to measure and then control the wafer temperature. Each and every tool and process requires new temperature measurement development. Development of temperature-measuring equipment is required for each tool because the conditions around the wafer make simple measurement difficult. There are issues which generally need to be solved or overcome to make acceptable measurements for each new technology. In most cases, contact cannot be made with the wafer and, even if contact is made, good thermal contact is almost impossible which results in optical noncontact methods being considered. The two largest issues with all pyrometric measurements are emissivity and background or stray light. The unknown and constantly changing wafer emissivity causes huge errors, and background light can render the measurement useless in a large number of processes. This stray light often comes from the plasma used in the process. The plasma light is frequently brighter than the emission from the wafer and will be collected in the pyrometer input optics. The measurement technique that is tool specific must deal with these light noise problems. One may consider looking at the backside of a wafer on a susceptor as a solution. However, the wafer is semitransparent and the transmission is both temperature and film dependent. In some situations the transmission can be used as a measure of temperature. The temporal signature and spectra of plasma radiation can be used to filter the noise from the desired pyrometric signal.

5. CALIBRATION

5.1. Lens system radiation thermometers

5.1.1. Lens collection
Most pyrometers use a lens (not a lightpipe) to collect emitted light from the hot object being measured. The lens allows smaller hot areas at greater distances to be measured than the lightpipe and therefore less range of angles. However, the lens allows the entire pyrometer to be placed farther from the hot and harsh conditions. The lens can collect the light and direct it to the photodetector or to a flexible fiber-optic cable which transmits the light to the photodetector.

5.1.2. Pyrometer systems in the semiconductor industry
The most important feature of any pyrometer used in the semiconductor industry is repeatability. This industry produces wafers by millions and consistency is the key requirement. Pyrometers are now widely used in the semiconductor industry to measure and control actual wafer temperature throughout high-temperature processing. The widespread direct measurement of production wafers occurred in the 1990s as the wafer value grew

and fostered tighter process control. Companies like Accufiber (USA) and later CI Systems (Israel) filled this need for better process control and actual real-time wafer temperature measurement. In the early use of OFT, repeatability was the main issue, but things have changed and now fabs and their processes are spread worldwide and require internationally traceable temperature control [1,31,45].

Today most semiconductor pyrometers have some similar features including highly sensitive photodetectors (high normalized detectivity D^*) and advanced amplifier circuits. The collection optics are usually designed for the process tool that the pyrometer is used on. The collection optics can be a lens or a lightpipe system, designed to give repeatable spot size and measurement. Most of these pyrometer systems use NIR light, with higher temperature ($>250°C$) measurements using a narrow band at $0.9\,\mu m$ and lower temperatures ($<300°C$) using bands around $1.6\,\mu m$. A metal oxide narrow-pass interference filter is used with a silicon photodetector for the 0.9-μm light. Typically, an InGaAs detector is used for the lower temperature measurements; sometimes extended (longer) wavelength InGaAs detector is used to push the measurement to lower temperature although the long-wave transmission of the wafer can cause difficulties. The short-wavelength 0.9-μm light is used chiefly because the wafer is opaque at wavelengths less than the bandgap. The log amplifier has been replaced with the transimpedance amplifier and the integrating amplifier, both of which improved the methods of handling low-level photodetector signal and allow for wide dynamic range when incorporating modern gain switching techniques. The transimpedance amplifier circuit will cover the range from 1×10^{-12} to $1 \times 10^{-3}\,A$, and the integrating amplifier circuit will cover the range from 1×10^{-14} to $1 \times 10^{-3}\,A$.

5.2. Lightpipe radiation thermometers (cold calibration, hot calibration, *in situ* calibration)

5.2.1. Hot versus cold lightpipe calibration

All pyrometers used in the semiconductor industry are calibrated, including lightpipe systems. Pyrometer calibrations typically use a traceable blackbody furnace measurement standard to check and characterize the collection optics, wavelength selective filter, photodetector, amplifier, and algorithms.

Lightpipe systems require calibration to "match" the process tool conditions. This special requirement is needed because of two lightpipe features. First, the lightpipe has a very wide (about $180°$) collection angle and will collect any light within that large acceptance angle. Second, the lightpipe will operate differently if it is hot or cold. If the lightpipe is cool, it will only transmit the light that it collects minus the very small transmission losses. If the lightpipe is hot, it can have self-emission which will add to the collected light. The details of lightpipe technology were discussed in Section 2.

There are three basic types of lightpipe calibrations using blackbody sources [5,25,38,41,46]. The "hot calibration" is where the lightpipe is inserted into the hot blackbody furnace, and the tip of the lightpipe is heated to close to blackbody temperature. In this calibration the self-emission from the tip of the lightpipe in included in the measurement. This type of calibration is sensitive to the insertion length. The "cold calibration" is where the lightpipe is kept cool and it only looks into the blackbody cavity. Here there is no self-emission. In both of these calibration methods, hot and cold fingers can sometimes be used to either heat or cool longer lengths of the lightpipe in the hope that the real measurement conditions can be calibrated. The third method is *in situ* calibration using either an instrumented wafer or real process temperature scale such as oxide growth or implant anneal.

5.3. Blackbody simulators

The development of sophisticated and complex multizone pyrometers for control in RTP tools created the need for an *in situ* method of calibration that could be implemented routinely and efficiently. Removing the pyrometers for external calibration would shut down the tool from the productions line for an extended time. External calibration techniques do not include the important effects of emittance enhancement from the chamber environment. Traditional blackbody calibration furnaces for high-temperature calibration are large, contaminating, and slow to respond, making them impractical as a field calibration source. The use of *in situ* wafers instrumented with TCs cannot be implemented as the wafer rotates to improve uniformity and for the stability of the thermal controller. This need is met with the development of small portable IR sources which can be conveniently placed in the processing chamber (Figure 13). The challenges to construction of such calibration sources are the necessity of simulating the spectra of blackbody radiation within the wavelengths of the pyrometer bandpass, and assuring repeatability over time in an industrial context. Several versions have been developed and have proven to yield reliable calibrations over long periods of time.

One version developed specifically for narrow bandpass pyrometers employs a combination of LEDs, which when combined will render the blackbody spectra over the region of interest (Figure 14). It is important that the calibration source be nearly Lambertian in the distribution of output angles as LEDs have a narrow angular output compared to the wide range of optical acceptance for most lightpipe radiation thermometers (LPRTs) (Figure 15). Precautions must be taken in selection and

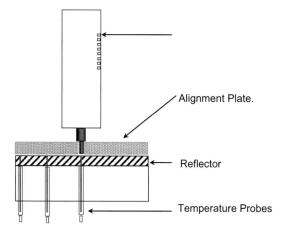

Figure 13   Photograph and schematic of LED blackbody for RTP tools.

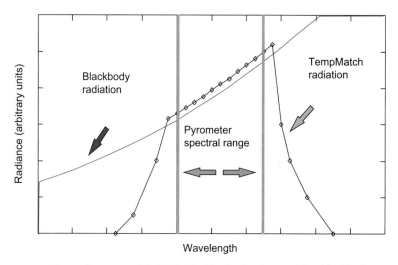

Figure 14   Comparing a portable LED blackbody simulator with a blackbody spectrum at the calibration temperature.

preconditioning the LEDs. Monitoring and real-time feedback control of the output is necessary to meet the stability and repeatability requirements.

An example of an error analysis with *in situ* calibration source was performed to look at the total error from the LPRT and the calibration source. Summing in quadrature, the instrument error of $\pm 1.31^\circ$C with the induced TempMatch™ error of $\pm 1.26^\circ$C gives a total system calibration error of $\pm 1.75^\circ$C (3σ). Errors introduced due to the transfer standard only increased the total error of the LPRT measurement from $\pm 1.31^\circ$C to

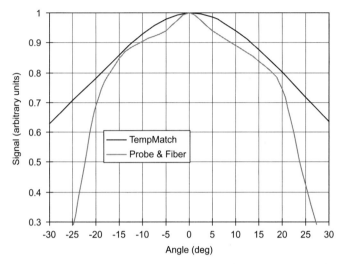

Figure 15 Illustration of the importance that the angular distribution of the LEDs simulates the Lambertian distribution of a blackbody over the NA of the pyrometer.

$\pm 1.75°\text{C}$. This performance is demonstrated to be sustained for a large population of instruments over a year of operation.

In order to meet the requirements of pyrometers with wide optical bandpass, an alternative design incorporated a very small IR thermal emitter which provided graybody radiation over the larger spectral region. LPRT calibration issues of sensor contamination, drift, and cable/coupler variability can be greatly reduced by *in situ* calibration. *In situ* calibration offers a shorter path of traceability and therefore reduces the error stack-up associated with traditional methods. *In situ* calibration tools have proven to be a practical realization of a portable transfer standard meeting the demanding requirements for use in semiconductor processing equipment [47–49].

5.4. Process repeatability standards (PRS) in RTP

As stated in Section 3.4 the requirements for repeatable temperature measurement and control in RTP are quite stringent. The high ramp rates and short processing times create a dynamic environment which does not obtain thermal equilibrium or even steady state. The intense background light and time-varying emittance make traditional radiation measurements uncertain. Under these conditions it is difficult to implement an accurate traceable path to an absolute standard.

As with many manufacturing technologies, repeatability is more critical than accuracy in RTP. Processes which can be empirically determined to produce an acceptable product and are repeatable may satisfy the needs of industry without absolute knowledge of the underlying fundamental parameters such as temperature, pressure, voltage, and current. Other processes, however, require accurate and traceable measurements and hence the NIST, and other national labs have put forth a substantial effort to continually improve traceability in RTP measurements over the years [41]. The lack of accuracy in temperature measurements became most apparent when looking at tool−to-tool variability, or the variation in processes between different tool vendors.

The one thing that has enabled RTP to develop as a successful technology has been the existence of several key PRS. These were developed back in the diffusion furnace days, when ramp rates were low enough for good traceable TC measurements and have been extendable for the modern processes. They are used worldwide every day, and without them RTP would not exist as a viable technology.

Two PRS, commonly referred to as monitors, that are the most widely used are the measurement of change in sheet resistance during RTA and the measurement of change in an oxide thickness in RTO. Both of these standards are based upon a high-purity single-crystal substrate, that is, a silicon wafer.

The RTA standard is created by implantation of silicon with ions of a dopant, usually boron, phosphorus, or arsenic. The dopant profile is primarily determined by the ion velocity distribution, while the dose is a function of the ion current over time (charge). When heated, several temperature-dependent changes will proceed. The implanted ions will move from interstitial sites to substitutional sites in the lattice (activation), and lattice damage left from the implant will anneal out (anneal). A function for the change in sheet resistance with time at process temperature can be developed which is very sensitive and highly repeatable. An example of an RTA standard: $1,050°C$, $20\,s$, $1 \times 10^{16}\,ions/cm^2$, $40\,KeV$, with 10% O_2, yields $-0.35\,\Omega/Sq/°C$. The sheet resistance is measured with a four-point probe which has become an industry standard.

The RTO standard is based on the measurement of the growth of an oxide given a specific humidity and oxygen concentration. As this is a diffusion-limited process under controlled conditions, it is very sensitive to temperature and highly repeatable. An example of an RTO standard: $1,100°C$, $60\,s$, yields $0.828 \pm 0.015\,A/°C$. The change in oxide thickness is measured by ellipsometry.

The advantages of PRS are the high degree of repeatability, composition from compatible materials, and potential of identifying nontemperature-related processing problems. The repeatability worldwide is within $\pm 5°C$ (3σ) and within $\pm 1°C$ (3σ) for a single vendor with a single fab. The disadvantages of PRS are that it is a nonreal-time process, results in only a single measurement point for the process cycle, is user dependent, expensive, and lacks ease in providing traceability to a freeze point standard. In spite of the disadvantages, process monitors are the final say for tool calibration, and they are the mainstay of the industry.

6. *IN SITU* CALIBRATION OF RADIATION THERMOMETERS IN RTP TOOLS

6.1. Proof wafer TC thermometry

TC-instrumented wafers [38,41,46] have been widely used in RTP systems to obtain wafer temperature distribution (static and dynamic temperature uniformity). They are useful for optimizing lamp irradiation patterns, and for in-chamber calibration of radiation thermometers (RTs). The purpose of such a calibration procedure is to transfer the TC temperature scale to the RT accounting for emissivity and stray radiation effects.

Three features need to be considered in assessing this approach. First, the reliability of type-K (Ni−Cr alloy vs. Ni−Al alloy) TC wire, especially in small gauges, is questionable for high-temperature applications requiring an uncertainty of $\pm 2°C$ at $1,000°C$ in the presence of a slightly oxidizing atmosphere. Second, the difference between the indicated TC temperature and the silicon wafer surface temperature is dependent upon the junction-wafer bonding structure, as well as upon effects caused by wire leads. Third, the temperature scale of the RT is appropriate only for the proof-wafer conditions and could be considerably in error when sighting on production wafers under different heating conditions or with different wafer emissivity.

A method to achieve high-accuracy, reliable *in-chamber* calibration of RTs can improve measurement practices. The instrumented proof-wafer concept is a viable approach. The above three features can be addressed as follows. Using the new platinum versus palladium (Pt/Pd) *wire* TCs made from pure metals, uncertainties for calibration *and* use of less than $\pm 1°C$ can be obtained at temperatures up to $1,100°C$ under ideal conditions [50]. Using Rh/Pt or Pt/Pd TFTCs, indicated temperatures will be those of the wafer surface, virtually unaffected by the presence of the thin films. The remaining challenge is to perform the calibration under conditions in which the radiation environment (emissivity variability and stray radiation) can be controlled and characterized.

6.1.1. Type-K wire TC thermometry

TCs with wires made of metal alloys are used extensively throughout the semiconductor industry. It is usually assumed that the TCs are stable with time and unaffected by any process conditions, and that the TC measures the difference in temperature between the measuring junction and the reference junction. In fact, the accuracy of a thermocouple depends on maintaining homogeneity of the thermoelements in the regions of temperature gradients [51]. For many commercial TCs, these assumptions may be poor due to phase changes in the alloy materials and to inhomogeneities, compositional or physical, along the length of the wires.

TC-instrumented silicon wafers are available from several commercial sources to use as calibration standards in RTP systems. Such calibration wafers typically have an array of TCs attached to one side of the wafer so that temperature uniformity can be measured. Type-K TCs made from 0.08 mm or 0.13-mm-diameter wires are currently the predominant choice for this purpose. The measuring junctions of the TCs are embedded within alumina-based cement in small cavities machined in the wafer surface. To minimize mechanical stresses on the embedded junctions in use, the TC wires are anchored near the periphery of the wafer by cementing them to ceramic posts. The wires are insulated electrically with braided silica sleeving that is suitable for use up to 1,100°C. Calibration wafers instrumented with *fine-wire* type-K TCs may be used up to 1,100°C. Above 500°C, however, these fine-wire TCs should be used only in inert gas atmospheres.

6.1.2. The new Pt/Pd wire TC

NIST has developed improved wire TCs consisting of high-purity Pt and Pd that minimize errors associated with instability and inhomogeneities [50]. Such TCs constructed using 0.5-mm-diameter wire have uncertainties less than 0.03°C below 960°C and less than 0.5°C at 1,500°C. Studies have been conducted to evaluate the effects of annealing, heat treatment, and oxidation in air. In collaboration with the Istituto di Metrologia "G. Colonnetti," the national metrology institute in Italy [50,52,53], NIST has produced an accurate Pt/Pd reference function for the range 0−1,500°C to facilitate the use of this new TC in industrial and scientific applications. As discussed subsequently, this new TC is used to determine the temperature of the proof wafer at the junction between the thin-film and wire leads of the new TFTCs.

6.1.3. The new Pt/Pd and Rh/Pt TFTCs

TFTCs have advantages over present wire TCs because they cause less disturbance of the heat transfer to and from the silicon wafer during high heat-flux loading. High-purity Pd, Rh, and Pt are attractive as TFTC materials for the RTP application because of their high stability, their good resistance to oxidation at high temperatures (up to 1,000°C), and their compatibility with Pt/Pd wire TCs. The stability of the interfaces between the films and the wafer are the most critical for thin films. At these high temperatures, TFTCs have advantages over thin-film resistance thermometers since they are not sensitive to mass changes which affect resistance, and the TC measuring junctions may be small. The key factor in TC stability (and also necessary for resistor stability) is uniform composition of the electrically conducting phase. Stress level and other metallurgical factors such as grain size and defect population have smaller effects on the Seebeck coefficient when compared with corresponding effects on the thermal coefficient of resistance. Reactions with oxygen from the atmosphere or solid phases, compound formation with nearby solid constituents, or diffusion to and from nearby phases, all can cause drift in the thermoelectric output and the Seebeck coefficient of the thermoelements and can have high vapor pressures of oxides. Because thin films inherently have short diffusion distances, these reactions cause instabilities at lower temperatures than the instabilities observed with wire TCs. At 1,000°C, the use of thin-film alloy TCs is practically precluded because of preferential reactions of their constituents. For example, TCs with thermoelements of platinum–rhodium alloys exhibit appreciable drift due to preferential oxidation of rhodium at 700–900°C [54,55].

The usefulness of high-temperature materials for TFTCs on silicon wafers has been described by Kreider and Gillen [56]. In summary, Pt/Pd TFTCs on silicon wafers have a high thermoelectric output (about 16 μV/K), but if the films are less than 1-μm thick, the practical limit for their use as an RTP calibration wafer is approximately 850°C, with brief excursions up to 900°C. The Pd leg is more sensitive to degradation by pore growth, coalescence, and delamination. Pt thin films have similar problems at 950–1,000°C. Some modifications would make the films more refractory: for example, increasing the thickness, adding a grain growth inhibitor such as an insoluble precipitate, or covering with an oxide can retard coalescence.

Both Rh and Pt thin films appear to be very stable in this application at temperatures up to 1,000°C. Rh thin films have very low hysteresis versus Pt and excellent repeatability in TFTC calibration tests. Optical microscopy, scanning electron microscopy, secondary ion mass spectrometry (SIMS) depth profiling, and stability of resistance of the films confirmed their suitability for this application. The absence of evidence of silicide reactions with any of the sensor materials indicated that 300 nm of thermal oxide on the silicon was an adequate diffusion barrier.

6.1.4. The TFTC proof wafer

In order to compare the performance of the commercial wire TC proof wafer to one with the thin-film differential TCs, a calibration wafer was prepared with both systems. The thin-film differential thermocouple measures the difference in temperatures between the LPRT target on the wafer and the Pt/Pd thermocouple junction at the periphery of the 100-mm wafer. Since this difference is generally less than 10°C, the measurement is not very sensitive to the accuracy of the TFTC calibration. The TFTC silicon proof wafer shown in Figure 16 is instrumented with conventional fine-wire (0.08-mm diameter) type-K TCs, the new Rh/Pt TFTCs, and the new *wire* Pt/Pd TCs. The region at the center of the

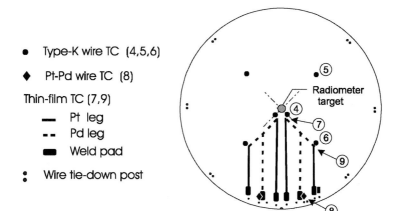

- • Type-K wire TC (4,5,6)
- ◆ Pt-Pd wire TC (8)

Thin-film TC (7,9)
- ━ Pt leg
- ▪▪ Pd leg
- ▬ Weld pad
- ⦙ Wire tie-down post

Figure 16 Proof wafer for comparing temperature measurements from type-K and Pt/Pd wire TCs, Pt/Pd TFTCs, and RTs.

wafer in close proximity to type-K TCs is intended as the primary target for sighting of an LPRT.

The 200-mm diameter silicon wafers purchased from the Semiconductor Manufacturing Technology (SEMATECH) consortium were thermally oxidized to an oxide thickness of 300−1,000 nm. These wafers were alcohol rinsed, dried with nitrogen, and ultraviolet/ozone irradiated to remove adsorbed hydrocarbons. The thin films were applied at NIST using physical masks prepared from 3-mm-thick aluminum alloy plates with beveled 0.5-mm patterns for the sputter deposition of Pt and Pd as shown in Figure 16. This array provides two banks of TFTC junctions with two junctions near the center where the RT measurement is made and two at positions halfway to the wafer edge. The sputter deposition of 1-μm-thick films from 99.99% Pt and 99.97% Rh with an 8−10-nm-thick Ti bond coat is described in more detail in Kreider and DiMeo [55].

The thin-film pattern included welding pads 10 mm from the edge of the wafer and the Rh pad was covered with Pt to improve its welding characteristics. Those pads connected with the "T" of the Pd thin films also accommodate a welded connection with Pt wire which completes a Pt/Pd wire TC junction on the pad. These TCs were constructed using fine (0.1-mm and 0.25-mm diameter) Pt and Pd wires that had been calibrated by a comparison calibration method [52]. Welded connections to the thin-film pads were accomplished with a Hughes[1] VTA 90 parallel gap welder [46] using 0.12 mm × 0.2 mm welding tips. The weld joint was made

[1] Certain commercial equipments, instruments, or materials are identified in this paper to foster understanding. Such identification does not imply recommendation or endorsement by the NIST, nor does it imply that the material or equipment are necessarily the best available for the purpose.

between the 1-μm-thick Pt thin film and a 0.1-mm diameter extension wire of Pt or Pd (10-mm long) which was butt welded to calibrated Pt and Pd wires of 0.25-mm diameter. These wires were anchored to the wafer with alumina cement in alumina tubes mounted in the wafer by SensArray. The wafer also had six conventional type-K wire TCs installed by SensArray in the locations shown in Figure 16. This instrumented wafer permits a comparison of temperature measurements from the RT, wire TCs (K and Rh/Pt), and Pt/Pd TFTCs. The difference in temperature measured on the wafer center between surface-mounted type-K wire thermocouples and the thin-film Rh/Pt thermocouples was about 2°C between 750°C and 875°C [57].

6.2. Calibration of TFTCs on silicon wafers

TFTCs have been used on silicon process-type wafers for the calibration of radiation temperature measurements in an RTP tool. TFTCs had been used in a broad range of applications for measuring surface temperatures [58,59]. The advantages of TFTCs include their small size (10^{-6}-mm^3 junctions), fast response, low cost, and flexibility in design and materials. This fast response time was demonstrated with an excimer laser to be faster than 1 μs [60,61]. Each application of TFTCs requires specific considerations of the materials of the thin films, their fabrication parameters, the substrate material, electrical insulation and connections, and the thermal and chemical environment of the measurement. The measurement of temperature using TFTCs also requires a calibration under the conditions present during service. Although standard reference tables [62] have been defined for the electrical response of wire TCs, it was determined from the earliest research [63] that TFTCs would require a specific calibration reflecting the relationship between the substrate material and the thin film and its physical, chemical, and metallurgical condition.

TFTCs can be fabricated from a much broader range of electrical conductors than wire TCs [64−66], but even with the same nominal composition as the wire TC, the TFTCs may have different thermoelectric coefficients. Also, the TFTCs are used to measure temperature differentials across short distances on the substrate and cannot be calibrated by standard procedures. NIST [67] has developed an improved method to calibrate TFTCs. The calibration test method for TFTCs uses 10 mm × 50 mm silicon wafer substrates using the comparison method up to 950°C with NIST-calibrated, highly accurate, Pt/Pd wire TCs as the reference thermometers. The resulting expanded uncertainty ($k = 2$) for the method for calibration of the TFTCs was less than 0.2% of the electromotive force (EMF) at 800°C.

The calibration system for TFTCs on silicon wafers was designed to obtain the thermoelectric output of a TFTC on a silicon wafer compared to a pure Pt wire. The design (Figure 17) includes a water-cooled clamp at

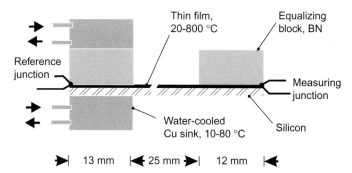

Figure 17   Calibration apparatus for TFTCs.

the cold junction with an electrical contact made by the platinum wire of the Pt/Pd reference TC. With effective water cooling, we could insert this water-cooled block with insulation into a tube furnace maintained at temperatures up to 1,000°C. The calibration test coupons were 50 mm × 10 mm wide for ease of fabrication and their ability to simulate severe thermal gradients of various applications. These coupons were coated at the same time as the silicon wafer to insure their representation of the wafer films. A Pt/Pd reference-wire TC was clamped to the hot junction of the thin film, creating an electrical contact and measuring junction. We used Pt/Pd reference TCs which were calibrated at NIST with an expanded uncertainty ($k = 2$) of 0.2°C at 1,000°C. The clamp for the hot junction/electrical contact was a 0.5-mm Pt wire wrapped around the semicylindrical boron nitride heat sink and the silicon substrate (Figure 17). This whole assembly was inserted into a tube furnace lined with a fused-quartz tube that permitted an N_2 purge. Typically the test was run at mole fraction of 0.02% O_2.

The instrumentation for these tests included a Keithley 7001 scanner, a Fluke 85056a voltmeter, and a PC for data acquisition. The thermocouple reference junctions were maintained at 0°C in a distilled-water/ice bath. The calibrated TCs were terminated in a distilled ice-water mixture to establish the reference junctions. Meter uncertainties were minimized by reading the digital voltmeter to a resolution of 0.1 μV, averaging four readings, and using standard precautions in handling low-voltage signals, such as shielding and correcting for the voltmeter zero offset. The sequence of EMF measurements was designed to compensate for temperature drifts that were linear in time. Using the mean deviation data and checks of the voltmeter calibration, an estimate of the EMF standard uncertainty is $u(E) = 0.4 \times 10^{-4}\ E + 0.1\ \mu V$, which is equivalent to 0.03°C at 800°C for a Pt/Pd TC.

To test the performance of the apparatus, platinum foil (99.99%, 0.025-mm thick) was tested with both welded (w) and clamped (c) hot junctions as described above. The measuring junction of a wire welded directly to a foil is located at the foil surface. In contrast, when a thermocouple is clamped to a foil or thin film, the measuring junction is located approximately one-half of the bead diameter from the film or foil surface. The observed difference between clamped and welded geometries, ΔE_{cw}, equals the measurement error caused by use of the clamped junction. The data were obtained from six runs with quasi-steady-states near 700°C and 900°C. Figure 18 shows the output of the welded Pt wire/Pt foil circuit, and the clamped Pt wire/Pt foil circuit on a typical thermal cycle with steady-state conditions at 900°C and 700°C. From these tests, the average values ± the standard deviations (SD) of ΔE_{cw} were $-1.0\,\mu V \pm 2.1$ μV and $-2.5\,\mu V \pm 2.5\,\mu V$ at 700°C and 900°C, respectively.

Similar comparison tests were run with palladium foil (99.95%, 0.012-mm thick). As was mentioned before, the small numbers for ΔE_{C-W} can be obscured if the hot junction temperatures are changing too rapidly with time (>5°C/min). Data were used from periods when the hot junctions are at nearly constant temperatures. The difference in temperature between the two hot junctions was 3–4°C for both the 700°C and 900°C quasi-steady-states. The corrected difference for ΔE_{C-W} averaged 10.9 μV (SD = 2.4 μV) for three runs at 900°C and 2.4 μV (SD = 2.0 μV) for four runs at 700°C. These tests indicate that clamped junctions can be used with less than a 1°C temperature differential between the foil (or film) and the measuring junction of the TC.

Using both the welded and clamped junctions, we used six comparison measurements to calculate the average Seebeck coefficient between 700°C and 900°C for both methods. We found the average Seebeck coefficient to be 16.4 μV/K ($u = 0.25\,\mu V/K$) for all 12 tests. There was better agreement between the welded and clamped junctions of a single run than among the separate runs.

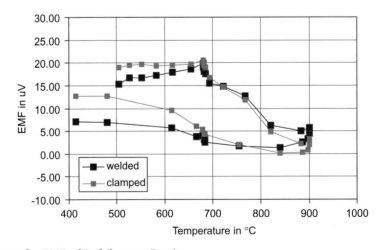

Figure 18 EMF of Pt foil versus Pt wire.

The primary source of the uncertainty in the calibration apparatus was found to be temperature differences between the measuring junction of the calibrated TC and the thin film itself. This uncertainty in the measurement of the thin-film temperature leads to an expanded uncertainty of 0.9°C at 950°C which is intended as the highest temperature of the calibration. This TFTC application has been coupled with Pt/Pd wire TCs for temperature measurement of calibration wafers in RTP tools and has enabled temperature measurements of the wafer with expanded uncertainties of less than 1°C at 900°C.

Figure 19 is a plot of the results of testing a sputtered platinum thin film on an oxidized silicon wafer. The oxide is 310-nm thick and the Pt film is 700-nm thick. Both the heating (above) and cooling curves (connecting the points) are included from points taken at intervals of 360 s and demonstrate the stability of the film and test apparatus. In Figures 19 and 20, the EMF of

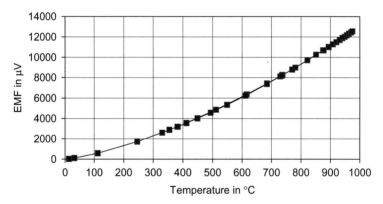

Figure 19   EMF of Pt film versus Pt wire.

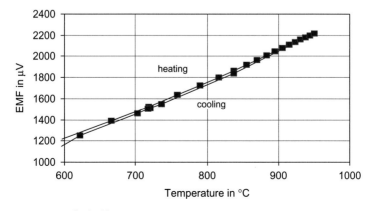

Figure 20   EMF of Rh film versus Pt wire.

the thin films versus the Pt wire [$E_{TF}/_{Pt}$ (t)] is plotted. From these data we derived the EMF and Seebeck coefficient of the film versus a Pt wire at temperatures up to 950°C. Since we are primarily interested in the Seebeck coefficient for this application between 700°C and 950°C, we solve for the slope of this plot. The average Seebeck coefficient of this film relative to TC grade Pt wire in this temperature range is 2.8 µV/K ($u = 0.25$ µV/K). This result may relate to impurities in the Pt film such as the Ti bond coat, surface effects of the thin film, or substrate-generated stresses. Both the heating and cooling curves are presented, and it can be seen that Rh thin films have been determined previously [68].

6.3. Calibration of RTs on RTP tools using silicon wafers with TFTCs

A TFTC wafer that enables more accurate calibration of the LPRTs was reported by Kreider et al. [39]. The NIST TFTC calibration wafer uses Pt/Pd wire TCs welded to Rh/Pt TFTCs to reduce the uncertainty of the wafer temperature measurement *in situ*. They presented the results of testing these TFTC calibration wafers in the NIST RTP test bed at temperatures ranging from 650°C to 830°C together with a discussion of the material's limitations and capabilities. The difference between the TC junction temperatures and the radiance temperatures indicated by the blackbody-calibrated LPRT can be attributed to the effective emissivity of the wafer, the parameter that accounts for the geometry and radiative properties of the wafer-chamber configuration. An analysis of the uncertainty, $u = 1.3$ K ($k = 1$), of the wafer surface temperature measurements in the NIST RTP test bed is presented below.

The wafer pattern shown in Figure 21 has 1.5-mm-wide weld pads and 0.4-mm-wide TFTC traces leading to the measuring junctions. The LPRT targets were in the center of the calibration wafer and at three of the four compass points, 54 mm from the center. Each target at 12-mm axial distance from the lightpipe has a view with 90% of the energy within a 10-mm-diameter circle. The sensitivity is heavily weighted near the center of the target. Pt (TC grade) and Pd (99.997% mass, nominal) 0.1-mm-diameter wires were welded to the weld pads and butt welded to 0.25-mm-diameter TC wires (Pt and Pd) that were secured to the wafer with alumina cement in Al_2O_3 tubes.

The NIST RTP test bed (Figure 22) [55] was used to simulate industrial wafer processing conditions as well as calibration of LPRTs. As described above, a bank of high-intensity lamps with a cold, highly reflective chamber (shown in Figure 22) is used to heat the wafer, and up to four 2-mm commercial lightpipes measure the bottom side of the wafer. The authors used a "shading" wafer, 150 mm in diameter, to reduce the thermal gradients in this static system. This wafer was 10 mm above the test wafer

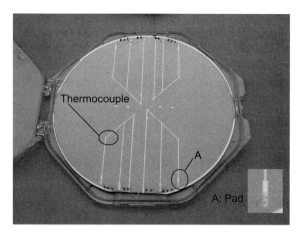

Figure 21 NIST calibration wafer with 8 TFTCs (angled junctions) and 12 pads for 16 Pt/Pd wire welds.

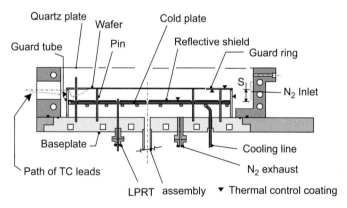

Figure 22 NIST RTP test bed wafer chamber.

and below the fused quartz plate which separates the light box from the test chamber. All tests were run in a purged atmosphere of N_2 with up to 0.1 mL/L O_2.

The 0.25-mm-diameter Pt/Pd TCs were calibrated by comparison with a Au/Pt wire TC, with a standard uncertainty ($k = 1$) of 5 mK at 950°C, in Na heat-pipe blackbody. Between 700°C and 950°C, the 0.25-mm-diameter Pt/Pd TCs agreed with the Au/Pt TC to within the equivalent of 0.1°C.

The Rh and Pt TFTCs were calibrated according to the procedure described above using the proof coupons deposited simultaneously with the calibration wafer. The Rh films had Seebeck coefficients of $(0.0144\ T+6.8)$ µV/K where T is the temperature in degree celcius. Typical calibration

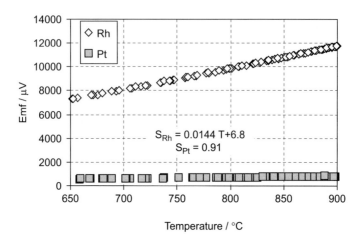

Figure 23 EMF of Rh and Pt thin films versus Pt wire as a function of temperature. Thermoelectric output of the Rh thin film was nearly the same as bulk Rh.

curves generated for the Rh and Pt thin films used on one of the tested
wafers are shown in Figure 23. The test results on Rh thin films have
consistently duplicated results for testing bulk Rh [68] with a standard
relative uncertainty for the Seebeck coefficient of 3%. The Pt films average
$0.9\ \mu V/K$ versus pure Pt reference wire. There is also some hysteresis on the
heating/cooling cycle with the Pt films which is probably due to plastic
deformation in the Pt film caused by the mismatch of thermal expansion
coefficient with the silicon wafer.

In situ calibration of the LPRTs was made with two TFTC calibration
wafers as described above after a 0.5-h anneal at 900°C. Typical calibration
curves for the three LPRTs are presented in Figure 24. This figure
compares the center lightpipe to the average of the TFTC junctions nearest
to the center as a function of the wafer temperature. This difference or
correction indicated for the center LPRT ranges from about 5°C at 660°C
to about 8°C at 830°C. We would expect this correction to be zero if the
reflecting shield had perfect reflectivity and there were no light leaks. This
case corresponds to an effective emissivity of one. The error correction is a
consequence of the effective wafer emissivity being less than unity in the
NIST test bed. Two additional corrections were applied to refine the
calibration. The first relates to the depression in temperature of the wafer
due to the LPRT, and the second relates to the increase in temperature due
to the low emissivity of the TFTC junction itself [39,69]. In the NIST
RTP test bed, the reflectivity of the lightpipe tip is 0.1, compared to that of
the cold shield ($\rho_{sh} = 0.95$).

A parallel-plate finite-element model was used to investigate the effects
of the lightpipe tip on the temperature distribution of the wafer in the

highly reflective RTP chamber. The wafer temperature differences $(T_{w,lp}-T_{w,nlp})$ between the with-lightpipe (lp) and the no-lightpipe (nlp) conditions are shown in Figure 25 for a wafer with an emissivity of 0.65 and a shield reflectivity of 0.95. At 800°C, the LPRT target temperature was depressed approximately 3.2°C, whereas the average temperature of the

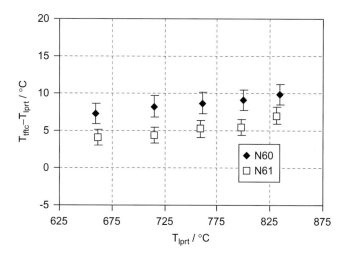

Figure 24 LPRT calibration $(T_{tftc}-T_{lprt})$ performed in RTP tool with NIST calibration wafer.

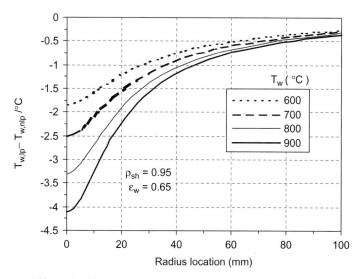

Figure 25 Effect of cold LPRT at 12-mm distance on temperature of calibration wafer.

TFTC junctions (at 10-mm radial distance) was depressed 2.7°C. At lower temperatures this effect is smaller. The resulting correction was 0.3−0.5°C for these experiments. Values depend on tool geometry and optical properties of the materials, but they are higher for LPRTs located closer to the wafer.

For the test wafer with Pt/Rh TFTC, the difference between the emissivities of the thin film ($\varepsilon_t = 0.25$) and the wafer ($\varepsilon_w = 0.65$) caused a small disturbance to the wafer temperature. A finite-element chamber model was used to investigate the temperature disturbance caused by a single TC with a width of 0.4 mm and a conjunction angle of 45°. The temperature corrections at the junction of the TFTC were predicted to be 0.14°C and 0.2°C for wafer temperatures of 700°C and 800°C, respectively. This correction for the NIST test bed is primarily controlled by the highly reflective ($\rho_{sh} = 0.95$) cold shield located 12 mm below the wafer. The greater the shield reflectivity becomes, the smaller the correction.

6.3.1. Uncertainty analysis

The temperature at the TFTC junction of the NIST calibration wafer was determined from the sum of the temperature at the periphery of the wafer, (Pt/ Pd wire TCs) and the temperature difference between the wafer periphery and the wafer center, given by the readings of the thin-film TC. The uncertainty of this measurement was a result of imperfections in these two TC circuits and imprecise knowledge of certain small correction terms. The equation for the temperature at the center TFTC of the wafer, T_C, is:

$$T_C = f^{-1}[f[T_{RJ}] + E_{Pt/Pd}] + \frac{E_{TF}}{S_{TF-Rh} - S_{TF-Pt}}$$
$$+ \frac{S_{TF-Pt}}{S_{TF-Rh} - S_{TF-Pt}}(T_{Rh} - T_{Pt}) + \frac{S_{WP} - S_{Pd}}{S_{Pd}}(T_{Pd} - T_{Rh}) \qquad (1)$$
$$+ \frac{S_{WP} - S_{TF-Rh}}{S_{TF-Rh} - S_{TF-Pt}}(T_{WP} - T_{Rh})$$

where $f[T]$ is the Pt/Pd TC reference function, T_{RJ} the ambient reference junction temperature, T_{Pt} the temperature of the junction between the Pt wire and the Pt thin film, T_{Rh} the temperature of the junction between the Pd wire and the Rh thin-film weld pad, T_{WP} the temperature at the edge of the Pt overcoating of the Rh weld pad, T_{Pd} the temperature of the junction between the palladium wire and the Rh weld pad, $E_{Pt/Pd}$ the EMF measured across the Pt/Pd TC, E_{TF} the EMF measured across the Pt extension wires attached to the Pt/Rh thin film TC, and S_{Pd}, S_{TF-Pt}, S_{TF-Rh}, and S_{TF-WP} are the Seebeck coefficients of the Pd wire, the Pt thin film, the Rh thin film, and the Pt-overcoated weld pad, respectively, with respect to pure Pt. The uncertainty budget for determination of temperature at the center of the NIST calibration wafer, in the NIST RTP test bed at a temperature of 800°C, is given in Table 2. Uncertainties in the reproducibility of the 0.1-mm Pd wire, weld pad effects, and thin-film TCs dominate the overall uncertainty.

A series of measurements comparing the EMF outputs of Pt and Pd wires in the severe thermal gradients of the NIST RTP chamber revealed that the uncertainty of the Pt/Pd wire TCs is primarily a result of inhomogeneities in the short sections of 0.1-mm-diameter Pd wire that run from the weld pads to a butt weld with 0.25-mm-diameter Pd wire. By welding two separate Pd wires to weld pads on a number of wafers, the standard uncertainty of the Pt/Pd TCs was found to be 0.89 K at 800°C. The 0.3-K calibration standard uncertainty of the 0.25-mm-diameter wire was included in the 0.94-K overall uncertainty.

The uncertainty of calibration of the thin films on the calibration coupons was confirmed by analyzing the repeatability of the calibrations of six thin films of Rh and nine thin films of Pt. The resulting Seebeck coefficient of the thin film pair, with its standard uncertainty, is $(17.9 \pm 0.5)\,\mu V/K$ at 800°C. Additional uncertainties result from temperature nonuniformities along the weld pads. For the Pt/Pd TCs used to determine the temperature of thin-film circuit at the edge of the wafer, the short section of weld pad between welds of the Pt and Pd wires introduces an additional EMF into the circuit. This additional EMF is proportional to the temperature difference between the welds and to the Seebeck coefficient of the weld pad relative to the Pd wire. At 800°C, $S_{WP} - S_{Pd}/_{Pt} = 35.3\,\mu V/K$, which is substantially larger than the Seebeck coefficient of the Pt/Pd TC. Consequently, the circuit is sensitive to temperature differences between the weld junctions of the Pt and Pd wires. TC readings at the wafer center and at the periphery of the wafer were combined with LPRT readings at the center of the wafer and at mid-radius

Table 2 Standard uncertainty (u) components for determination of the wafer temperature at a single thin-film junction on the NIST calibration wafer, at a temperature of 800°C.

Component	$u/°C$
Reproducibility/calibration of thin films	0.53
Reproducibility/calibration of wire TCs	0.94
Weld pad effects	0.64
Pt junction temperature, T_{Pt}	0.21
Reference junction temperature, T_{RJ}	0.18
Combined standard uncertainty	1.3

to estimate values for the temperature gradients along the weld pads. This was done after normalizing the LPRT readings to the TC readings at the center of the wafer. For each wafer, a quadratic function was fit to this temperature versus distance data. Evaluation of the slope of the function at the periphery of the wafer gives a temperature gradient at the weld pads of (0.3 ± 0.1) K/mm.

The temperature of the junction between the Pt wire and the Pt thin film was determined by interpolation of the nearby measurements at the weld pads for the Rh thin films. The resulting large uncertainty of the Pt weld pad temperature, 3 K, has a small effect on the wafer temperature measurement because of the small Seebeck coefficient of the Pt thin film relative to the Pt wire. A variety of other uncertainty components were calculated and found to be negligible, including calibration of the voltmeter and reference junction compensation of the Pt/Pd wire TCs.

The difference in location of the TFTC junction and the LPRT also generates an uncertainty of the LPRT calibration. The TC junctions were 4 and 14 mm from the center of the LPRT target. Applying the quadratic fit used to determine the temperature gradient across the weld pad, the temperature gradient near the center was found to be 0.1°C/mm. This gradient is not uniform, and its impact on the uncertainty is partially offset by the LPRT cooling effect discussed above. The estimate of the net effect at 15-mm radius led to a total combined standard uncertainty $u_c = 1.6°C$ and 1.4°C at 4 mm for LPRT calibration in the NIST RTP test bed. The values in Figure 24 for two wafers are expected to be within 2.2°C ($k = 1$). The observed values are 3.0−3.5°C apart. Measurement of the center zone of the wafer with four independent TC circuits gave an average reading that was less sensitive to temperature gradients, providing a more stringent evaluation of the uncertainty budget in Table 2.

In a commercial RTP apparatus, the temperature nonuniformities along the plane of the Si wafer are typically significantly smaller ($6 \times$) than the nonuniformities in the NIST RTP test bed. The uncertainty components corresponding to temperature nonuniformities at the weld pads and of the thin-film calibrations are approximately proportional to the temperature nonuniformities and would be reduced for a commercial RTP apparatus. With a maximum temperature nonuniformity of 5 K, the standard uncertainty of the NIST calibration wafer is expected to be 1.0 K or less.

The primary causes of the uncertainty were related to temperature differences in the locations of the TFTC junction and the LPRT target, the inhomogeneity of the 0.1-mm Pd wire used in the TC and temperature gradients on the weld pad. Reduction of thermal gradients in commercial tools, relative to the NIST RTP test bed, provides a solution to the first problem. Better compositional control on the Pd wire would reduce the second problem.

6.4. Transient thermal response of TFTCs

TFTCs do not significantly perturb the convective heat transfer to their substrate, and their low thermal mass enables rapid thermal response as a sensor. An NIST study [62] was pursued to evaluate the utility of TFTCs to determine their transport response and compare their response to those predicted using existing thermal property data. Platinum, platinum-10% rhodium, and gold-sputtered films were used as the sensors. The substrates included high-purity alumina; plasma-sprayed, partially stabilized zirconia; and space shuttle type tile material MinK2000.

An ArF excimer laser (193 nm) with a pulse width (FWHM) of roughly 12 ns was used as a pulsed heat source. The laser beam was apertured and focused to obtain $10-300\,mJ/cm^2$ energy density incident in a $4-7$-mm-diameter spot centered on the TFTC junction. The spatial profile of the apertured spot was relatively uniform and larger than the junction region. Laser intensities were used that were sufficient to obtain reproducible TC outputs with good signal-to-noise ratios, yet not enough to cause damage to the TFTC.

For all the measurements of the TFTCs, the rise time of the TC was $<500\,ns$. This fast response is due to the heat transfer directly by photon absorption within the extinction length of the Pt ($<1\,\mu m$). The characteristic thermal times of the TFTCs on alumina substrates were found to range from 5 to 7 μs for thin (0.4 μm) TFTC to $5-30\,\mu s$ for the thick (4 μm) TFTC.

6.5. Transient response of sensors in PEB process

Recent studies on dynamic temperature profiling and lithographic performance modeling of the post-exposure bake (PEB) process have demonstrated that the rate of heating and cooling may have an important influence on resist lithographic response [70]. Measuring the transient surface temperature during the heating or cooling process with such accuracy can only be assured if the sensors embedded in or attached to the test wafer do not affect the temperature distribution in the bare wafer. NIST scientists reported on an experimental and analytical study [70] to compare the transient responses of embedded platinum resistance thermometer (PRT) sensors with surface-deposited TFTCs. The TFTCs on silicon wafers were developed at NIST to measure wafer temperatures in other semiconductor thermal processes. Experiments were performed on a test bed built from a commercial, fab-qualified module with hot and chill plates using wafers that had been instrumented with calibrated type-E (NiCr/CuNi) TFTCs and commercial PRTs. Time constants were determined from an energy-balance analysis fitting the temperature$-$time derivative to the wafer temperature during the heating and cooling processes. The time constants for instrumented wafers ranged from 4.6 to 5.1 s on heating for both the TFTC and PRT sensors, with an average difference less than 0.1 s between the TFTCs and PRTs and slightly greater differences on cooling.

The 200-mm test wafers were designed to have four pairs of corresponding TFTC junctions and matching commercial PRTs (see Figure 26). The TFTC junctions are the obtuse and acute angles and the PRTs are the double white spots. These sensor pairs were placed close to each other (6−7 mm) in order to measure the 200-mm wafer under nearly identical thermal conditions. Two wafers with dual instrumentation were designed and fabricated. One wafer had type-E (Ni/Cr versus Cu/Ni) TFTCs on the wafer surface connected to type-E wires, as well as a set of commercial embedded PRT sensors. Type-E TCs have high output (50−70 μV/K) and permit lower measurement uncertainty at a high speed of data acquisition. The second dual sensor wafer had TFTCs and wire TCs of Pt and Pd (5−6 μV/K). Both wafers had embedded thin-film PRTs on 0.25-mm-thick alumina substrates secured to the Si wafer with polyimide in a 2.74 mm × 1.43 mm oval-shaped hole in the silicon. The PRT sensors were also covered with an AlN-filled polyimide and had four-wire Pt foil leads.

This design permitted the metallurgically bonded TFTC junction to be essentially massless because it is 1-μm thick compared to the

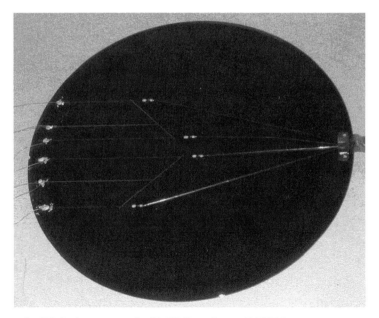

Figure 26 Wafer instrumented with PRTs and type-E TFTCs.

0.76-mm-thick wafer. The response times of the thin-film junctions were measured [60,61] and found to be < 5 μs. The TFTC is used as a differential TC between the measuring junction and the interface with the wire TC. TFTCs were calibrated using the comparison method described in Ref. [52]. Since this temperature difference was less than 5°C, the measurement was not very sensitive to the calibration accuracy of the TFTC. The wafers instrumented with both PRTs and TFTCs were designed to compare the transient responses of the two types of sensors and not their absolute temperature measurements. In fact, the PRTs have smaller temperature measurement uncertainties than the TFTCs.

Cross-sections of two of the PRTs on the 200-mm instrumented wafer were cut. Figure 27 is an electron microscope image of the cavity in the Si wafer (above the white line) containing an Al_2O_3 substrate which holds the thin-film PRT and a "dome" of high thermal conductivity polyimide cover. The assembly was cemented into the drilled pit in the Si wafer and covered with a filled polyimide (white color). The gap between the Al_2O_3 substrate and the Si wafer is approximately 25-μm wide and can be measured using the scale on the figure. The Si pit also had a dip where the gap was larger. The electron microscope image of the second sensor displayed a slightly larger gap.

The tests were planned to investigate the effect of various temperatures (100−150°C), various rotational positions of the test wafer, various sensor locations (near center and at mid-radius), various wafer to hotplate gaps (0.10 and 0.13 mm), and two different TCs (type-E and Pt/Pd). The tests also included some redundancies on different days.

A typical thermal cycle is depicted in Figure 28, where the differential TFTC output is indicated by the open diamonds, the output of the TFTC and the wire TCs is indicated by the dashed thin black line, and the PRT

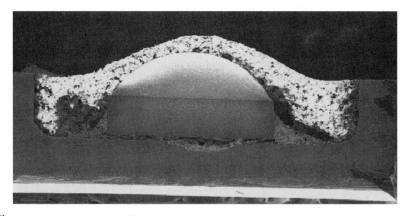

Figure 27   Cross section of a PRT as seen using an electron microscope.

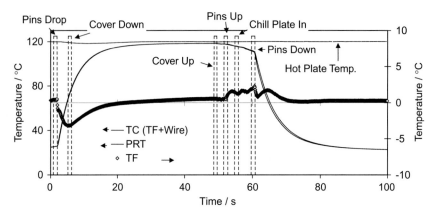

Figure 28 Wafer instrumented with type-E thin-film TCs and PRT and heated using the PEB bed hot plate set at 120°C.

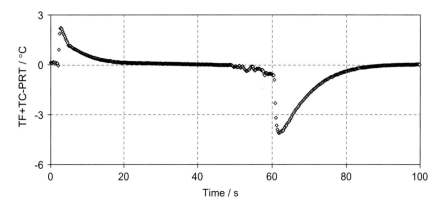

Figure 29 The difference between the temperature of the sum of the thin film plus wire TC and the PRT during the thermal cycle shown in Figure 28.

temperature is indicated by the thin, solid black line. The activities of the PEB module pneumatic actuators are indicated by the dashed line pillars, and the uppermost solid line indicates the temperature of the hotplate. The TFTC output is indicated by the *y*-axis to the right. The left *y*-axis indicates the temperature of the PRT, the hotplate, and the sum of the wire and thin–film TCs.

The difference in response time of the sensors with the wafer heating and cooling processes is indicated in Figure 29 for a typical run. In both heating and cooling the thin films reacted more quickly to their surrounding thermal environment than the PRTs. The difference between the response times of the sensors was greater in cooling than in heating.

7. EMISSIVITY METHODS

7.1. Problem of surface finish

The process side of wafers has patterns and film stacks which are continually changing over their development as an IC. For this reason many tools have been designed with the pyrometers viewing the back side (non process side) of the wafer. For wafers 200 mm and smaller, the back sides were often intentionally damaged by bead blasting or a caustic etch to create dangling bonds to trap defects. Various films were sometimes used to passivate the wafer and prevent auto-doping. Figure 30 illustrates a typical 200-mm wafer backside bi-directional reflectivity function (BDRF) for a wafer with a caustic etch. The BDRF quantifies the scattered light in the incident plane as a function of reflected angle. Different wafers can have very different BDRFs depending on their manufacture. Table 3 shows modeling results for temperature errors with pyrometers of varying field of views. These data clearly illustrate the advantages of the wide field of view of a sapphire LPRT when measuring rough surfaces or film stacks. Figure 31 shows the results of computer simulations for a wide variety of materials used in the semiconductor industry as viewed with pyrometers of differing NAs. The optical characteristic of the target materials was determined by actual measurements of the BDRF for these materials. It becomes clear from this

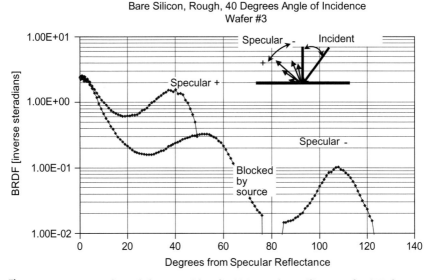

Figure 30 BDRF of rough bottom side of a 200-mm bare silicon wafer (Wafer #3) at $40°$ angle of incidence. These wafers are often roughened to help trap defects.

Table 3 This data modeled by Monte Carlo ray tracing demonstrate the large degree of sensitivity that pyrometers have to surface roughness as a function of view angle. It shows the temperature change (in °C) for a shift from totally diffuse to totally specular.

Pyrometer half angle	Nitride $\varepsilon = 0.95$	Silicon $\varepsilon = 0.69$	Polysilicon $\varepsilon = 0.4$
$40°$	0.05	0.39	1.18
$10°$	1.69	10.55	20.89
$2.5°$	3.67	24.29	51.32

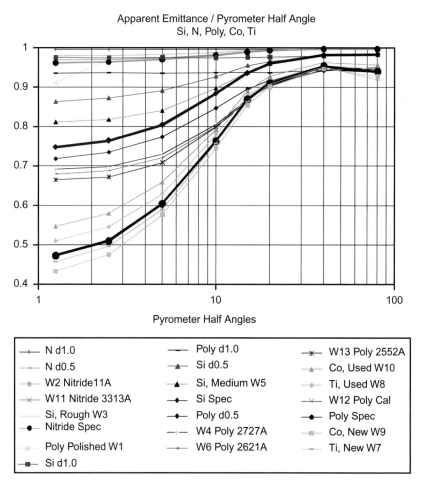

Figure 31 Results of computer simulations for pyrometers with various NAs. The surface characteristics for the different target materials were taken from measured BDRFs.

study the important of pyrometer view angle in diminishing the dependence on surface emittance. The roughness issue has been nearly eliminated for 300-mm silicon wafers as they have been standardized to being high polished on both sides [31,69].

7.2. Use of virtual blackbodies

With the difficult constraints limiting the use of contact thermometers in many semiconductor applications, combined with the complex and variable thermo-optical nature of semiconductor materials, methods of constructing virtual blackbodies have proven very important [5,16,31,44]. This approach allows the designer to construct a chamber where multiple reflections of thermally emitted radiation within the chamber will result in an increase in the apparent emissivity. The underlying principle is Kirchhoff's Law of thermal radiation which states that under equilibrium conditions, the emissivity of a body (or surface) equals its absorptivity, that is, good absorbers are good emitters. Consider a closed chamber with the exception of an entrance aperture which allows for viewing the interior walls which are in thermal equilibrium. For opaque surfaces, the absorbance is one minus the reflectance. The apparent absorption of the aperture will therefore be related to the probability that a given photon which enters through it will be reflected back out. Successful designs have chamber wall that have a high view factor of their surrounding surfaces, and keep the reflected energy directed away from the entrance aperture until it is absorbed by multiple reflections. A good example of the concept is the long-tapered cones commonly made of graphite used in high-temperature blackbody calibration furnaces. The aspect ratio of the taper is large enough to keep the incident rays always directed toward the apex of the taper until they are absorbed. Cavities with very high effective emissivities have also been developed using highly specular surfaces, which eliminate the backscatter which is associated with diffuse surfaces. The international standards laboratories, and universities, have investigated the optimum chamber design for high-emissivity blackbody calibration furnaces [71−73].

A second approach to create a virtual blackbody can be used where not all of the chamber walls are in thermal equilibrium. In this scenario, one or more of the chamber walls are cold, but highly reflective. This "mirrored" wall creates a virtual image of the hot surfaces and is optically equivalent to an environment in thermal equilibrium. This is the method which has been very successful in RTP chambers with cold reflective walls, or a reflector plate located near the wafer.

Very simple virtual blackbodies can sometimes be effectively implemented by drilling a small hole part way into a susceptor and viewing its entrance aperture with a pyrometer or lightpipe. Even a shallow hole with a 4:1 aspect ratio can effectively enhance the emittance and greatly reduce the dependence of contamination on the walls. Where the surface is rotating, as is common in many process applications, a sharp "V" groove can sometimes be fabricated into the moving wafer susceptor which is viewed by the immobile lightpipe or pyrometer.

7.3. Pyrometric sensitivity and difficulties with ratio pyrometry

It is critical in the design of a pyrometer to optimize the sensitivity to change in temperature if high-resolution measurements are to be made. What is desirable is to have a high change in radiant power for a given change in temperature. This pyrometric sensitivity

$$S_P \equiv \frac{T}{L} \frac{\Delta L}{\Delta T} \approx \frac{c_2}{\lambda T} \tag{2}$$

for a spectral pyrometer operating at a single wavelength λ can be calculated by differentiating the Planck equation with respect to temperature and substituting the differentials into the definition of sensitivity. The sensitivity equation is appropriate when

$$\exp\left(\frac{c_2}{\lambda T}\right) \gg 1 \tag{3}$$

Equation (2) clearly demonstrates that sensitivity is inversely proportional to the wavelength, so the shortest practical bandpass will yield the best sensitivity. Historically, industry has often chosen the pyrometer bandpass to correspond to the peak intensity of thermal emission for the temperature of interest. This is a logical choice if the pyrometer is limited by the signal-to-noise ratio of the detector and related electronics. The current generation of low-noise detectors and ICs have allowed for the measurement of temperatures down to 200°C on silicon wafers at wavelengths shorter than 1 μm. This is an ideal region as it is shorter than the bandgap of silicon where the emissivity and opacity are high and nearly independent of temperature as shown in Figure 6.

Choosing the optimum pyrometer bandpass does not solve all the emissivity problems encountered in a processing chamber. Dual wavelength (two-color) or ratio pyrometry, often marketed as a cure for the uncertainty in emittance, is based on the concept of computing the ratio of thermal emission at two different wavelengths then relating this ratio to temperature. If the emittance, view factor, atmospheric absorption, or any measurement artifact are the same at both wavelengths, then they will cancel out in the ratio and not influence the calculated temperature. This standard industrial technique has been around for a long time and has found success in a limited number of applications where its assumptions are met. Applications where the target does not completely fill the pyrometers field of view, or measuring through an obscure viewport, or graybody objects are all good candidates for the ratio technique. The optical properties of most semiconductor materials have a very strong dependence on wavelength and are therefore inappropriate objects for this method. Additionally, by taking the ratio of the radiances, there is an associated loss in pyrometric sensitivity. Calculating sensitivity as in Equation (2) is simplified for the ratio case by using Wien's approximation

$$L(\lambda, T) = \frac{c_{1L}}{\lambda^5(e^{c_2/\lambda T} - 1)} \approx \frac{c_{1L}}{\lambda^5 e^{c_2/\lambda T}} \tag{4}$$

Taking the ratio of the radiances at the two different wavelengths λ_1 and λ_2 yields

$$R_{1,2} \approx \left(\frac{\lambda_2}{\lambda_1}\right)^5 e^{((1/\lambda_2)-(1/\lambda_1))(c_2/T)} \tag{5}$$

Taking the partial derivative with respect to temperature T results in

$$\frac{\partial R_{1,2}}{\partial T} \approx -R_{1,2}\left(\frac{1}{\lambda_2} - \frac{1}{\lambda_1}\right)\frac{c_2}{T^2} \tag{6}$$

Rearranging and substituting Equations (5) and (6) gives the relationship of pyrometric sensitivity $S_{1,2}$ as a function of wavelengths and temperature for the ratio method.

$$S_{1,2} \equiv \frac{T}{R_{1,2}}\frac{\partial R_{1,2}}{\partial T} \approx \left(\frac{1}{\lambda_1} - \frac{1}{\lambda_2}\right)\frac{c_2}{T} \tag{7}$$

Equation (7) clearly shows that as the two wavelengths are chosen to be closer together, the sensitivity goes to zero, and the further away they are from each other the more they approach the single bandpass case. For the semiconductor industry, the region shorter than the bandgap of silicon and the noise limits imposed from the detector-electronics is only a few hundred nanometers at best, and even in this "sweet zone", there is significant wavelength dependency in emissivity. For most materials commonly used in the semiconductor industry, the measurement uncertainty will increase from the loss of sensitivity over the uncertainty introduced from emittance errors.

7.4. RTP reflector plate enhancement theory

A very successful method of temperature measurement in an RTP chamber was developed at Applied Materials and is based on the principles of emissivity enhancement by the chamber itself. The phenomenon of an apparent or effective increase in the emissivity of a surface as a result of multiple reflections from the surrounding surfaces was described by Gouffé

[74] and developed by many others [71,75,76]. This field of study has gained greatly with the development of high-speed computer models with the use of Monte–Carlo ray tracing techniques, where it has been extensively used to design high-emittance blackbody cavities for calibration furnaces.

Placing a mirror near the surface of a wafer creates a virtual image of the wafer, forming a chamber which appears to be isothermal. Inside a completely isothermal cavity, the thermal radiation will be independent of emissivity and thus will be blackbody radiation. The closer one can realize a closed isothermal cavity, the closer will the radiation in the cavity approximate an ideal blackbody. Many practical concerns prevent one from realizing a completely emissivity-independent cavity. Perfect reflectors do not exist in practice, and they must have an edge and contain holes for lift pins, gas purges, and an aperture to monitor the radiation. The use of sapphire lightpipes has proven to be an effective way to sample thermal radiation from a wide range of angles with only a minimal intrusion into the system. Figure 32 shows a reflector plate with a LPRT near the wafer surface to create a virtual blackbody cavity. The arrows represent the first three paths for thermal emission to reach the detector. They correspond to the first three terms of the infinite series,

$$L_T = \varepsilon L_0 + \varepsilon L_0 R(1 - \varepsilon) + \varepsilon L_0 [R(1 - \varepsilon)]^2 + \varepsilon L_0 [R(1 - \varepsilon)]^3 + \ldots$$

$$= \varepsilon L_0 \sum_{i=0}^{n} [R(1 - \varepsilon)]^i \tag{8}$$

where R is the plate reflectivity and ε is the wafer emissivity. In the limit the series converges to

$$L_0 \varepsilon \sum_{i=0}^{n} [R(1 - \varepsilon)]^i = \frac{L_0 \varepsilon}{1 - R(1 - \varepsilon)} = \varepsilon_A L_0 \tag{9}$$

but the terms modifying L_0 must be equal to ε_A, the apparent or effective emittance of the wafer. The degree of emissivity enhancement one would

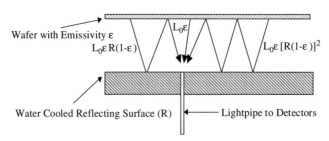

Figure 32 Schematic of reflector plate with LPRT near the wafer surface to create a virtual blackbody cavity.

expect to observe for any given plate reflectivity under ideal conditions is

$$\varepsilon_A = \frac{\varepsilon}{1 - R(1 - \varepsilon)} \tag{10}$$

7.5. RTP *in situ* emissometer application

With the use of thin-layer optical coatings, reflective surfaces can be manufactured with reflectivities very close to unity over the pyrometer bandpass. In a practical application, an effective reflectance of 0.91 is more realizable when considering the edge effects, holes, and limited NA of the pyrometer.

Figure 33 shows the temperature error as a function of wafer emissivity in free space without any reflector plate enhancement, that is, the plate reflectivity is zero. With a pyrometer bandpass of about 0.95 μm, Figure 34 demonstrates the reduction of temperature error from the effects of a plate enhancement with a practical reflectance of 0.91. As can be seen by comparing Figures 33 and 34, the unenhanced emissivity for a silicon wafer at 0.69 will create a temperature measurement error of almost 40°C, but would be reduced to about 4°C for a plate reflectivity of 0.91. This nearly one order of magnitude reduction in emissivity dependence is very good,

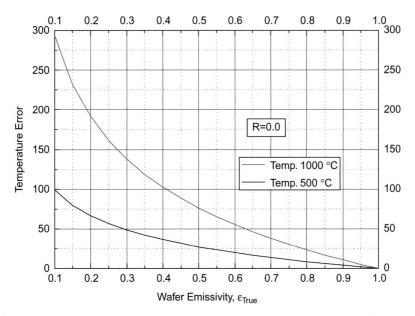

Figure 33   Graph of temperature error as a function of wafer emissivity in free space without any reflector plate enhancement with a plate reflectivity of zero. The pyrometer bandpass is about 0.95 μm.

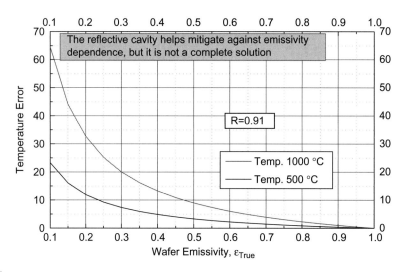

Figure 34 Illustration of the reduction of temperature errors for the effects of a plate enhancement from a practical reflectance of 0.91.

but when considering some polysilicon or metal film stacks with emissivities of 0.2, it is not good enough. In this case the temperature error would be reduced from 185°C to 33°C, which is still too large.

If two different lightpipes are inserted into the reflector plate, but the effective reflectance around the second one is reduced by making the area near its aperture more absorbing, two independent equations of the form of Equation (10) are defined. Figure 35 shows a plot of two different lightpipe configurations, one with a plate enhancement of 0.91 and the other at 0.7. Figure 35 shows the influence of different plate reflectance on the enhanced emittance of the wafer.

Combining these equations with Equation (4) gives us the two equations with two unknowns which are necessary to solve for emissivity. Calibration can be accomplished by measuring well-characterized wafers with a wide range of emissivities on the pyrometer surface, and determining the temperature with a PRS as defined in Section 5.4. Figure 36 shows the excellent correspondence between measured and predicted values of temperature correction for various emissivity wafers.

The two lightpipe methods can be integrated into a single-probe pyrometer—emissometer combination. In this design the effective emissivity is different for two detectors as they have very different NAs. One detector samples the radiation that comes through the lightpipe nearly parallel to its axis and has primarily been emitted from the wafer with little interaction from the reflector plate. The other detector receives radiation from a wide range of angles, where most of the energy has been reflected between the plate and the wafer several times. This change in effective

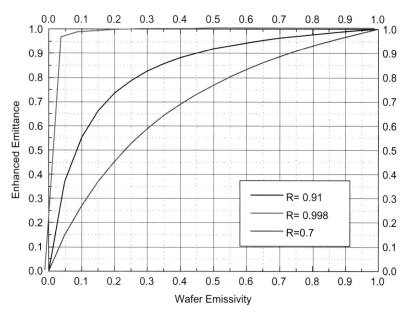

Figure 35 A graphical representation of the influence of different plate reflectance on the enhanced emittance of the wafer.

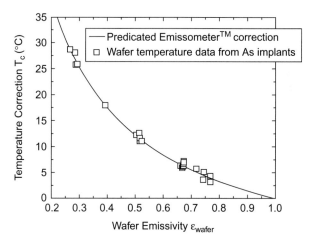

Figure 36 Shown here is the excellent correspondence between measured and predicted values of temperature correction for various emissivity wafers.

emittance can be mathematically treated as a change in effective reflectance, and Equation (10) can be used as in the two-probe system.

A schematic diagram of the integrated pyrometer–emissometer is drawn in Figure 37, while Figure 38 depicts a schematic drawing of an

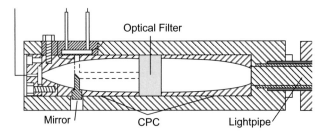

Figure 37 Schematic diagram of the integrated pyrometer—emissometer.

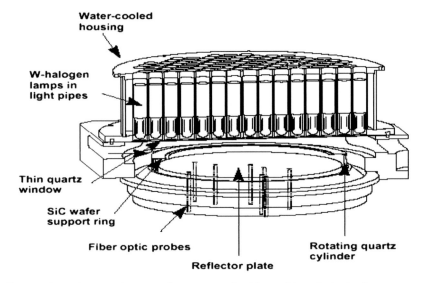

Figure 38 Schematic drawing of an RTP tool with sapphire LPRTs/emissometers with reflector plate. The honeycomb lamphead heats the process side of a rotating wafer, and each pyrometer monitors the temperature of a small area of the multi-zone controls surface.

RTP tool with sapphire LPRTs that serve as emissometers with a reflector plate. The honeycomb lamphead heats the process side of a rotating wafer, and each pyrometer monitors the temperature of a small area of the multizone controls surface. Figure 39 shows a histogram with the repeatability of an RTP tool equipped with LPRTs and emissometers for RTO and RTA. These methods were calibrated to PRS as discussed in Section 5.4. For the oxidation the temperature was related to film growth determined by ellipsometry, and the anneal temperature determined by sheet resistance as measured by a four-point probe.

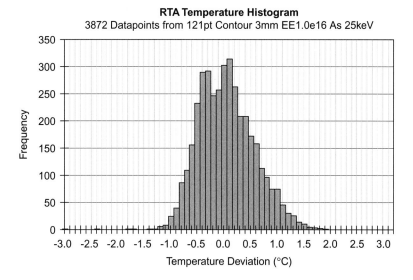

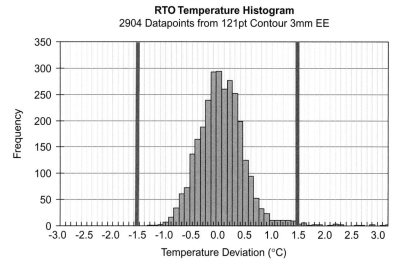

Figure 39 Histogram showing the repeatability of an RTP tool equipped with LPRTs and emissometers for RTO and RTA. These methods were calibrated to PRS as discussed in Section 5.4. For the oxidation the temperature was related to film growth determined by ellipsometry, and the anneal temperature determined by sheet resistance as measured by a four-point probe.

7.6. Low-temperature measurements in RTP

As IC processing has advanced, the thermal budgets have progressively been reduced, increasing the requirement to measure and control temperature at lower and lower temperatures. Radiation thermometry in RTP is constrained by a number of factors which limit the lowest temperature that can be usefully measured. These factors are the bandgap of silicon which limits the pyrometer bandpass to wavelengths shorter than $1\,\mu m$, the bandwidth of the controller requiring at least $100\,Hz$ for adequate control, the noise limits of cost-effective detectors, optical throughput required to access the chamber, methods to filter out the intense lamp radiation. Some success with longer wavelength measurements beyond the silicon bandgap have been demonstrated [77], and methods independent of thermal radiation have been employed, as described in Section 7.10.

7.7. Ripple Technique theory

The Ripple Technique™ was developed specifically to measure wafer surface temperature in an RTP chamber and derives its name from the small changes in radiant output that results from the AC of the electrical input. This method may be used to measure and control temperature in most applications where an object is heated by incandescent lamp radiation. For some semiconductor manufacturing steps, it is necessary to keep the time at processing temperature short, so ramp rates of $150°C/s$ are common. The extremely intense emission from the heating lamps ($10^6\,W/m^2$) will override the thermally emitted radiation from the wafer sometimes by orders of magnitude, making a direct thermal radiation measurement impossible. The ideal wavelength for a pyrometer to measure a silicon wafer is in the $900-950\,nm$ range where the emissivity and opacity are high and almost independent of temperature. This wavelength band is near the peak output of the heaters which can be used to measure the hemispherical reflectivity of the wafer. To accomplish this, the AC component of the spectral radiosity from the wafer is divided by the AC component of the spectral irradiance on to the wafer. It is important to point out that a measurement of the radiosity will consist of both the emitted and reflected components from the wafer, but the wafer has little change in thermal signature from the 120-Hz ripple.[2] The ratio of the AC components will represent only the reflected and not the emitted power, and yields the hemispherical spectral reflectance of the wafer by use of the bi-directional reflectivity theorem. The conditions for use of this theorem are met as the chamber interior is highly reflective to give a uniform irradiance onto the wafer from many banks of lamps with high efficiency. The concept is to create a photon box where the major absorbing surface area is the wafer, and the photons will bounce around until they eventually reach it. The lamp filament area is small

[2] For 60 cycle line frequency the current stops 120 times a second, creating a 120-Hz ripple signature of a few percent. Lower frequencies can be used to gain larger ripple amplitudes, and corrective terms can be developed for the influence of changing wafer temperature.

compared to the wafer, and the chamber walls are highly reflective and water cooled so they do not emit a significant amount of thermal radiation. The hemispherical spectral reflectance of the wafer is

$$\rho \approx \frac{\Delta M_W}{\Delta M_L} \tag{11}$$

where ΔM_W represents the AC component of the spectral radiosity and ΔM_L represents the AC component of spectral irradiance onto the wafer from the lamps.

From the conservation of energy, there are only three different events that can occur to a photon when it impinges onto the wafer surface. Either it penetrates through it, is absorbed by it, or reflected from it. Therefore

$$\tau + \alpha + \rho = 1 \tag{12}$$

where τ is the transmittance, α the absorbance, and ρ the reflectance. Invoking Kirchhoff's Law allows the substitution of emittance (ε) for absorbance (α), and for wavelengths where the wafer is opaque the transmittance is zero.[3] Thus

$$\varepsilon = 1 - \rho = 1 - \frac{\Delta M_W}{\Delta M_L} \tag{13}$$

The reflection component is the irradiance time the reflectance (ρM_L). If the M_L is known, the exitance (M_T) from the wafer can determined:

$$M_T = M_W - \rho M_L = M_W - M_L \left(\frac{\Delta M_W}{\Delta M_L} \right) \tag{14}$$

With the wafer exitance and emittance known from Equations (13) and (14), it is possible to solve for temperature using Planck's equation (assuming diffuse so that $M_T = \pi L_T$).

$$T = \frac{c_2}{\lambda \ln(1 + (\varepsilon c_{1L}/\lambda^5 L_T))} \tag{15}$$

Thus, given the irradiance and radiosity of a wafer with adequate temporal resolution to quantify the lamp ripple, it is possible to derive the *in situ* emittance and temperature of the wafer real time in an intense radiation-heated environment.

---

[3] Systems have been developed to use pyrometers at wavelengths longer than the bandgap of silicon where it becomes transparent. This allows Ripple pyrometry to be extended to lower temperature measurements. This required additional terms [64].

7.8. Ripple Technique application

In practice, the detector and electronics should support 10 kHz bandwidth and have 16 bit resolution for a 120-Hz ripple system. As in conventional radiation thermometry the photodetector current I_d is a function of the emittance and temperature of the target material, the throughput of the optics τ in square metre per steradian (m^2 sr), and the detector responsivity R_d in Ampere per Watt (A/W).

$$I_d = \frac{R_d \tau \varepsilon c_{1L}}{\lambda^5 (e^{c_2/\lambda T} - 1)} \tag{16}$$

The throughput is the area of the limiting aperture of the optics times the solid angle defined by the marginal rays in that plain. When calculating the wafer low temperature measurement limit, I_d must be greater than the electronics noise floor.

There are many possible optical configurations depending on the requirements of the measurement and limits imposed by the chamber design. The original work was using sapphire lightpipes with a right angle bend near the end. In this configuration, the probes could be mounted on the side wall of the chamber and extend over to the center of the wafer, one probe measuring the irradiance into the wafer from the lamps, and the other collecting the radiosity from the wafer. Later, more practical designs evolved in which the two lightpipes were removed from a position directly over the wafer, to the edge of the lamp reflectors. One of the lightpipes was designed to have a narrow view factor and was positioned to measure only radiation from the wafer by looking between adjacent lamps. The other lightpipe was designed to have a large view factor and was positioned to receive emission primarily from the lamps. This sensor design and position formed characteristic points in the chamber. That is, although not directly measuring the radiation near the surface of the wafer, it measures radiation that is proportional to the irradiance. Many designs have evolved over the years to meet specific chamber requirements [5,27,44,77−83], which is also reflected in the partial list of patents referenced in [18,84−92]. It has been shown that a collection angle of about 45° will satisfy Kirchhoff's Law and allow accurate wafer temperature measurements using the Ripple Technique [93].

Calibration techniques vary by chamber design and application, but they can be divided into two areas: the use of process repeatability scales (see Section 5.4), or the use of instrumented wafers. High-temperature emissivity standard such as graphite, platinum, and bare silicon to cross-check the emissometry has been very helpful, but is limited by the lack of accurate understanding of emissivity as a function of temperature for a wide variety of suitable high-temperature materials.

Even though the Ripple Technique was invented and demonstrated at Accufiber Inc., it was first implemented in a practical form by Tony Fiory at Bell Labs when he reduced to commercial practice the concepts to retrofit all of his in-house RTP tools with improved temperature measurement and control. Dr. Fiory compared the temperature measurement of the two standard pyrometers with the upgraded ripple pyrometer by using a process repeatability standard for a control (see Section 5.4). Sheet resistance was measured with splits of 60 boron implanted, and 178 arsenic implanted wafers. Oxide thickness growth for 954 wafers was also measured for control by all three pyrometers. The results of these repeatability tests in a production environment are shown in Table 4 [94]. Figure 40 is a photograph of Ripple pyrometers which were retrofitted into a commercially available RTP tool.

The last three rows of Table 4 present the three sigma variability for the pyrometers, and demonstrate over a 20-fold improvement with the Ripple pyrometer for oxide growth. Following the success at Lucent Technologies, several RTP tool manufactures have continued to evolve Ripple pyrometry for various chamber designs and extend its capabilities to measure temperature at lower temperature where the wafer is transparent to longer wavelength pyrometers.

Ripple pyrometry was used extensively in the early years of RTP to measure filmed and device wafer emissivity at high resolution during heating and at process temperatures. *In situ* real-time surface emissivity can be measured to a resolution of ± 0.001 a hundred times per second with Ripple pyrometers. Ripple pyrometers were also successfully applied to a number of applications which were not semiconductor related, including spacecraft surface temperature measurements at NASA and the European Space Agency.

7.9. Other methods in lamp-heated systems

Over the years there have been many ingenious techniques investigated to develop an emissivity-independent measurement of wafer temperature in a lamp-heated environment. National laboratories, universities, pyrometer companies, and tool manufactures have worked on this important and complex problem. All of these methods were able to demonstrate a measurement which compensated for surface emittance, but were never implemented into a successful tool design for a variety of reasons. They were either too complex, fragile, slow, and expensive; they couldn't be integrated into existing chamber designs; or they were not developed for business decisions unrelated to technology. The authors reference them here as they may offer solutions in other lamp-heated processes, or they may have become practical from currently advanced technology, that is, faster computers or inexpensive powerful diode lasers.

Table 4   Process results data.

Process test name	B-RTA	As-RTA	RTO
Number of 150-mm wafers in test	60	178	954
Wafer measurement metric for inferred process temperature	Sheet resistance	Sheet resistance	Oxide thickness
Measurement tool	Prometrix	Prometrix	Thermawave
Number of sites wafer map	121 pt polar	49 pt polar	52 pt area
Edge exclusion	4 mm	7.5 mm	5 mm
Units of measurements	Ohms/Sq.	Ohms/Sq.	Angstrom
Sensitivity coefficient	$-0.424°C/\Omega$	$-2.78°C/\Omega$	$1.95°C/A$
Mean of all wafer measurements, 3σ	$240.0\,\Omega$	$64.47\,\Omega$	$70.84\,A$
All wafers all sites, measurement, 3σ	$40.1\,\Omega$	$4.19\,\Omega$	$6.82\,A$
All wafers all sites, temperature, 3σ	$17.0°C$	$11.65°C$	$13.3°C$
Wafer-to-wafer mean, measurements, 3σ	$22.0\,\Omega$	$3.06\,\Omega$	$5.93\,A$
Wafer-to-wafer mean, temperature, 3σ	$9.33°C$	$8.50°C$	$11.5°C$
Within wafer uniformity, measurement, 3σ	$33.5\,\Omega$	$2.86\,\Omega$	$3.37\,A$
Within wafer uniformity, temperature, 3σ	$14.2°C$	$7.95°C$	$6.57°C$
Lucent Ripple sensor rel. to wafer means, 3σ	$8.83°C$	$8.82°C$	$11.7°C$
AG DTC, passive, rel. to wafer means, 3σ	$76.5°C$	$65.4°C$	$53.5°C$
AG ERP, passive, rel. to wafer means, 3σ	$137.0°C$	$242.1°C$	$280.0°C$

Note: Variations are expressed as 3 standard deviations [94].

Several RTP manufacturers developed methods to derive an average bulk wafer temperature based on the thermal expansion of silicon. Snow used a very sensitive optical micrometer to monitor the diameter of the wafer during processing [95,96], Loewenstein used a Michelson interferometer to measure the diameter [97], and Moslehi measured the change in optical scattering from surface roughness due to thermal expansion [98]. Brueck demonstrated the use of a tiny diffraction grating etched onto the wafer [99] and Burckel used speckle interferometry to measure the change in temperature from thermal expansion [100].

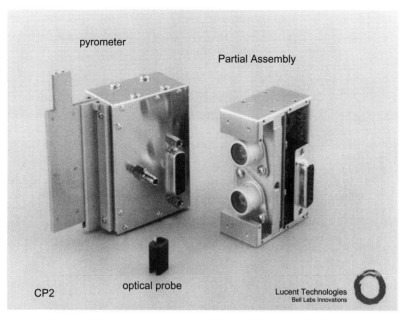

Figure 40   Lucent Ripple pyrometer type-CP2 for AG Heatpulse (model numbers 4100/2146).

Wafer temperature measurement by the use of ellipsometry, which exploits the temperature dependence of the optical values of n and k for silicon, was reported by Massoud. Many of the limiting factors of computational speed and chamber access could be solved today by current computer and fiber-optic technologies [101].

Acoustic thermometry for wafer temperature measurement was demonstrated by Lee. In this method a laser pulse on the surface of the wafer creates a shock wave. The Lamb wave propagation rate is temperature dependent and can be measured by the deflection of a second laser beam located a known distance from the origin. The time of flight can then be related to temperature. The influence of surface film stacks complicates the analysis, but it is independent of the thermal emissivity [102].

Several versions of laser thermometry have been suggested for use in the semiconductor industry. In one of these techniques the wafer surface temperature is excited with a laser pulse to create a small change in temperature. The change in radiance is detected at two different wavelengths. The ratio of the emissivity at the two monitor wavelengths can be determined which allows one to accurately employ classical dual-wavelength pyrometry for an absolute temperature measurement. This system was designed with fiber-optic cables which facilitate chamber compatibility [103].

A more comprehensive review of surface temperature measurements using optical techniques was done by Zhang, and contains many of the technical highlights of the applications mentioned above [104].

7.10. Laser methods

As the size of transistors on ICs has become progressively smaller over time, the most advanced circuits require processing times shorter than the thermal relaxation time of the lamp filaments used in traditional RTP systems. To achieve the required thermal impulse, the surface is rapidly heated by a laser or flash lamps, and quickly cooled by conduction into the bulk of the substrate. The resulting thermal pulse will have a duration from tens of nanoseconds to milliseconds depending on the technology. These new tools offer an abundance of new challenges with peak temperature ramp rates up to 10^{11}°C/s and intense irradiative environments from $10^8 - 10^{11}$ W/m^2.

7.10.1. Dynamic surface anneal (DSA) pyrometry

DSA pyrometry is a laser-diode-based technology where approximately a kilowatt of optical power is focused onto a line about 50-μm wide and 10-mm long. This line is scanned across the wafer surface at several hundred millimeters per second to create a thermal impulse in the millisecond range along a 10-mm-wide strip. The scans are repeated until the complete wafer surface has been annealed.

The DSA pyrometer design is shown schematically in Figure 41. The DSA pyrometer shares the same optics train used to homogenize and focus

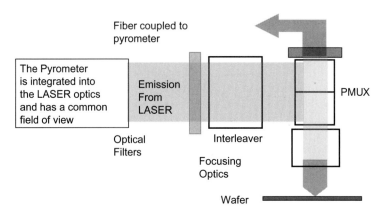

Figure 41 The DSA pyrometer shares the same optics that is used to focus the laser onto the wafer. This method enables the pyrometer to monitor the wafer under the same field of view as the laser [105].

the laser diodes. The pyrometer field of view is therefore identical to the scanning beam. A dichroic mirror is used to separate the pyrometer and laser wavelengths. The laser absorption and emittance variations on the wafer surface are compensated for by either the use of an absorbing layer on the wafer surface or the use of dummy structures in the IC layout. Calibration is accomplished by the use of process repeatability standards as described in Section 6.7 and by noting the abrupt change in emittance when silicon melts at $1,414°C$ [105,106].

7.10.2. Laser spike annealing (LSA) pyrometry

LSA is a technology where a wafer is scanned beneath an elliptical spot of energy ($2.5\,mm \times 95\,\mu m$) focused from a CO_2 laser ($10.6\,\mu m$). The laser beam is polarized and directed to incident the wafer at Brewster's angle so as to be highly absorbed onto the surface. The pyrometer is also directed to view the wafer at Brewster's angle to create a target with minimal emissivity dependence. Figure 42 shows a schematic layout of the LTA system with associated temperature measurement and control [107].

7.10.3. Flash rapid thermal annealing (fRTA) pyrometry

fRTA is a flash-lamp-based technology where the full surface of the wafer is preheated with conventional quartz halogen lamps to a base temperature then flashed with a high-powered discharge lamp to obtain a millisecond heat pulse. High-speed pyrometers independently monitor both stages for thermal control [108].

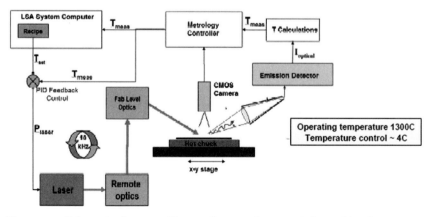

Figure 42 Schematic diagram of laser spike anneal system (Ultratech). The substrate temperature is measured during anneal, using optical pyrometry at three wavelengths. The feedback control bandwidth is 10 kHz.

8. PROBLEMS TO SOLVE

The largest problem in pyrometry, for many applications, continues to be unknown and changing emissivity. There are some excellent emissivity measurement and correction techniques, but they are often sensitive to the measurement environment and therefore require significant development and testing to assure the emissivity is measured and accounted for correctly in each environment. Some of the techniques developed to slove specific needs in the semiconductor industry could be commercialized to address a more general market, both in and out of the semiconductor industry. The Ripple Technique™ and portable temperature calibration sources are both examples to this nature.

The second largest measurement problem is background light. The background light can come from any number of sources, but most often comes from either room light or hot sources near the object being measured. In the semiconductor industry a wafer is typically being measured, and the stray or reflected problem light usually comes from the heater which is generally much hotter than the wafer. The fact that the heater is hotter than the object being measured and hence more radiative can lead to errors even though only a small percentage of the heater light may be reflected into the collection system. Semiconductor processes use plasma techniques, including etch, ash, and plasma-assisted depositions. All of these processes have significant amounts of stray light from the plasma which can cause large errors in the pyrometer reading. The problem is not just reflected light but also transmitted light which proceeds through the wafer and into the pyrometer. Many of these processes are lower temperature, requiring slightly longer wavelength pyrometers (InGaAs photodetectors) and the wafer is often semitransparent and has a film stack, doping and temperature-dependent transmission. The wafer's temperature-dependent transmission can be used to make these low-temperature measurements, but the stray light from the plasma is still problematic.

9. SUMMARY

The successful and widespread use of radiation thermometry in the manufacture of semiconductor devices is the result of careful consideration of the physics within the measurement environments. The technical hurdles continue to be dominated by:

1. stray light,
2. unknown and changing emissivity,
3. lack of thermal radiation at short wavelengths, and
4. training of the end user.

The most robust wafer temperature measurements quantify and minimize the effects of sources of light other than direct wafer emission as well as minimizing emissivity errors.

This chapter reviews wafer temperature measurement with the inclusion of a brief history of the subject with many important references cited. Much of the chapter is devoted to one of the key technologies, namely the sapphire lightpipe which forms the core of OFT methods. OFT forms the foundation for many of today's solutions for both measurement and control of wafer process temperature. As examples we describe the measurement methods used in a few important wafer processes. The techniques and methods used in RTP are described in detail. RTP is a difficult environment to make optical measurements; however, appropriate temperature measurements are being used to control manufacturing of every IC today. The sections on TFTCs form a body of work developed to establish better absolute temperature traceability to the ITS-90 in an RTP environment for pryometric techniques. These techniques should have applicability in areas other than semiconductor manufacturing.

As semiconductor devices continue to scale to smaller and smaller dimensions, the processing times will become shorter, and the thermal budgets lower. Laser processing may be required to meet these challenges and that will drive innovation to measure and control temperature on the millisecond timeframe. We would expect the next few decades to be as fertile of ground for developments in radiations thermometry due to the demands of the manufacturing industry as it has been over the last 20 years.

ACKNOWLEDGMENTS

The authors give special thanks to one of the brightest scientists solving today's semiconductor temperature control problems using advanced pyrometry, Jason Mershon of Portland, Oregon. The authors also express gratitude to Aaron Hunter for his insightful comments and review of the manuscript.

REFERENCES

[1] A. E. Braun, "Optical Pyrometry Begins to Fulfill its Promise", Semiconductor International, March (1998).
[2] R. R. Dils, "Optical Fiber Thermometer," US Patent 4,576,486 (1986).
[3] C. W. Schietinger and B. A. Adams, "Reliability Improvement Methods for Sapphire Fiber Temperature Sensors," in Selected Papers on Optical Techniques for Industrial Inspection, edited by P. Cielo and B. Thompson (1997), DOI: 10.1117/12.24716.
[4] C. Schietinger, "Temperature measurements using optical fibers," Electronics Cooling 3(1) (1997), Technical Brief, CRC Press. Available at: http://electronics-cooling.com/articles/1997/jan/jan97_tb.php.

[5]   B. A. Adams and C. W. Schietinger, "*In-Situ* Optical Wafer Temperature Measurement," in *Temperature: Its Measurement and Control in Science and Industry*, edited by D. C. Ripple, B. C. Johnson, C. W. Meyer, R. D. Saunders, G. F. Strouse, W. L. Tew, B. K. Tsai, and H. W. Yoon, AIP, New York, pp. 1081–1086 (2002).

[6]   C. T. Wallis and R. C. Schwarz, "Radiation Pyrometer," US Patent 1,318,516 (1919).

[7]   J. L. Schueler and C. A. Kellogg, "Apparatus for Measuring High Temperatures," US Patent 1,475,365 (1923).

[8]   S. Ruben, "Pyrometer," US Patent 1,639,534 (1927).

[9]   F. S. Marcellus, "Temperature Measurement," US Patent 1,894,109 (1933).

[10] W. J. Bohnet, "Apparatus for Transmitting Radiant Heat for Temperature Measurement," US Patent 2,709,367 (1955).

[11] J. E. Stewart, D. E. Evans, and G. D. Larson, "Remote Radiation Temperature Sensor," US Patent 3,626,758 (1971).

[12] K. A. Wickersheim, "Optical Temperature Measurement Technique Utilizing Phosphors," US Patent 4,075,493 (1978).

[13] T. Ito, "Charging Current Integrating Type Photodetectors," US Patent 4,348,110 (1982).

[14] R. R. Dils, "Blackbody Radiation Sensing Optical Fiber Thermometer System," US Patent 4,750,139 (1988).

[15] R. R. Dils, "Method and Apparatus for Determining Temperature in a Blackbody Radiation Sensing System," US Patent 4,845,647 (1989).

[16] P. Vandenabeele and K. Maex, "Temperature measurement during implantation at elevated temperatures ($300-500°C$)," J. Vac. Sci. Technol. B **9**, 2784–2787 (1991).

[17] M. M. Moslehi, "Method and Apparatus for Real-Time Wafer Temperature Measurement Using infrared Pyrometry in Advanced Lamp-Heated Rapid Thermal Processors," US Patent 4,956,538 (1990).

[18] C. W. Schietinger and B. E. Adams, "Non-contact Techniques for Measuring Temperature or Radiation-Heated Objects," US Patent 5,154,512 (1992).

[19] E. M. Jensen, "Electro-Optical Board Assembly for Measuring the Temperature of an Object Surface from Infra-Red Emissions Thereof, Including an Automatic Gain Control Therefore," US Patent 5,717,608 (1998).

[20] T. Heinke, J. Ysaguirre, S. King, and P. Carlson, "Ratio Type Infrared Thermometer," US Patent 5,815,410 (1998).

[21] E. M. Jensen, "Electro Optical Board Assembly for Measuring the Temperature of an Object Surface from Infra Red Emissions Thereof Including an Automatic Gain Control Therefore," US Patent 5,897,610 (1999).

[22] M. Yam and A. M. Hunter, "Apparatus and Method for Measuring Substrate Temperature," US Patent 6,007,241 (1999).

[23] C. W. Schietinger and R. A. Palfenier, "Wafer temperature Measurement Method for Plasma Environments," US Patent 6,799,137 (2004).

[24] J. P. Kottmann and C. Stenzel, "Characterization of Flexible Sapphire Fibers in High-Temperature Pyrometers," Sens. Mater. **11**, 233–246 (1999).

[25] R. R. Dils, J. Geist, and M. L. Reilly, "Measurement of Silver Freezing Point with Optical Fiber Thermometer: Proof of Concept," Journal of Applied Physics **59**, 1005–1012 (1986).

[26] B. E. Adams, "Investigation of a Slightly Cooled Integrating Detector (SCID) for Low Level Radiometric Measurements," Master's Thesis, Oregon Graduate Institute of Science and Technology (1998).

[27] C. W. Schietinger, B. E. Adams, and C. Yarling, "Ripple Technique: A Novel Non-Contact Wafer Emissivity and Temperature Measurement Method for RTP,"

in *Materials Research Society Symposium Proceedings Volume 224*, edited by J. C. Gelpey, M. L. Green, R. Singh, and J. J. Wortman, Materials Research Society, Pittsburgh, pp. 23–31 (1991).

[28] Y. Nishi and R. Doering (eds.), *Handbook of Semiconductor Manufacturing Technology*, CRC Press, Boca Raton, FL, p. 141 (2000).

[29] P. J. Timans, "The Thermal Radiative Properties of Semiconductors," in *Advances in Rapid Thermal and Integrated Processing, NATO ASI Series*, edited by F. Roozeboom, Kluwer Academic Publishers, The Netherlands, pp. 35–101 (1996).

[30] T. Sato, "Spectral Emissivity of Silicon," Jap. J. Appl. Phys. **6**, 339–347 (1967).

[31] B. E. Adams, A. Hunter, M. Yam, and B. Peuse, "Determining the Uncertainty of Wafer Temperature Measurements Induced by Variations in the Optical Properties of Common Semiconductor Materials," in *The Electrochemical Society Proceedings Volume 2000-9*, pp. 363–374 (2000).

[32] B. K. Tsai, F. J. Lovas, D. P. DeWitt, K. G. Kreider, G. W. Burns, and D. W. Allen, "In-Chamber Calibration using a Silicon Proof-Wafer," in *Proceedings of the 5th IEEE International Conference on Advanced Thermal Processing – RTP 1997*, edited by R. B. Fair, M. L. Green, B. Lojek, and R. P. S. Thakur, New Orleans, LA, pp. 340–346 (1997).

[33] K. G. Kreider, D. P. DeWitt, B. K. Tsai, F. J. Lovas, and D. W. Allen, "Calibration Wafer for Temperature Measurements in RTP Tools," in *Characterization and Metrology for ULSI Technology*, edited by D. G. Seiler, A. C. Diebold, W. M. Bullis, T. J. Shaffner, R. McDonald, and E. J. Walters, American Institute of Physics, Woodbury, NY, Vol. 449, pp. 303–309 (1998).

[34] Y. H. Zhou, Y. J. Shen, Z. M. Zhang, B. K. Tsai, and D. P. DeWitt, "A Monte Carlo Model for Predicting the Effective Emissivity of the Silicon Wafer in Rapid Thermal Processing Furnaces," Int. J. Heat Mass Transf. **45**, 1945–1949 (2002).

[35] B. K. Tsai and D. P. DeWitt, "ITS-90 Calibration of Radiometers using Wire/Thin-Film Thermocouples in the NIST RTP Tool: Effective Emissivity Modeling," in *Proceedings of the 7th IEEE International Conference on Advanced Thermal Processing – RTP 1999*, edited by D. P. DeWitt, J. Kowalski, B. Lojek, and A. Tillmann, Gaithersburg, MD, pp. 125–135 (1999).

[36] B. K. Tsai, D. P. DeWitt, F. J. Lovas, K. G. Kreider, C. W. Meyer, and D. W. Allen, "Chamber Radiation Effects on Calibration of Radiation Thermometers with a Thin-Film Thermocouple Test Wafer," in *Proceedings of TEMPMEKO '99, 7th International Symposium on Temperature and Thermal Measurements in Industry and Science*, edited by J. F. Dubbeldam and M. J. de Groot, Edauw & Johannissen, Delft, Netherlands, pp. 726–731 (1999).

[37] C. W. Meyer, D. P. DeWitt, K. G. Kreider, F. J. Lovas, and B. K. Tsai, "ITS-90 Calibration of Radiation Thermometers for RTP Using Wire/thin-film Thermocouples on a Wafer," in *Characterization and Metrology for ULSI Technology*, edited by D. G. Seiler, A. C. Diebold, T. J. Shaffner, R. McDonald, W. M. Bullis, P. J. Smith, and E. M. Secula, American Institute of Physics, Woodbury, NY, Vol. 550, pp. 254–258 (2000).

[38] B. K. Tsai, C. W. Meyer, and F. J. Lovas, "Characterization of Lightpipe Radiation Thermometers for the NIST Test Bed," in *Proceedings of the 8th IEEE International Conference on Advanced Thermal Processing – RTP 2000*, edited by D. P. DeWitt, J. Kowalski, B. Lojek, and A. Tillmann, Gaithersburg, MD, pp. 83–93 (2000).

[39] K. G. Kreider, W. A. Kimes, C. W. Meyer, D. C. Ripple, B. K. Tsai, D. C. Chen, and D. P. DeWitt, "Calibration of Radiation Thermometers in Rapid Thermal Processing Tools Using Si Wafers with Thin-Film Thermocouples," in *Temperature: Its Measurement and Control in Science and Industry*, edited by D. C. Ripple, B. C. Johnson, C. W. Meyer, R. D. Saunders, G. F. Strouse, W. L. Tew, B. K. Tsai, and H. W. Yoon, AIP, New York, pp. 1087–1092 (2002).

[40] B. J. Lee, Z. M. Zhang, E. A. Early, D. P. DeWitt, and B. K. Tsai, "Modeling Radiative Properties of Silicon with Coatings and Comparison with Reflectance Measurements," J. Thermophys. Heat Transf. **19**, 558–569 (2005).

[41] B. K. Tsai, "A Summary of Lightpipe Radiation Thermometry Research at NIST," J. Res. NIST **111**, 9–30 (2006).

[42] R. J. Schreuutelkamp, P. Vandenabeele, B. Deweerdt, and W. Coppye, "*In-situ* Emissivity Measurements to Probe the Phase Transformations during Rapid Thermal Processing Cosilicidation," Appl. Phys. Lett. **61**, 2296–2298 (1992).

[43] SVT Associates, Inc., "In-Situ 4000 White Paper: Solving the Problems of Pyrometry and Thickness Measurement during MBE and MOCVD," www.svta.com/products/monitoring/IS4K_White%20Paper.pdf (2009).

[44] C. W. Schietinger, "Wafer Temperature Measurement in RTP, and Wafer Emissivity in RTP," in *Advances in Rapid Thermal and Integrated Processing*, edited by F. Roozeboom, Kluwer Academic Publishers, The Netherlands, Chapters 3 and 4 (1996).

[45] C. Schietinger, "Closed loop process control," *Electronic News* **45**(37), 46 (1999), Capital Equipment Section, Reed Business Information, Inc. ©1999.

[46] W. A. Kimes, K. G. Kreider, D. C. Ripple, and B. K. Tsai, "Emissivity Compensated Pyrometry for Specular Silicon Surfaces on the NIST RTP Test Bed," in *Proceedings of the 12th IEEE International Conference on Advanced Thermal Processing – RTP 2004*, edited by J. Gelpey, B. Lojek, Z. Nenyei, and R. Singh, Portland, OR, pp. 156–161 (2004).

[47] M. Yam, A. Rubinchik, and B. Peuse, "Novel Approach to the Temperature Calibration in the Centura," in *5th International Conference on Advanced Thermal Processing of Semiconductors – RTP'97*, RTP'97, Round Rock, TX, pp. 102–104 (1997).

[48] M. Yam, A. Rubinchik, A. Hunter, B. Adams, and R. Dils, "Performance of Tempmatch (TM) Transfer Standards for In Situ Calibration of Optical Fiber Thermometers," in *7th International Conference on Advanced Thermal Processing of Semiconductors – RTP'99*, RTP-Conference, pp. 160–162 (1999).

[49] A. Hunter, B. Adams, A. Rubinchik, and G. Pham, "A Novel in-Situ Lightpipe Pyrometer Calibration Technique," in *9th International Conference on Advanced Thermal Processing of Semiconductors – RTP'2001*, RTP'97, RTP-Conference, pp. 169–172 (2001).

[50] G. W. Burns and D. C. Ripple, "Techniques for Fabricating and Annealing Pt/Pd Thermocouples for Accurate Measurements in the Range 0°C to 1300°C," in *Proceedings of TEMPMEKO '96, 6th International Symposium on Temperature and Thermal Measurements in Industry and Science*, Torino, Italy (1996).

[51] R. E. Bentley, "The Theory and Practice of Thermoelectric Thermometry," *Handbook of Temperature Measurement*, Vol. 3, Springer (1998).

[52] ASTM International, *Manual on the Use of Thermocouples in Temperature Measurement*, *ASTM Manual Series*, ASM, Philadelphia (1993).

[53] M. Battuello, K. Ali, and F. Girard, "Characterization of Pt/Pd Thermocouples from 800°C to 1300°C by Comparison with a Standard Radiation Thermometer," in *Proceedings of TEMPMEKO '96, 6th International Symposium on Temperature and Thermal Measurements in Industry and Science*, Torino, Italy, September, pp. 10–12 (1996).

[54] S. Knight and A. Settle-Raskin (eds.), "Project Portfolio FY 1997, The National Semiconductor Metrology Program," Publication NISTIR 5851, Office of Microelectronics Programs, NIST (1997).

[55] K. G. Kreider and F. DiMeo, "Platinum/Palladium Thin-Film Thermocouples for Temperature Measurement on Silicon Wafers," Sens. Actuators A **69**, 46–52 (1998).

[56] K. G. Kreider and G. J. Gillen, "High Temperature Materials for Thin-Film Thermocouples on Silicon Wafers," Thin Solid Films **376**, 32−37 (2000).

[57] K. G. Kreider, D. P. DeWitt, B. K. Tsai, F. J. Lovas, and D. W. Allen, "RTP Calibration Wafer Using Thin-Film Thermocouples," in *Materials Research Society Symposium Proceedings, Volume 525*, edited by J. C. Gelpey, M. L. Green, R. Singh, and J. J. Wortman, Materials Research Society, Pittsburgh, pp. 87−94 (1998).

[58] H. P. Grant and J. S. Przybyseweski, "Thin Film Temperature Sensors," J. Eng. Power **99**(4), 497 (1977).

[59] J. R. Kinard, D. X. Huang, and D. B. Novotny, "Performance of Multilayer Thin-Film Multijunction Thermal Convertors," IEEE Trans. Inst. Meas. **44**(2), 383−386 (1995).

[60] D. Burgess, M. Yust, and K. G. Kreider, "Transient Thermal Response of Plasma-Sprayed Zirconia Measured with Thin-Film Thermocouples," Sens. Actuators A **24**, 155−161 (1990).

[61] M. Yust and K. G. Kreider, "Transparent Thin Film Thermocouple," Thin Solid Films **176**, 73−78 (1989).

[62] G. W. Burns, M. G. Scroger, G. F. Strouse, M. C. Croakin, and W. F. Guthrie, *Temperature-Electromotive Force Reference Functions*, NIST Monograph, Vol. 175, U.S. Government Printing Office, Washington (1993).

[63] K. G. Kreider, "Thin-Film Thermocouples for Internal Combustion Engines," J. Vac. Sci. Technol. A **4−6**, 2618−2623 (1986).

[64] M. Yust and K. G. Kreider, "Transparent Thin Film Thermocouple," Thin Solid Films **176**, 73−78 (1989).

[65] K. G. Kreider, "Thin Film Ruthenium Oxide-Iridium Oxide Thermocouples," in *Materials Research Society Symposium Proceedings Volume 234*, edited by D. Allred and G. Slack, Pittsburgh, PA, pp. 205−211 (1991).

[66] K. G. Kreider, "High Temperature Silicide Thin-film Thermocouples," in *Materials Research Society Symposium Proceedings Volume 322*, edited by D. Allred and G. Slack, Pittsburgh, PA, pp. 285−290 (1994).

[67] K. G. Kreider, D. C. Ripple, and D. P. DeWitt, "Calibration of Thin-Film Thermocouples on Silicon Wafers," in *Proceedings of the 7th International Symposium on Temperature and Thermal Measurements in Industry and Science*, edited by J. F. Dubbeldam and M. J. de Groot, VSL, Delft, pp. 286−291 (1999).

[68] K. G. Kreider, "Sputtered High Temperature Thin-Film Thermocouples," J. Vac. Sci. Technol. A **11**(4), 1401−1405 (1993).

[69] K. G. Kreider, D. H. Chen, D. P. DeWitt, W. A. Kimes, and B. K. Tsai, "Effects of Lightpipe Proximity on Si Wafer Temperature in Rapid Thermal Processing Tools," in *Characterization and Metrology for ULSI Technology*, edited by D. Seiler, et al., AIP, New York, pp. 200−204 (2003).

[70] K. G. Kreider, D. P. DeWitt, J. B. Fowler, J. E. Proctor, W. A. Kimes, D. C. Ripple, and B. K. Tsai, "Comparing the Transient Response of a Resistive-Type Sensor with a Thin-Film Thermocouple during the Post-Exposure Bake Process," in *Proceedings Data Analysis and Modeling for Process Control*, edited by K. W. Tobin Jr., Bellingham, WA, pp. 81, 92 (2004).

[71] M. Groll and G. Neuer, "A New Graphite Cavity Radiator as Blackbody for High Temperatures," in *Temperature: Its Measurement and Control in Science and Industry*, edited by H. H. Plumb, AIP, New York, pp. 449−456 (1972).

[72] H. W. Yoon, C. E. Gibson, and J. L. Gardner, "Spectral radiance comparisons of two high-temperature blackbodies with temperatures determined using absolute detectors and ITS-90 techniques," in *Temperature: Its Measurement and Control in Science and Industry*, edited by D. C. Ripple, B. C. Johnson, C. W. Meyer, R. D.

Saunders, G. F. Strouse, W. L. Tew, B. K. Tsai, and H. W. Yoon, AIP, New York, pp. 601–606 (2002).

[73] H. W. Yoon, C. E. Gibson, and B. C. Johnson, "The Determination of the Emissivity of the Variable-Temperature Blackbody Used in the Dissemination of the US National Scale of Radiance Temperature," in *Proceedings of the 8th International Symposium on Temperature and Thermal Measurements in Industry and Science* (TEMPMEKO '01), edited by B. Fellmuth, J. Seidel, and G. Scholz, VDE Verlag Gmbh, Berlin, pp. 221–226 (2001).

[74] A. Gouffé, "Corrections: d'ouverture des Corps-noirs Artificiels Compte Tenu des Diffusions Multiples Internes," Rev. d'Optique **24**, 1 (1945).

[75] R. E. Bedford, "Effective Emissivities of Blackbody Cavities — A Review," in *Temperature: Its Measurement and Control in Science and Industry*, edited by H. H. Plumb, ISA, Pittsburgh, PA, p. 425 (1972).

[76] T. Iuchi and R. Kusaka, "Two Methods for Simultaneous Measurement of Temperature and Emittance Using Multiple Reflection and Specular Reflection, and Their Application to Industrial Processes," in *Temperature: Its Measurement and Control in Science and Industry*, edited by J. F. Schooley, AIP, New York, p. 491 (1982).

[77] M. Oh, B. Nguyenphu, and A. T. Fiory, "In-Line Temperature Monitoring of Rapid Thermal Annealing Processes," in *Materials Research Society Symposium Proceedings Volume 387*, edited by S. R. J. Brueck, J. C. Gelpey, A. Kermani, J. L. Regolini, and J. C. Sturm, Materials Research Society, Pittsburgh, pp. 131–136 (1995).

[78] C. W. Schietinger and B. A. Adams, "A Review of Wafer Temperature Measurement Using Optical Fibers and Ripple Pyrometry," in *Proceedings of the 5th IEEE International Conference on Advanced Thermal Processing — RTP 1997*, edited by R. B. Fair, M. L. Green, B. Lojek, and R. P. S. Thakur, New Orleans, LA, pp. 335–339 (1997).

[79] C. W. Schietinger and E. Jensen, "Wafer Temperature Measurement: Status Utilizing Optical Fibers," in *Materials Research Society Symposium Proceedings, Volume 429*, edited by J. C. Gelpey, M. C. Öztürk, R. P. S. Thakur, A. T. Fiory, and F. Roozeboom, Materials Research Society, Pittsburgh, pp. 283–290 (1996).

[80] C. Schietinger and B. Peuse, "Wafer Emissivity for RTP-Modeled and Measured," in *Proceedings of the 3rd IEEE International Conference on Advanced Thermal Processing — RTP 1995*, edited by B. Lojek, Amsterdam, The Netherlands (1995).

[81] Z. H. Wang, C. S. Schietinger, and M. Sun, "A New Fiber Optic Array Sensor System for Wafer Temperature Measurement in a Multi-zone and Multiphase RTP Furnace," in *Proceedings of the 3rd IEEE International Conference on Advanced Thermal Processing — RTP 1995*, edited by B. Lojek, Amsterdam, The Netherlands (1995).

[82] Z. H. Wang, C. W. Schietinger, and B. E. Adams, "Emittance Compensated RTP Temperature Measurement Principle and Practice," in *Proceedings of the 1st IEEE International Conference on Advanced Thermal Processing — RTP 1993*, edited by B. Lojek, Scottsdale, AZ (1993).

[83] A. T. Fiory, C. W. Schietinger, B. E. Adams, and F. G. Tinsley, "Optical Fiber Pyrometry with in-situ Detection of Wafer Radiance-Accufiber's Ripple Method," in *Materials Research Society Symposium Proceedings, Volume 303*, edited by J. C. Gelpey, J. K. Elliott, J. J. Wortman, and A. Ajmera, Materials Research Society, Pittsburgh, pp. 139–145 (1993).

[84] C. W. Schietinger, A. N. Hoang, and D. V. Bakin, "Optical Techniques for Measuring Layer Thicknesses and Other Surface Characteristics of Objects Such as Semiconductor Wafers," US Patent 7,042,581 (2006).

[85] C. W. Schietinger, A. N. Hoang, and D. V. Bakin, "Optical Techniques for Measuring Layer Thicknesses and Other Surface Characteristics of Objects Such as Semiconductor Wafers," US Patent 6,934,040 (2005).

[86] C. W. Schietinger, A. N. Hoang, and D. V. Bakin, "Optical Techniques for Measuring Layer Thicknesses and Other Surface Characteristics of Objects Such as Semiconductor Wafers," US Patent 6,654,132 (2003).

[87] C. W. Schietinger and A. N. Hoang, "Optical Techniques for Measuring Layer Thicknesses and Other Surface Characteristics of Objects Such as Semiconductor Wafers," US Patent 6,570,662 (2003).

[88] C. W. Schietinger and B. E. Adams, "Non-contact Optical Techniques for Measuring Surface Conditions," US Patent 5,769,540 (1998).

[89] C. W. Schietinger and B. E. Adams, "Non-contact Optical Techniques for Measuring Surface Conditions," US Patent 5,490,728 (1996).

[90] C. W. Schietinger and B. E. Adams, "Non-contact Optical Techniques for Measuring Temperature of Radiation-Heated Objects," US Patent 5,318,362 (1994).

[91] C. W. Schietinger and B. E. Adams, "Non-contact Optical Techniques for Measuring Surface Conditions," US Patent 5,310,260 (1994).

[92] C. W. Schietinger and B. E. Adams, "Techniques for Measuring the Thickness of a Film Formed on a Substrate," US Patent 5,166,080 (1992).

[93] H. Xu and J. C. Sturm, "Emissivity of Rough Silicon Surfaces: Measurement and Calculations," in Materials Research Society Symposium Proceedings, Volume 387, edited by S. R. J. Brueck, J. C. Gelpey, A. Kermani, J. L. Regolini, and J. C. Sturm, Materials Research Society, Pittsburgh, pp. 29–34 (1995).

[94] A. T. Fiory, "Rapid Thermal Annealing and Oxidation of Silicon Wafers with Back-Side Films," in Materials Research Society Symposium Proceedings, Volume 470, edited by T. J. Riley, J. C. Gelpey, F. Roozeboom, and S. Saito, Materials Research Society, Pittsburgh, pp. 50–56 (1997).

[95] K. A. Snow, "Method and Apparatus for Temperature Measurement Using Thermal Expansion," US Patent 5,221,142 (1993).

[96] B. Peuse and A. Rosekrans, "In-situ Temperature Control for RTP via Thermal Expansion Measurement," in Materials Research Society Symposium Proceedings, Voume 303, edited by J. C. Gelpey, J. K. Elliott, J. J. Wortman, and A. Ajmera, Materials Research Society, Pittsburgh, pp. 125–131 (1993).

[97] L. M. Loewenstein, J. D. Lawrence, W. G. Fisher, and C. J. Davis, "Semiconductor Wafer Temperature Measurement System and Method," US Patent 5,102,231 (1992).

[98] M. M. Moslehi, "Method for Real-Time Semiconductor Wafer Temperature Measurement Based on a Surface Roughness Characteristic of the Wafer," US Patent 5,474,381 (1995).

[99] S. R. J. Brueck, S. H. Zaidi, and M. K. Lang, "Temperature Measurement for RTP," in Materials Research Society Symposium Proceedings, Volume 303, edited by J. C. Gelpey, J. K. Elliott, J. J. Wortman, and A. Ajmera, Materials Research Society, Pittsburgh, pp. 117–123 (1993).

[100] D. Burckel, S. H. Zaidi, and S. R. J. Brueck, "Sub-feature Speckle Interferometry: A New Approach to Temperature Measurement," in Materials Research Society Symposium Proceedings, Volume 387, edited by S. R. J. Brueck, J. C. Gelpey, A. Kermani, J. L. Regolini, and J. C. Sturm, Materials Research Society, Pittsburgh, pp. 125–130 (1995).

[101] H. Z. Massoud, R. K. Sampson, K. A. Conrad, Y. Z. Hu, and E. A. Irene, "Applications of in-situ Ellipsometry in RTP Temperature Measurement and Process Control," in Materials Research Society Symposium Proceedings, Volume 224,

edited by J. C. Gelpey, M. L. Green, R. Singh, and J. J. Wortman, Materials Research Society, Pittsburgh, pp. 17−22 (1991).

[102]   Y. J. Lee, B. T. Khuri-Yakub, and K. C. Sarawat, "Temperature Measurement in Rapid Thermal Processing using Acoustic Techniques," Rev. Sci. Instrum. **65**, 974−976 (1994).

[103]   G. J. Edwards, A. P. Levick, and Z. Xie, "Laser Emissivity Free Thermometry (Left)," in *Proceedings of the 7th International Symposium on Temperature and Thermal Measurements in Industry and Science*, edited by P. Marcarino, Levrotto & Bella, Torino, pp. 383−388 (1997).

[104]   Z. M. Zhang, "Surface Temperature Measurements using Optical Techniques," Annu. Rev. Heat Transf. **11**, 351−411 (2000).

[105]   B. Adams, A. Mayur, A. Hunter, D. Jennings, and R. Ramanujam, "Pyrometry for Laser Annealing" in *Proceedings of the 13th IEEE International Conference on Advanced Thermal Processing − RTP 2005*, Santa Barbara, CA (2005).

[106]   D. Jennings, A. Mayur, V. Parihar, H. Liang, R. Mcintosh, B. Adams, T. Thomas, J. Ranish, A. Hunter, T. Trowbridge, R. Achurtharaman, and R. Thakur, "Dynamic Surface Anneal: Activation Without Diffusion." in *Proceedings of the 12th IEEE International Conference on Advanced Thermal Processing − RTP 2004*, edited by J. Gelpey, B. Lojek, Z. Nenyei, and R. Singh, Portland, OR, pp. 47−51 (2004).

[107]   J. T. McWhirter, D. Gaines, and P. Zambon, "Emission Feedback Control System for Sub-millisecond Laser Spike Anneal," in *Proceedings of the 16th IEEE International Conference on Advanced Thermal Processing − RTP 2008*, Las Vegas, NV (2008).

[108]   P. J. Timans, Yao Zhi Hu, J. Gelpey, S. McCoy, W. Lerch, S. Paul, D. Bolze, H. Kheyrandish, J. Reyes, and S. Prussin, "Optimization of Diffusion, Activation and Damage Annealing in Millisecond Annealing," in *Proceedings of the 16th IEEE International Conference on Advanced Thermal Processing − RTP 2008*, Las Vegas, NV (2008).

CHAPTER 4

THERMOMETRY IN STEEL PRODUCTION

Tohru Iuchi[1], Yoshiro Yamada[2], Masato Sugiura[3] *and*
Akira Torao[4]

Contents

1. Introduction   217
2. Review of the Steel Production Process   219
   2.1. Smelting and refining processes   220
   2.2. Hot-rolling processes   221
   2.3. Annealing processes   222
   2.4. Coating processes   222
   2.5. Stainless steel and electrical steel processes   223
3. Characteristics of Radiometric Temperature Measurement in
   Steel Processes   224
   3.1. Passive methods   226
   3.2. Active methods   227
4. Applications   228
   4.1. Notable technical developments in the past and present   228
   4.2. Molten iron and steel   237
   4.3. Continuous heat-treatment lines   245
   4.4. Coating lines   261
5. Summary   270
References   271

1. INTRODUCTION

By the first half of the 19th century, the Industrial Revolution that
began in England was completed. The invention of the steam engine had

[1] Department of Mechanical Engineering, Toyo University, 2100 Kujirai, Kawagoe Saitama 350-8585, Japan
[2] National Metrology Institute of Japan, National Institute of Advanced Industrial Science and Technology, 1-1-1
Umezono, Tsukuba, Ibaraki 305-8563, Japan
[3] Environment & Process Technology Center, Nippon Steel Corporation, 20-1 Shintomi, Futtsu, Chiba 293-8511,
Japan
[4] New Products Planning & Development Department, JFE Advantech Co. Ltd., 3-48 Takahata-cho, Nishinomiya,
Hyogo 663-8202, Japan

Experimental Methods in the Physical Sciences, Volume 43   © 2010 Elsevier Inc.
ISSN 1079-4042, DOI 10.1016/S1079-4042(09)04304-5   All rights reserved.

drastically changed the industrial structure of the world. In this period, scientific understanding of heat processes in connection with the heat efficiency of steam engines contributed to the foundation of modern thermodynamics. The study of the most efficient conversion of heat to power, the "Carnot Cycle" introduced by Sadi Carnot, laid the path to the formulation of the key concept of the physical quantity of thermodynamic temperature by Lord Kelvin. These developments led, in the latter half of the 19th century, to a transformation of manufacturing and in particular the production of iron and steel. Here, temperature measurement of molten iron and steel was one of the most crucial tasks that needed to be performed and experiments into radiation pyrometry began at this time. So, both the establishment of thermodynamic temperature and the development of radiometric temperature measurement were closely related to the rise of heavy industries and are the result of successful interaction between the industrial and the academic world.

The pressing thermometry requirements of the steel industry and the development of radiometric temperature measurement were inextricably linked in the early stage of industrialization of iron and steel production. However, radiometric temperature measurement was not necessarily indispensable for the production of steel, until various innovations took place which transformed the steel processes from batch to continuous production. New processes with rapidly running steel sheets in heat-treatment furnaces increased enormously the importance of non-contact temperature measurement because the large temperature differences between the surroundings and the steel sheet inevitably forced direct measurement of the temperature of the steel sheet itself.

In this chapter, various examples of radiometric temperature measurement used in a modern integrated steel plant with the blast furnace (BF) at its core are described. In order to apply radiometric temperature measurement systems to steel processes, well-devised measurement principles are inevitably required that may require additional equipment to accommodate the adverse environment. Moreover, *in situ* calibration systems are often necessary. Practical examples of such systems are discussed.

First, in Section 2 the various manufacturing processes of a modern integrated steel industry and their products (in connection with radiation thermometry) are summarized. In Section 3, the importance of radiometric temperature measurement and the difficulties in its implementation in the steel industry are outlined and a classification of the different methods of radiation thermometry implemented in the steel industry is given. In Section 4, more sophisticated methods of radiometric temperature measurements applied to steel processes are described in detail. This section is divided into four: some pioneering achievements as well as unique methods that are of interest for the future are described in Section 4.1, successful applications of radiation thermometry for molten iron and steel, continuous heat-treatment lines and coating lines are detailed in Sections 4.2, 4.3 and 4.4, respectively. Concluding remarks with some future prospects for radiometric temperature measurement in steel industry are presented in Section 5. Throughout this chapter, radiance, emissivity, absorptance and reflectance are treated as *spectral* quantities, even though in the notations dependence on the wavelength is omitted for brevity. Unless otherwise stated, the surfaces are assumed to be opaque.

2. REVIEW OF THE STEEL PRODUCTION PROCESS

The most common form of production of iron and steel around the world is through an integrated process consisting of the BF and the basic oxygen furnace (BOF). The World Steel Association [1] has indicated that world crude steel output reached about 1,343 million metric tons (mmt) (1 mt = 1,000 kg) for the year 2007, an increase of 7.5% from 2006. BF–BOF process accounts for almost half of the total amount of world steel production.

Figure 1 shows a flowchart of the integrated manufacturing process for iron and steel [2,3] and describes the process from raw materials to various

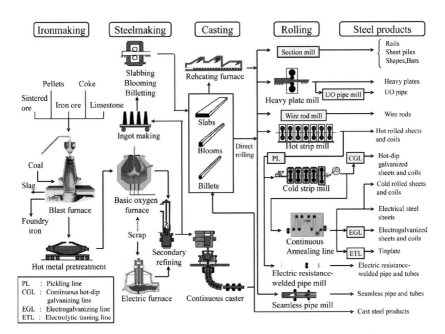

Figure 1   Iron and steelmaking process.

steel products. The smelting and refining process for iron and steel in the BF—BOF process involves the use of carbon to reduce iron ore (Fe_2O_3) in the BF to make molten iron [4], and then the molten iron is decarbonized in the BOF to make molten steel [5]. In this sense, the BOF is often called a "converter". Besides BF—BOF process, there is the electric furnace (EF) process which utilizes mainly scrap iron and steel in combination with direct reduced iron (when necessary) to produce molten steel. The molten steel from the BOF and the EF is then deoxidized and alloying elements are added in prescribed amounts to achieve the target composition and temperature.

After this procedure for converting iron to steel, the molten steel (temperature range: 1,500—1,600°C) is then cast in continuous casting machines to produce slabs, blooms and billets, and the castings obtained are cut into the required lengths [6]. After heating to the rolling temperature in a reheat furnace, the castings are hot-worked to the required product shapes on various types of mills: bars and wire rods are worked on section mills, bar mills and wire-rod mills, plates are processed by reversing mills, while hot-rolled steel sheets are rolled by hot strip mills [7].

After pickling (continuous dipping process in liquid acid) to remove oxidation layers from the surface, the hot-rolled steel sheets are worked to cold-rolled steel sheets on reversing mills or tandem rolling mills, and the cold-rolled steel sheets are tinned or galvanized as required to produce various surface-treated steel-sheet products.

Steel pipe is produced by forming and welding steel sheets or plates, or by piercing a billet and rolling to the final dimensions to produce pipes without a seam [8]. Other major steel products are stainless and electrical steel sheets, which are produced by special types of rolling mills and heating processes.

2.1. Smelting and refining processes

The largest BFs at present are about 100 m in total height, with a maximum internal diameter of about 16 m, and have an internal volume of over 5,500 m^3. A BF of this size can produce around 12,000 t of molten iron (also called hot metal or pig iron) per day. The temperature inside this huge facility exceeds 2,000°C where hot air is introduced to come into contact with the molten iron. The pig iron temperature at the taphole is around 1,500—1,550°C. To monitor the heat level inside the BF, a large number of thermocouples are used: in the furnace refractory, in lances inserted from the BF sides or occasionally through cables introduced from the BF top so as to slowly sink with the raw material. Tapped molten iron temperature measurement is described in detail in Section 4.2.

The main reaction in the BOF process is the oxidization of the carbon in the hot metal by the introduction of oxygen gas (O_2) into the furnace.

The residual oxygen, after contributing to this decarbonization reaction, remains in the molten steel. This oxygen is fixed and removed by deoxidation reagents such as silicon and aluminum which forms SiO_2 and Al_2O_3 or is removed as carbon monoxide gas in a subsequent vacuum degassing process (secondary refining). In the BOF, the temperature of the molten metal rises from around 1,400°C when freshly charged to above 1,600°C upon tapping in a single batch operation lasting approximately 20 min. The temperature is commonly measured by thermocouple probes mounted on a lifter called a sublance from the furnace top opening. For BF−BOF and EF processes, precise temperature control of molten iron or steel is necessary to produce high-quality products and to maintain stable operation of facilities.

2.2. Hot-rolling processes

When steel is heat treated, changes occur not only to the crystal structure and grain size, but also to the state of the foreign atoms present in the steel, namely they form solid solutions or precipitate as compounds according to their concentrations.

In the hot-rolling process, the strength of the steel can be increased by precipitating fine particles and by reducing the size of the crystal grains. This is accomplished by controlling the rolling condition and the cooling rate. Different properties can be obtained in steel of the same composition by controlling the crystal structure, the size and distribution of the precipitated particles. This means that a wide range of steel properties can be obtained by an effective and optimum heat treatment. This feature makes it possible to produce steel materials with diverse characteristics, and thus to make them suitable for specific applications in various fields.

A typical hot-rolling process is now described. A slab of steel about 250 mm in thickness is heated to about 1,200°C in the reheating furnace, and then delivered for hot rolling at the hot strip mill which contains both roughing mills and finishing mills. The thickness of strips rolled on the hot strip mill ranges roughly from 0.8 to 25.4 mm, the maximum strip width from 1.3 to 2.2 m and the rolling speed of the final stand is about 22 m/s. Throughout the hot-rolling process, the temperature of the slab and the strip is measured after each process and temperature control applied to ensure that the rolling, cooling and coiling are undertaken at the target temperature (500−700°C). A hot strip mill alone is generally equipped with more than 20 radiation thermometers of various kinds [9−11].

In the traditional hot-rolling process, the batch rolling process, the rolling of sheet bars and coiling of strips are processed individually to the finish rolling. In a new process, the so-called "endless hot strip rolling", the sheet bars are joined and continuously rolled. For joining the head and tail ends of sheet bars, induction heating with upset welding [12] and laser welding [13] are adopted. Finishing hot strip rolling is performed under an uninterrupted tension between finish rolling stands. The result of using the continuous method is that product quality, such as small deviations of strip thickness, as well as productivity and stability of rolling, is greatly improved. In addition, an ultra-thin strip almost as thin as a cold-rolled strip can be produced at the strip mill, bypassing the cold-rolling process.

2.3. Annealing processes

Steel hardens after cold rolling due to the dislocation tangling formed by plastic deformation. Annealing is therefore subsequently carried out to soften the material. The annealing process is comprised of heating, holding the material at an elevated temperature (soaking) for a pre-determined time and cooling of the material. Heating facilitates the movement of iron atoms, resulting in the disappearance of tangled dislocations, and stimulates the nucleation and growth of grains of various sizes, which depend on the heating and soaking conditions.

The annealing of cold-rolled coils used to be performed as coil stacks in a bell-type furnace. This process was known as batch annealing. However, a continuous annealing furnace (CAF) is now more commonly used. This type of annealing involves uncoiling, welding of strips' ends, passing the welded strips continuously through a heating furnace, and then cutting the welds and recoiling the strips [14,15]. In this process, the heating cycle applied to strips differs from product to product and has to be controlled according to the required mechanical properties such as hardness and tensile strength for each product. It is consequently important to measure the temperature of the strip while it is traveling through the heating furnace in order to control the operating conditions [16].

2.4. Coating processes

Flat sheet products for which corrosion resistance is essential are coated after annealing. Coating processes are broadly divided into hot dipping and electroplating. The hot-dip process is more suitable for heavy coating weights, and electroplating for lighter coatings. Typical hot-dip-coated products include galvanized strips for automobiles, construction materials and home electrical appliances, while tin- and chrome-plated strips for food and beverage cans and other containers are typical products of electroplating. Continuous coating of strip is more common than coating of cut sheets due to its increased productivity.

In recent times, zinc-coated steel sheets for use in automobile manufacture are mainly produced in a continuous galvanizing line (CGL).

In this method, the strip of steel is initially degreased, pickled and cleansed, and it is then passed through the annealing furnace and into a pot containing molten zinc. The annealing furnace is used to apply the necessary heat cycle to obtain the required mechanical properties and to activate the surface with a reducing gas, which makes it easier for zinc to wet the strip surface. The cross-section of a galvanized strip is composed of the steel substrate, iron−zinc alloy layers and a zinc layer on the surface. As basic properties like paint adhesion and weldability of steel strips are essential for automobile use, a galvannealing process has been developed to improve these properties. In this process, the zinc-coated strip emerges from the pot and is heated in a galvannealing furnace, forming an iron−zinc alloy layer by interdiffusion of iron and zinc, so that the surface of the zinc layer also contains some iron.

For both galvanizing and galvannealing processes, accurate temperature measurement of the strips is very important to achieve the required precise process control that enables stable quality products.

2.5. Stainless steel and electrical steel processes

As stainless steel has a lot of outstanding characteristics such as high corrosion resistance, good press formability, high oxidation resistance, high temperature strength and good appearance (high glossiness or brightness), it has consequently a large number of application fields including construc-tion, automobile, home electrical appliances, food production equipment and so on [17]. Electrical (or silicon) steel is manufactured so as to possess superior magnetic properties such as high permeability, low core loss and low magnetostriction. Therefore, it is commonly used for rotating machines and transformers [18,19]. Manufacturing facilities for these steels include a heating furnace, rolling mill and annealing line, and in the case of electrical steel complicated coating processes are also essential to obtain good magnetic properties. As optimum control of the crystal structure and size through heat treatment is a key factor to achieve the material properties necessary for their application, temperature control in the heating and annealing processes plays a critical role. As a result, accurate temperature measurement is necessary at several specific positions in the furnace.

As shown, a wide variety of steel products are produced by various processes in the steel industry. One prominent feature is that major processes such as casting, rolling, annealing and coating are operated continuously. Another is high-speed, automated operation of facilities to produce high-quality sheet products efficiently. Most new facilities are now highly automated by use of the latest electronic devices, computer, network, mechanical systems and, in particular, a large number of process control sensors [20]. These trends are based not only on recent progress in electronics, information and mechatronics technologies, but also on research and development activities in such fields as: online measurement and inspection for surface or internal qualitative properties and various process variables, process control and scheduling technology, process analysis using mathematical statistics and numerical simulation, practical application of quality control and chemical analysis, etc. [21].

Many practical examples of radiation thermometry in steel industry are well reviewed in Refs. [22−26]. A comprehensive steel database is given in Ref. [27].

3. CHARACTERISTICS OF RADIOMETRIC TEMPERATURE MEASUREMENT IN STEEL PROCESSES

Heating and cooling in a controlled manner are the defining process features in iron and steel manufacture as described in Section 2. Therefore, from the very beginning of steel production to the current highly sophisticated modern steel processes, there has been an unchanging strong demand for reliable temperature measurements in process control, in order to produce high-quality products, to maintain process stability and to optimize energy use.

Both contact and non-contact temperature measurements play very important roles in the steel production process, because there are certain processes where the contact method is favorable, while there are others where the non-contact method is better suited. For a batch process like a reheating furnace, objects generally have large volumes and thermal equilibrium of the object with the furnace is required in order to eliminate temperature gradients inside the object. In this situation, temperature measurement by thermocouples inserted into the furnace is sufficient for its operation. On the other hand, in the case of continuous processes like a continuous annealing line, there is a large difference between the temperatures of the surrounding furnace and the object, and therefore temperature measurement of the object itself is mandatory for the control of product quality. Radiometric temperature measurement is indispensable for these processes. In the course of time, most batch processes in steel production have been converted to continuous ones and the demand for radiometric temperature measurement has rapidly increased. Even in batch processes such as slab reheating furnaces, demand for improved steel quality has placed stronger demands on temperature control, and the importance of direct temperature measurement of the objects before reaching thermal equilibrium has increased. More recently, with the advent of new steel products such as new types of coated steel sheets to meet increasing customer demands, steel production depends even more on critical process temperature control. Detailed examples of radiometric temperature measurements that were developed to meet such demands are described in Section 4.

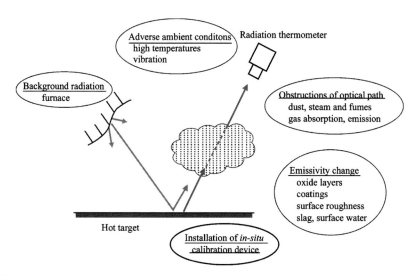

Figure 2   Difficulties in radiometric temperature measurement in steel processes.

Radiometric temperature measurement encounters numerous difficulties when implemented in steel processes, as summarized in Figure 2. First, variation in surface emissivity and existence of background radiation are inherent problems. In the case of steel, emissivity is never close to unity for any choice of wavelength and depends on surface coatings, surface oxide layers, surface roughness, material composition and temperature. For radiometric temperature measurement of an object in a furnace, the effect of background radiation from heat sources in the furnace must be ameliorated before correcting for emissivity change. Second, most steel processes are far from clean, and temperature measurement must be conducted in harsh environments. Molten metals are mostly covered with slag. Many steel processes are accompanied by steam, fumes and dust which obscure the radiant flux from the measurement object. At a continuous line, the object runs at a high speed, causing vibration and tilting of the target, which often degrade the accuracy of the radiance measurement. Thus, radiometric temperature measurement systems must be supplemented with additional equipment to eliminate these external disturbances or in some cases to actively control them, resulting in a high-cost rugged apparatus.

In situ sensor calibration is an effective straightforward means to check and mitigate the effects of these sources of error [28,29]. The importance of a traceability system to ITS-90 within the steel plant for radiometric temperature measurement should also be recognized, especially because of the large number of radiation thermometers in use covering a wide temperature range [30]. Traceability of radiation thermometers is described in Chapter 2 in the companion volume of *Radiometric Temperature Measurements: I. Fundamentals*, Vol. 42 of this series.

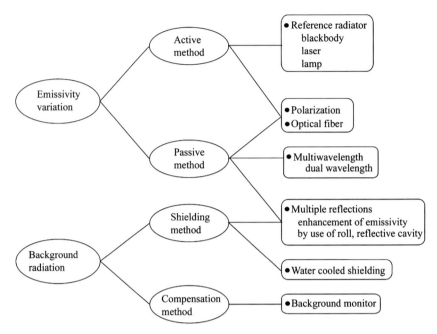

Figure 3 Classification of radiometric temperature measurement techniques in steel processes.

In the next section, radiation thermometry methods which enable measurements of objects with varying emissivity are treated extensively. Initially, offline-evaluated emissivity values were applied to online measurements [28,31−34]. Soon, various methods to enable radiation temperature measurement of objects with unknown or varying emissivity were devised. Such radiometric temperature measurement methods are classified into two categories: passive and active.

Figure 3 shows the classification of radiometric temperature measurement methods in steel processes.

3.1. Passive methods

The passive method requires no external radiant source other than the thermal radiation emitted from the measured object. The method is subdivided into two groups: one detects the radiance of the object and corrects with *a priori* knowledge of emissivity without additional apparatus [35−37]. The other makes use of geometric enhancements of the emissivity either within the process [29,38] or by using interaction of the radiance with an auxiliary apparatus [33,39−41].

The former determines the temperature from radiance measurements at two or more wavelengths (or alternatively at two polarizations).

The problems of conventional ratio pyrometry are discussed in Chapter 1 (this volume). Ways to overcome these shortcomings are described in Section 4.3.4. The latter group makes use of the emissivity enhancement through multiple reflections between the surfaces of the object and the auxiliary apparatus or other similar means within the process. If the number of reflections can be considered to be infinite, the detected radiance, L, is the sum of all the radiances originating from the object at temperature T_o and the apparatus at temperature T_r that reaches the detector, and therefore the following relation holds:

$$
\begin{aligned}
L &= \varepsilon_o L_b(T_o) + R_o \varepsilon_r L_b(T_r) + R_o R_r \varepsilon_o L_b(T_o) \\
&\quad + R_o^2 R_r \varepsilon_r L_b(T_r) + (R_o R_r)^2 \varepsilon_o L_b(T_o) + \cdots \\
&= \left\{ 1 + \sum_{i=1}^{\infty} (R_o R_r)^i \right\} \{ \varepsilon_o L_b(T_o) + R_o \varepsilon_r L_b(T_r) \} \\
&= \frac{\varepsilon_o L_b(T_o) + R_o \varepsilon_r L_b(T_r)}{1 - R_o R_r} = \frac{(1 - R_o) L_b(T_o) + R_o (1 - R_r) L_b(T_r)}{1 - R_o R_r}
\end{aligned}
\tag{1}
$$

here, R and ε are the reflectance and the emissivity, the subscripts o and r represent the object and the apparatus, and $L_b(T)$ is the blackbody radiance of an object at temperature T. To derive the last equality, the following relation based on the Kirchhoff's law (Section 3.3 of Chapter 3 in the companion volume of *Radiometric Temperature Measurements: I. Fundamentals*, Vol. 42 of this series) was applied:

$$
\varepsilon = 1 - R
\tag{2}
$$

Inspection of the last expression in Equation (1) tells us that $L = L_b(T_o)$ holds (in other words, the effective emissivity becomes one) for the following two extreme cases:

(1) $R_r = 1$ (i.e. the apparatus is a perfect mirror);
(2) $T_r = T_o$.

Neither of these conditions can be rigorously met in real situations, nor can the number of reflections reach infinity. Even so, the enhancement of the emissivity through multiple reflections has been effectively put to use as described in detail in Sections 4.3.2, 4.3.3 and 4.4.1.

3.2. Active methods

The active method requires the supply of additional radiant energy. One approach is to position a large blackbody radiation source, of the same temperature as the object, so that its radiation is reflected from the object surface [42,43]. This is equivalent to putting $R_r = 0$ and $T_o = T_r$ at the same time in Equation (1). It is easy to verify that under this condition,

$L = L_{\mathrm{b}}(T_{\mathrm{o}})$ holds. Another approach is to measure the reflectance by an external radiance source incident on the object surface and apply Equation (2) to obtain the emissivity. However, both of these approaches become impractical for real surfaces with roughness and variable tilt. A method to overcome these problems is described in Section 4.4.2 [44]. A third approach is to apply a high power laser beam to thermally modulate the object surface. The detected modulated thermal radiation signal then carries information of the absorptance of the object surface, which is equal to its emissivity [45,46]. These are described in Section 4.1.4.

Finally, it should be emphasized that understanding the behavior of the emissivity of the object, that is, of steels in the current case, from both theoretical [47−54] and experimental aspects [55−64], is very helpful for developing radiometric temperature measurement system for specific applications.

4. APPLICATIONS

In this section, practical examples of radiometric temperature measurement methods as applied to the production and processing of steel are described in detail. This section consists of four parts; the first reviews pioneering achievements of the past and introduces some unique and attractive techniques for the future. The rest of the section deals with practical applications of radiation thermometry to various types of steel production.

4.1. Notable technical developments in the past and present

4.1.1. Radiation thermometry of steel sheets in furnaces [31,32]

The radiometric temperature measurement of objects in a furnace is one of the most important tasks in the production of steel. A major difficulty in using the technique is the large temperature difference between the surroundings (i.e. the furnace) and the object. The background thermal radiation from the furnace is in most cases much larger than the thermal radiation originating from the object itself. A proposed solution to this problem is shown in Figure 4. This is known as the furnace load temperature measurement proposed by Barber [31].

Two radiometers are used in this method: radiometer 1 measures the radiance of the object and radiometer 2 detects the radiance of the furnace wall. The radiance signals L_1 and L_2 detected by the two radiometers 1 and 2 can be written as shown in Equations (3) and (4), respectively:

$$L_1 = \varepsilon_{\mathrm{o}} L_{\mathrm{b}}(T_{\mathrm{o}}) + (1 - \varepsilon_{\mathrm{o}}) L_{\mathrm{b}}(T_{\mathrm{e}}) \qquad (3)$$

$$L_2 = L_{\mathrm{b}}(T_{\mathrm{e}}) \qquad (4)$$

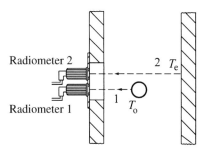

Figure 4   Furnace load temperature measurement. (Original figure from Ref. [31].)

where T_o and T_e are the temperatures of the object and the environment (surrounding furnace wall), respectively, ε_o the object emissivity and $L_b(T)$ the blackbody radiance signal at temperature T detected by the radiometers. The following equation is derived from the above two equations and from this T_o is obtained through an *a priori* determination of ε_o:

$$L_b(T_o) = \frac{L_1 - (1 - \varepsilon_o)L_2}{\varepsilon_o} \tag{5}$$

The temperature measurement of the object via the above equation is based on the assumption that ε_o is constant with temperature during processing, the environment temperature T_e is uniform and the related surfaces are diffusely reflecting. In steel processing, these conditions are rarely met, so the application of this method to steel processes is limited [65,66].

Barber contributed to another method of radiometric temperature measurement in a continuous heat-treatment furnace [31]. The method consists of having a miniature radiation pyrometer mounted in a water-cooled jacket which shields the thermometer from the furnace background radiation. The pyrometer observes the strip through a hole in a flange that shields the strip from incident ambient radiation. A major difficulty with the method is the provision of an adequate shield. He considers the ratio, d/D, which can be interpreted as the normalized distance, where d is the shielding flange−object distance and D the shielding flange diameter. When the ratio is small, the background radiation is sufficiently suppressed so that the only unknown variable in the measurement is the emissivity. He derives maximum values for this ratio that enable temperature measurement with accuracy within a certain range for various reflecting properties of the object surface.

These methods played a pioneering role in developing radiometric temperature measurement of objects in a furnace, where contamination of emitted radiance with background radiation is an unavoidable difficulty.

Figure 5 shows the arrangement of a radiation thermometry system developed by Murray, which was applied to the continuous annealing line of cold-rolled steels at United States Steel Corp. [32]. In this temperature

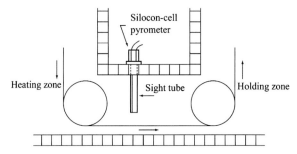

Figure 5 A pyrometer arrangement on continuous annealing line. (Reproduced from Ref. [32] with permission.)

measurement system, Murray introduced several techniques that are commonly practiced today.

A continuous annealing line is composed of several heat zones having different temperatures for metallurgical treatment of a steel strip. Murray's measurement system was mounted in the roof of the tunnel between the heat and hold zones, looking down on the strip as it passed between these zones. The tunnel location was a suitable place for the thermometer as there is much reduced background radiation emitted at this point of the process.

He used a silicon detector-based radiation thermometer responsive from 0.8 to 1.0 µm for measurement in the temperature range of $500-800°C$. Due to the relatively short operating wavelength compared to thermometers based on thermal detectors, he anticipated enhanced performance [67]. However, he recognized that some background radiation was present in the readings of the pyrometer, even though the installation of the pyrometer in the tunnel was chosen to minimize this problem. To mitigate the effect of residual background radiation, he introduces a second pyrometer with an identical optical system but with a lead sulfide detector sensitive at 2.2 µm, for which the surface becomes more specular with longer wavelengths when viewing from an inclined direction and therefore is less sensitive to background radiation from the hot surroundings. He compares the two readings and argues that temperature measurement with the silicon-based pyrometer system is possible within errors of $\pm 14°C$ at about $500°C$ by adjusting the emissivity setting to 0.6 to account for the background radiation instead of 0.4, the true emissivity of the cold-rolled steel at wavelength 0.9 µm.

To verify this argument, an apparatus to separately measure the emissivity of a cold-rolled steel sheet was designed and constructed by Murray. A strip specimen was heated without oxidation in a small chamber in an atmosphere of hydrogen−nitrogen gas and the emissivities of the strip at 0.9 and 2.2 µm were determined *a priori* to be 0.40 and 0.25, respectively. These emissivity values are used to derive the optimum emissivity setting of 0.6 at the online system.

The emergence of these integrated measurement systems in continuous annealing lines was one of the epoch-making developments in radiometric temperature measurement. Various temperature measurement systems in continuous annealing lines have been developed since, some of which are comprehensively described in Section 4.3 [28,29,35−37,40,41,68]. For further discussions on reflection errors in furnaces, interested readers are referred to a comprehensive textbook on radiation thermometry with focus on this topic [69].

4.1.2. Pyrometer with hemispherical reflector [33]

Drury et al. developed a pyrometer which is highly acclaimed for its pioneering achievement as a passive radiometric method for temperature measurement [33] that formed the basis of many successively developed passive methods for dealing with unknown emissivities [29,38,40,41]. Figure 6 shows the photograph of this type of pyrometer head which consists of a silicon detector that receives radiance through a small hole at the top of a highly reflecting hemisphere. This method is widely known as the "gold-cup" technique. When the reflecting hemisphere briefly makes contact with the measurement object surface, or covers it with only a small gap, the apparent radiance through the small hole increases by multiple reflections between the hemisphere and the object surface causing the effective emissivity of the object to approach 1, that is, equivalent to that of a blackbody. This results in improved temperature measurement in spite of emissivity variations of the object.

Under ideal conditions, the effective emissivity ε_{eff} of the object using the hemisphere method is obtained from Equation (1) by putting $L_b(T_r) = 0$ and is shown in the following equation:

$$\varepsilon_{eff} = \frac{L}{L_b(T_o)} = \frac{1 - R_o}{1 - R_o R_r} = \frac{\varepsilon_o}{1 - (1 - \varepsilon_o)R_r} \tag{6}$$

where ε_o is the emissivity of the object and R_o and R_r the reflectances of the object and the inside surface of the hemisphere, respectively. Here Equation (2) was applied to replace R_o with ε_o, and ε_{eff} was calculated to be 0.91 by Equation (6) when ε_o and R_r are 0.5 and 0.9, respectively. In practice, the system reaches pseudo-blackbody condition when ε_o becomes about 0.8.

A pyrometer with a hemispherical reflector and a silicon detector sensitive at relatively short wavelengths is widely applicable to measurements in the high temperature range above 500°C. This type of pyrometer was later applied in-process for obtaining a reference temperature for a CAF thermometer (cf. Section 4.3.2) [29,70].

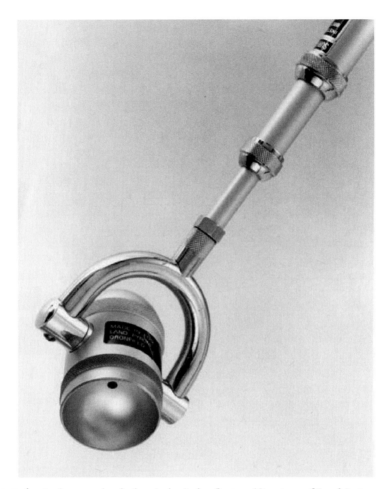

Figure 6 A photograph of a hemispherical reflector. (Courtesy of Land Instruments International Ltd.)

4.1.3. Radiation thermometry based on polarization method [43]

Murray proposed a radiation thermometer named the "polaradiometer", which utilizes the optical properties of reflected polarized light and that of thermally emitted light by the metal surfaces [43,71]. The principle of this method is essentially based on the approach described in Tingwaldt [72]. The "polaradiometer" is an emissivity-free method of radiometric temperature measurement based on a null-balanced active method.

Figure 7 shows the principle of the polaradiometer which consists of examining the sum of polarized radiances of the emitted components from the measurement object and the components reflected at the surface of the object that originates from a reference blackbody. The radiance emitted by

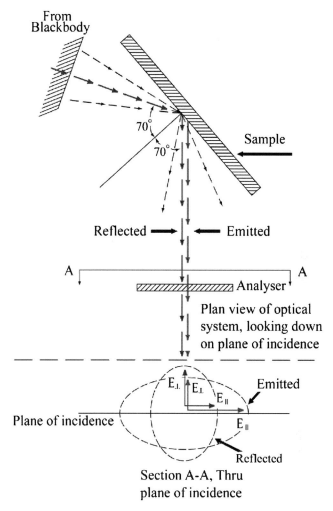

Figure 7 Polarization of emitted and reflected beams. (Reproduced from Ref. [43] with permission.)

the metal surface consists of a p-polarized (i.e. in the plane of incidence) major component and an s-polarized (i.e. perpendicular to the plane of incidence) minor component. Since Kirchhoff's law holds for each component, the originally unpolarized blackbody radiance, when reflected on the metal surface, has the s-polarized as the dominant component and the p-polarized as the lesser component. The sum of the emitted and reflected radiances is observed to be the resultant of two elliptically polarized radiances when viewed through a rotating analyzer, as indicated in Figure 7.

The following equation is obtained for the radiance signal I as seen through the rotating analyzer (the theoretical analysis and the derivation of Equation (7) are described in detail in Ref. [43]):

$$I = \frac{1}{4} \left[E_{br}^2 + \left(\varepsilon_p + \varepsilon_s \right) \left(E_{bo}^2 - E_{br}^2 \right) + \left(\varepsilon_p - \varepsilon_s \right) \left(E_{bo}^2 - E_{br}^2 \right) \cos 2\phi \right] \qquad (7)$$

where E_{bo} and E_{br} are the amplitudes of the blackbody radiation emitted at the object temperature T_o and the reference blackbody temperature T_r, respectively, ε_p and ε_s the emissivities of the object for radiation polarized parallel and perpendicular to the plane of incidence, respectively, and ϕ the time-varying azimuth angle of the analyzer. The first two terms of Equation (7) represent the DC components and the third term is the alternating signal. The measurement method relies only on measurement of the alternating component. The third term depends equally on the radiance of the object and that of the blackbody. This term disappears when the object and blackbody temperatures are equal, *irrespective* of the emissivity of the object. Therefore, the temperature of the object is obtained from the blackbody temperature by null balancing the AC component signal by varying the temperature of the reference blackbody.

Field tests were conducted with polaradiometers in several plants of the US Steel Corporation. However, it was found that in practice the method was not completely emissivity-free. Murray states that the failure of the polaradiometer to be completely emissivity-free is related to the geometry of the system. For diffuse samples such as MgO-coated steel, a larger blackbody is required and the object-to-blackbody spacing becomes impracticably small. Further investigation is necessary to determine whether this limitation can be overcome by changes in the optics of the system. He also states that the use of smaller acceptance angles, requiring a smaller angle to be filled by the blackbody, may be helpful.

Murray, moreover, argues that one of the obstacles to the development of the method is the widespread misconception that polarization techniques can be applied only to optically smooth, clean, metallic surfaces. This argument may be true to some extent, but one must also understand that there are *significant* problems associated with the polarization technique. It is strongly recommended for anyone applying this technique to study beforehand its feasibility by evaluating the polarization characteristics of the real target material, because these are highly dependent on its surface condition. One concrete achievement of radiometric temperature measurements by a polarization technique, using a different approach, in steel industry is described in Section 4.3.4 [35]. A comprehensive comparison between simulated and experimentally obtained polarized emissivity of oxidized steel surfaces is reported in Ref. [63].

4.1.4. Laser absorption radiation thermometry (LART) [45,46,73]

4.1.4.1. Three-wavelength LART [45,73].

A method of actively negating the effect of emissivity using laser sources was proposed by DeWitt and Kunz in the 1970s [56]. This regained strong attention in the past decade with the advances in solid-state laser and photodetector technologies [45,73]. The method is an extension of two-color ratio pyrometry, where, instead of assuming *a priori* the emissivity ratio of the object, it is determined *in situ* from the modulated signal of the thermal radiation induced by laser heating.

In this technique, the measurement object at temperature T is heated by laser pulses at two wavelengths, λ_1 and λ_2. The pulsed temperature rises ΔT_1 and ΔT_2 are proportional to the incident laser powers $P(\lambda_1)$ and $P(\lambda_2)$ and the absorptance $\alpha(\lambda_1)$ and $\alpha(\lambda_2)$ of the object, that is

$$\Delta T_1 = c'\alpha(\lambda_1)P(\lambda_1), \qquad \Delta T_2 = c'\alpha(\lambda_2)P(\lambda_2) \tag{8}$$

where c' is the proportionality constant which is assumed to be equal for both conditions.

If ΔT_1 and ΔT_2 ($\ll T$) are monitored at a third wavelength, λ_3, different from the laser wavelengths, then the modulated radiance amplitudes $\Delta L(\lambda_3, \lambda_1)$ and $\Delta L(\lambda_3, \lambda_2)$ in the radiance $L_{\lambda_3}(T)$, induced by the incident lasers at wavelengths λ_1 and λ_2, respectively, are given by the following:

$$\Delta L(\lambda_3, \lambda_1) = \left(\frac{\partial L}{\partial T}\right)_{\lambda_3} \Delta T_1, \qquad \Delta L(\lambda_3, \lambda_2) = \left(\frac{\partial L}{\partial T}\right)_{\lambda_3} \Delta T_2 \tag{9}$$

If $P(\lambda_1)$ and $P(\lambda_2)$ are monitored, then from the above equation, the absorptance ratio $\alpha(\lambda_1)/\alpha(\lambda_2)$ is determined from the following:

$$\frac{\alpha(\lambda_1)}{\alpha(\lambda_2)} = \frac{(\Delta L(\lambda_3, \lambda_1))/P(\lambda_1)}{(\Delta L(\lambda_3, \lambda_2))/P(\lambda_2)} = \frac{\varepsilon(\lambda_1)}{\varepsilon(\lambda_2)}$$

The last equality comes from the Kirchhoff's theorem (cf. Section 3.3 of Chapter 3 in the companion volume of *Radiometric Temperature Measurements: I. Fundamentals*, Vol. 42 of this series), which tells us the (directional spectral) absorptance α is equal to the (directional spectral) emissivity ε. The determined emissivity ratio can be applied to two-color pyrometry at the same two wavelengths λ_1 and λ_2 to obtain a temperature measurement by the following relation:

$$\frac{L_{b,\lambda_1}(T)}{L_{b,\lambda_2}(T)} = \frac{\Delta L(\lambda_3, \lambda_2)}{\Delta L(\lambda_3, \lambda_1)} \frac{P(\lambda_1)}{P(\lambda_2)} \frac{L_{\lambda_1}(T)}{L_{\lambda_2}(T)} \tag{10}$$

where $L_{\lambda_1}(T)$ and $L_{\lambda_2}(T)$ are thermal radiances of the object detected at wavelengths λ_1 and λ_2, respectively, and $L_{b,\lambda}(T)$ the blackbody radiance at wavelength λ and temperature T. Note that the right-hand side of

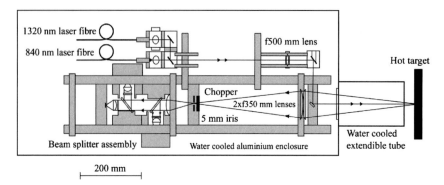

Figure 8 Schematic plan of LART optical head. (Reproduced from Ref. [73] with permission of IMEKO Secretariat.)

Equation (10) consists of six measured values. Therefore, solving this equation for temperature T gives an emissivity-free temperature measurement.

A schematic diagram of the LART optical head is shown in Figure 8. Two optical fibers provide 1 W CW radiation from a solid-state laser and a diode pumped Nd:YAG laser, amplitude modulated at 44 and 23 Hz, respectively, at wavelengths 840 and 1,320 nm. A silicon photodiode detects the radiation at 840 nm, while two InGaAs photodiodes are used at wavelengths 1,320 and 1,550 nm.

4.1.4.2. Two-wavelength LART [46].

An alternative approach, instead of measuring the temperature rises ΔT_1 and ΔT_2 induced by laser pulses at wavelengths λ_1 and λ_2, respectively, at a third wavelength λ_3, is to measure them with the same wavelength as the incident laser, that is, to measure at wavelength λ_2 the temperature modulation ΔT_1, and *vice versa*. In this case, the measurement can be simplified as follows:

$$\Delta L(\lambda_2, \lambda_1) = \left(\frac{\partial L}{\partial T}\right)_{\lambda_2} \Delta T_1, \qquad \Delta L(\lambda_1, \lambda_2) = \left(\frac{\partial L}{\partial T}\right)_{\lambda_1} \Delta T_2 \qquad (11)$$

Taking the ratio and combining with Equation (8) gives

$$\frac{\Delta L(\lambda_1, \lambda_2)}{\Delta L(\lambda_2, \lambda_1)} = \frac{\varepsilon(\lambda_1)}{\varepsilon(\lambda_2)} \frac{((\partial L_{b,\lambda}(T))/\partial T)_{\lambda_1}}{((\partial L_{b,\lambda}(T))/\partial T)_{\lambda_2}} \frac{\alpha(\lambda_2)}{\alpha(\lambda_1)} \frac{P(\lambda_2)}{P(\lambda_1)} \qquad (12)$$

Therefore

$$\frac{((\partial L_{b,\lambda}(T))/\partial T)_{\lambda_1}}{((\partial L_{b,\lambda}(T))/\partial T)_{\lambda_2}} = \frac{\Delta L(\lambda_1, \lambda_2)}{\Delta L(\lambda_2, \lambda_1)} \frac{P(\lambda_1)}{P(\lambda_2)} \qquad (13)$$

Now the right-hand side of the above equation consists of four measured quantities, and from this T can be determined without the effect of emissivity.

As compared to the three-wavelength LART, the two-wavelength LART has the advantage that, since only the AC modulated part of the radiance detection is used, the measurement is free from the effects of DC background radiation. This is similar to temperature measurement by the modulated photothermal effect [74,75], which is free from reflected background radiation, with the added advantage of being free of the effect of emissivity and absorption in the optical path.

The three-wavelength LART method has reportedly been field-tested in the steel plant though difficulties in measurement at the plant precluded reporting the results [76]. Improvements in instrumentation and further trials in the field [77,78] may transform the system to become a viable tool for application in steel production.

4.2. Molten iron and steel

Until the 1930s, temperature measurement of molten iron and steel relied on optical pyrometry based on the experienced human eye. Goggles of cobalt glass were used, which has transmission bands at both ends of the visible spectrum, and thus forms a sort of two-color pyrometer. Use of such goggles rendered the eye very sensitive to changes in temperature [79]. Correspondingly, an instrument based on optical pyrometry was not a practical proposition, due to the variation in the emissivity of molten metal, as well as the adverse environment: that is, slag covering the surface of the molten metal, dust and fumes in the optical path, and severe heat. In the open-hearth furnace (a shallow steel bath for melting and refining commonly in use prior to BOF), innovative ideas such as the "blow-tube pyrometer" were attempted [80,81]. In this method, the open end of an iron tube was inserted beyond the slag layer into the molten metal. Gas blown through the tube end maintains a well-stirred undulated bubble blackbody cavity in the bath, and the optical detector on the other end of the tube viewed the thermal radiation from the cavity. Despite this attempted innovation, the temperature measurement of liquid steel remained an unsolved problem [82].

This situation was changed when the "quick-immersion" thermo-couple was developed in the late 1930s by Schofield at the National Physical Laboratory in the UK [79,81,83]. The thermocouple tip on the end of a rod is inserted through the slag into the molten metal beneath. It is left in position until the temperature is recorded, and then is quickly withdrawn. By reducing the protective sheaths to the minimum possible, a reasonably reliable reading can be obtained in a few seconds and the thermocouple then withdrawn intact. After a certain number of immersions, the sensing thermocouple tip is discarded. After some years of development, the method came to be used in most steel plants worldwide [84,85]. The current situation is that accurate and reliable temperature measurement is possible by immersion thermocouples, but the cost of the disposable sensing cartridges and the difficulty in automating the measurement limit its application to infrequent batch measurement.

Recently, optical pyrometry is being reintroduced, replacing immersion thermocouples in processes in which (semi-)continuous temperature measurement of molten iron and steel is required [34,39,86,87]. Novel optical radiation thermometers utilizing modern optical methods while retaining the advantages of the classical "blow-tube" technique are described in this section.

4.2.1. Immersion-type optical-fiber radiation thermometry [39]

This thermometer is a contact measurement instrument that measures the temperature of the molten metal by thrusting the tip of an optical fiber below the metal surface. Thermal radiation from beneath the surface of the molten metal propagates along the fiber and is detected by the optical sensor at the other end. In Figure 9, the measurement sequence for an immersion-type optical-fiber radiation thermometer is shown. The measurement is rapid and is made before the immersed tip of the optical fiber is dissolved or torn away in the molten-metal flow. The fiber is retracted, and is thrust again when the next measurement is required. The greatest advantage of such a measurement method is that the immersed fiber tip, with its small diameter of 125 μm, forms a cylindrical glass cavity with an isothermal wall of molten metal whose effective emissivity is virtually equal to one. Other advantages are that the system is unaffected by the slag or powder that may cover the surface of the molten metal, and it is unaffected by the harsh environment (e.g. fumes, dust and extreme heat).

The method was applied to the continuous caster tundish. The tundish is a vessel placed above the continuous casting machine to allow the

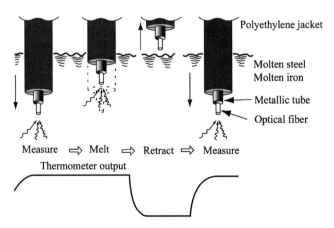

Figure 9 Measurement sequence by an immersion-type optical-fiber thermometer. (Reproduced from Ref. [39] with permission of IMEKO Secretariat.)

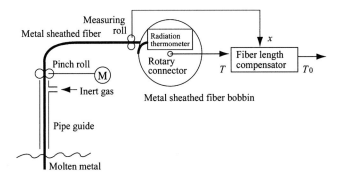

Figure 10 Measurement system configuration of tundish thermometer. (Reproduced from Ref. [39] with permission of IMEKO Secretariat.)

exchange of ladles carrying the molten steel without interrupting the casting. The knowledge of the molten steel temperature at the tundish is used to determine the operating conditions of the caster. The installed system is shown schematically in Figure 10. The fiber comes in a length of $100-500$ m wound around a rotating bobbin, with one end fed through a pipe guide that leads the fiber to the molten-metal surface. The other end of the fiber is connected to the radiation thermometer, which detects the transmitted radiation with a silicon photodiode with nominal wavelength of $0.9\,\mu m$. The system is equipped with a measuring roll that measures the consumed fiber length, and correction is made for the transmission loss through the fiber (approximately $0.08\,K/m$) according to the remaining length. The optical fiber is a SiO_2-graded index fiber of 50-μm core diameter and 125-μm cladding diameter, which is contained in a stainless-steel tube of outer diameter $1.2\,mm$. The metal sheathed fiber is produced using techniques for welded steel pipes. The metal sheath is then clad in a 4-mm diameter polyethylene jacket containing carbon particles, which carbonizes to form a protective layer when thrust into the molten steel. The fiber of this structure can be immersed 300-mm deep in the molten steel while only 40 mm of fiber is consumed. Comparison of results obtained by the method with immersion thermocouples is shown in Figure 11. The system results in a reduction in measurement cost by replacing the immersion thermocouples.

Another application of this measurement system has been made in the BF molten iron temperature measurement at the taphole [88,89]. Pig iron produced in a BF is tapped at a taphole of the furnace body and flows through a runner. The temperature at the taphole carries information of the thermal conditions inside the BF and is useful for optimizing the furnace control. However, the molten iron/slag mixture stream at the taphole has density and velocity of approximately $5{,}500\,kg/m^3$ and $10\,m/s$, respectively, and the flow is too rapid for immersion thermocouple measurements.

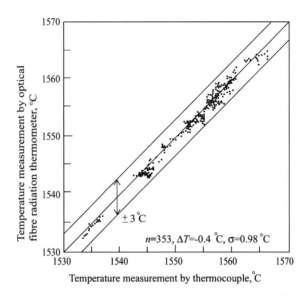

Figure 11 Comparison of measurement by immersion-type optical fiber radiation thermometer and immersion-type thermocouple. (Reproduced from Ref. [39] with permission of IMEKO Secretariat.)

Conventionally, a factory worker makes measurements at the slag skimmer, a dam located tens of meters downstream the runner from the taphole, where the slag is separated from the molten iron and the iron surface is exposed. However, the measured temperature is significantly less than that of the metal as it leaves the taphole due to heat loss to the runner refractory. In addition, because the frequency of the measurement is restricted for reasons of workload and thermocouple cost, it is difficult to identify rapid temperature changes.

To enable automated measurement at the taphole by applying the immersion-type optical fiber radiation thermometer, a double-metal-tube-sheathed fiber of outer diameter 3.6 mm was employed to achieve the rigidness required to realize sufficient immersion of the fiber tip into the rapid flow of dense liquid iron. The measurement system is depicted in Figure 12. An example of measurements with this system is shown in Figure 13 compared with measurements using an immersion thermocouple made at the skimmer. As can be seen, the temperature measured at the skimmer has a dead time of about 1.5 h during which heat is lost to the cold trough, whereas at the taphole the measured temperature is seen to vary continuously. The improved control of the furnace heat is reported to have contributed to an average decrease in molten iron temperature of approximately 20°C and hence to more efficient energy utilization and to better control of the pig iron composition [89].

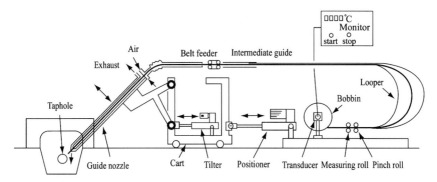

Figure 12   Blast furnace taphole molten iron measurement system. (Reproduced from Ref. [87] with permission.)

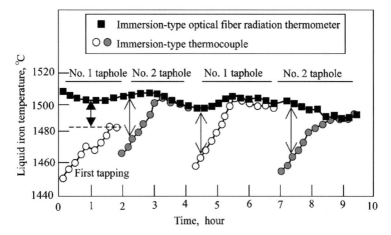

Figure 13   Measurement of molten iron temperature with an immersion-type optical-fiber radiation thermometer and with an immersion-type thermocouple. (Reproduced from Ref. [89] with permission.)

Another application of the optical fiber molten-metal thermometer is reported for continuous temperature measurement at a BOF [90]. Commercialized portable systems are also available which are applied to melting, casting or welding processes [91].

4.2.2. Two-dimensional radiation thermometry [34,87]

A two-dimensional radiation thermometry system has been developed, which automatically determines the measuring target in a thermal image and obtains its temperature. The system consists of a CCD camera, which is calibrated as a radiometer, and a high-speed image processor.

Molten iron at the BF is the object of the two-dimensional radiation thermometer [87]. If non-contact radiation thermometry can be applied for molten iron stream temperature measurement as it emerges from the taphole, it will provide a much simpler means to measure the temperature automatically and continuously compared with the immersion-type optical fiber radiation thermometer described in the previous section. One difficulty the method faces is that the molten iron stream contains slag, which is composed of various kinds of molten oxides, and the mixing rate of iron and slag varies during the tapping. Emissivity of slag is higher than that of molten iron. The effective emissivity is unknown in real time because we have no knowledge of the mixing rate. Error in the estimation of the emissivity will result in a large temperature error even if a short measurement wavelength is used. The two-dimensional observation method overcomes these difficulties in the following way.

A monochromatic camera, which is connected to an image processing system, observes the molten iron stream as shown in Figure 14. The camera is protected from the hostile environment by an air-cooled jacket and is placed at a distance of more than a few meters from the target to avoid the high ambient temperatures and splashing with molten iron. The stream moves at a high velocity of roughly 10 m/s, but the short exposure time of the camera enables image acquisition without blurring. As shown in the upper section of Figure 15, the thermal image has a marble pattern composed of two regions with different brightness. Temperature of the liquid should be uniform, so the difference in brightness is caused by the difference in emissivity of the two regions corresponding to slag and molten iron. The dark area of the pattern, that is, lower emissivity material, corresponds to the molten iron, and thermometry is carried out by targeting this area.

The radiance intensity of the molten iron is determined from the gray level histogram. As shown in the lower section of Figure 15, the radiance intensity distribution of the molten iron has a clear peak. The histogram shape varies with time, because the mixture rate of the molten iron and the slag varies gradually, but the intensity level of the molten iron peak depends

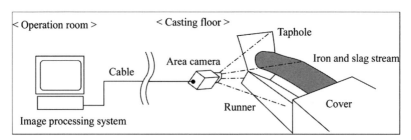

Figure 14 Schematic of tapped flow observation at a BF. (Original diagram taken from Ref. [87], reproduced with permission.)

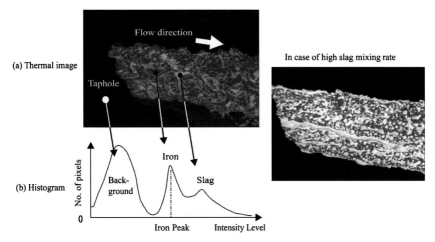

Figure 15   (a) Thermal image of the stream and (b) gray level histogram. (Original diagram taken from Ref. [87], reproduced with permission.)

only on its temperature. The image processing system finds this peak, obtains the radiance level and calculates the temperature assuming emissivity of molten iron of about 0.4 [92]. Field tests of the system show that the measured temperature obtained by this method agrees with discrete temperature measurements obtained with immersion thermocouples.

Molten steel temperature measurement at the BOF is another application of two-dimensional radiation thermometry [34]. When applying radiation thermometry to molten steel in a BOF, observation through the furnace top opening is impossible. This is because there are difficulties related to the measurement environment as follows:

- Large amounts of dust are generated in the space above the molten steel, so the thermal radiation from the molten steel is strongly attenuated.
- Slag floats on the molten steel surface. The emissivity and temperature of the slag is different from that of molten steel.
- High-temperature combustion gas formed by the top-blown oxygen gas is a source of an intense background radiation.

To overcome the difficulties above, the two-dimensional radiation thermometer system views through a window provided at the bottom of a furnace as shown in Figure 16. Pressurized inert gas is injected into the nozzle and prevents the molten steel from flowing out [93,94]. As with the "blow-tube pyrometer", the gas blown out from the nozzle forms a gas column in the molten steel which acts as a blackbody cavity. However, the gas adversely affects the field of view through the nozzle because solidified steel called a "mushroom" grows at the end of the nozzle where the cold inert gas comes in contact with the molten steel. To overcome this, the field

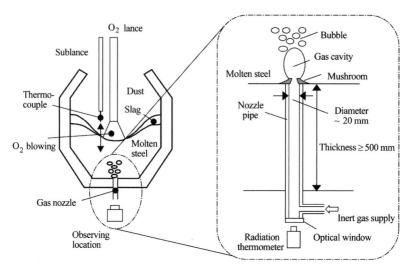

Figure 16 Schematic diagram of a BOF and observation at bottom gas nozzle. (Original diagram taken from Ref. [34], reproduced with permission.)

of view of the two-dimensional detector is adjusted to cover the whole inner diameter of the nozzle at its furnace inner end and, by doing so, we obtain a thermal image in which high-temperature molten steel appears as a high-intensity area against a dark area corresponding to low-temperature inner wall of the nozzle, as shown in Figure 17(a). Even if the optical axis shifts (Figure 17(b)) or the field of view narrows by a "mushroom" forming (Figure 17(c)), the radiance intensity level of the molten steel remains unchanged, though its relative position or area changes. Image processing then determines the temperature of the molten steel area.

Preliminary tests were carried out at a 1.5-t laboratory steel bath. Argon was used as the inert gas. Molten steel in the bath was heated by induction and cooled naturally. The temperature measured continuously by two-dimensional radiation thermometry agreed closely with that measured intermittently by thermocouples at 3-min interval. The agreement is an indication that the gas column formed a stable blackbody cavity.

Field tests were then conducted on a commercial 60-t stainless steel BOF. Furnace shell temperatures around the observation nozzle reach 400°C or higher in operation, which is intolerable for a CCD camera. To protect the CCD camera, a heat-resistant fiberscope was fitted to the nozzle and the image was fed to a CCD camera installed at a remote and cooler location. The measurement results by the two-dimensional radiation thermometry system agreed with the results of conventional immersion thermocouple measurement at the sublance, a lance lowered from the furnace top opening for measurement of molten steel temperature and composition.

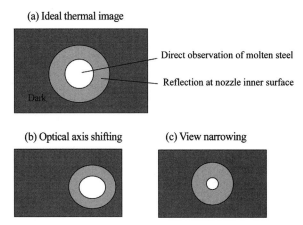

Figure 17 (a–c) Thermal images obtained by a camera at a nozzle. (Original diagram taken from Ref. [34], reproduced with permission.)

4.3. Continuous heat-treatment lines

Continuous heat-treatment lines for steel strips such as cold-rolled, stainless and silicon steels, in which steel sheets continuously run through furnaces at high speed, have superseded batch treatment processes. New CAFs employ rapid heating by combustion in the heating zone for increased thermal efficiency. However, the weak oxidizing atmosphere causes wide variations of the emissivity of the steel sheet, leading to difficulties in applying radiation thermometry. Another type of a continuous heat treatment is the galvannealed steel process, where the steel is alloyed with the zinc coating also resulting in rapid and large emissivity change.

In this section, radiometric temperature measurement methods developed to meet such emerging demands are described.

4.3.1. Temperature measurement system with a water-cooled shielding flange [28]

The first radiation thermometry system to be successfully applied online for the production of cold-rolled steel strips and silicon steel strips is described in Ref. [28]. The system consists of a radiation pyrometer in passive mode (i.e. without an external radiation source) with a water-cooled shielding flange and a contact thermometer for intermittent online calibration. The system is an evolved version of the methods by Barber and by Murray described in Section 4.1.1. A measurement model taking into account the reflections of thermal radiation originating from various surfaces is constructed to optimize the design of the radiation shield. The model also evaluates the effect of remaining reflected background radiation.

4.3.1.1. *Shielding effects.* Below a model is considered to simulate the system with a water-cooled shielding flange in a furnace as shown in Figure 18, where ε_o, ε_e and ε_s are emissivities of the object steel strip, environment furnace wall and shielding flange with diameter D, respectively. T_o, T_e and T_s are the temperatures of the above three surfaces. The distance between the shielding flange and the steel strip is given by d.

The effective radiance signal, L_i ($i =$ o, e, s), detected by a radiation pyrometer is written as follows:

$$L_i = \varepsilon_i L_b(T_i) + (1 - \varepsilon_i) \sum_k F_{i,k} L_k \qquad (14)$$

where $F_{i,k}$ represents the fraction of radiance leaving surface i and reaching surface k [53]. The subscripts (i, $k =$ o, e, s) represent the object steel strip, the environment (furnace wall) and the shielding flange surfaces, respectively. $L_b(T_i)$ detected by the radiation pyrometer is the blackbody radiance signal at temperature T_i. L_o, L_e and L_s are the effective radiances of the object, the furnace wall and the shielding flange, respectively.

The temperature reading T_a of the radiometer with emissivity setting ε_o is obtained from L_o/ε_o. The temperature difference, $\Delta T = T_a - T_o$, calculated from the model, shows the following behaviors:

(1) ΔT increases with increasing normalized distance, $\tilde{d} = 2d/D$.
(2) ΔT decreases rapidly with increasing ε_s.
(3) The measurement wavelength λ, which minimizes ΔT, shifts toward longer wavelength as the temperature difference, $T_e - T_o$, increases.

The temperature drop of the steel strip due to the heat loss because of the water-cooled flange was estimated using the model given in Figure 18 and

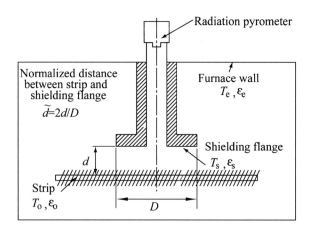

Figure 18 Simulation model of a radiometric temperature measurement system with a water-cooled shielding flange. (Reproduced from Ref. [28] with permission.)

the calculated value was confirmed experimentally using a contact thermometer. The results showed that the temperature drop is negligible and has no effect on the strip temperature under normal operating conditions (line speed ≥ 0.83 m/s) with $\tilde{d} \geq 0.2$. Based on the simulated results, a water-cooled shield was designed with blackened surface of emissivity $\varepsilon_s = 0.95$, so that the effect of radiation from the shield L_s and the reflection of the radiation originating from the object L_o could be neglected.

4.3.1.2. Background radiation factor, η.

The effect of the background radiation was considered and quantified through the introduction of the factor, η. η represents the ratio of the radiance reflected from the steel strip surface to the radiance emitted from the furnace walls and other heat sources. This factor enables a good estimate of the error in radiometric temperature measurement in a furnace to be calculated. The background radiance term is given by $\eta L_b(T_e)$. Using this term, Equation (14) is rewritten as follows:

$$L_o = \varepsilon_o L_b(T_o) + \eta L_b(T_e) \tag{15}$$

The output signal $L_b(T_a)$ of the radiometer after correcting for emissivity ε_o is given by the following equation:

$$L_b(T_a) = \frac{L_o}{\varepsilon_o} = L_b(T_o) + \frac{\eta}{\varepsilon_o} L_b(T_e) \tag{16}$$

Generally, η is related to ε_o, ε_s, $\tilde{d}$ and λ. To obtain η, the furnace temperature T_e is intentionally raised so that $T_o \ll T_e$, that is, $L_b(T_o) \ll L_b(T_e)$, holds. η is determined from Equation (15), as follows:

$$\eta = \frac{L_o}{L_b(T_e)} \tag{17}$$

Experimental value of η was found to clearly depend on the wavelength λ. In addition, a significant difference was observed for η between specular and diffuse surfaces. Cold-rolled steel and brightly annealed stainless steel belong to the former, while heavy steel plate belongs to the latter. Furthermore, an ε_s larger than 0.9 was required in order to reduce η to an acceptable level. The procedure to quantify η is detailed in Ref. [28]. Applying the Wien's approximation to $L_b(T)$, the equation estimating the correction term ΔT is given in the following equation:

$$\Delta T \equiv T_a - T_o = \frac{-(c_2/\lambda)}{\ln[\exp(-(c_2/(\lambda T_o))) + (\eta/\varepsilon_o) \exp(-(c_2/(\lambda T_e)))]} - T_o \tag{18}$$

where the temperatures are expressed in Kelvin.

The correction term can be estimated accurately enough if η and ε_o are obtained experimentally, and if T_o and T_e in real operating conditions in a CAF are known. It can be shown from Equation (18) that with values of η

and ε_o to match real situations, if $T_e \sim T_o$, $\lambda = 1\,\mu m$ is more suitable than $\lambda = 2\,\mu m$ in order to reduce ΔT, while in the case where $T_e - T_o \geq 200°C$, $\lambda = 2\,\mu m$ is better than $\lambda = 1\,\mu m$. In other words, the optimum measurement wavelength moves to longer wavelengths with increasing difference, $T_e - T_o$. Interested readers are guided to a generalized, comprehensive description of strategies for wavelength choice in Ref. [69]. With a suitable choice of wavelength and optimized shielding flange design, namely by taking $\lambda = 1\,\mu m$ and $\tilde{d} \leq 0.35$, ΔT can be kept within $\pm 4°C$ at $800°C$ in the case where $T_e - T_o \leq 200°C$. The factor η experimentally derived from the above procedure is quite useful when designing a water-cooled shielding flange to be installed in a newly constructed CAF. Once η is determined, ΔT is easily calculated by Equation (18). The emissivity ε_o of the steel strip can be obtained in advance by a laboratory experiment, but by use of the contact thermometer described below it can also be directly measured in the production line.

4.3.1.3. In situ measurement and calibration.
Additional factors need to be considered for online use, such as absorption loss by the atmosphere, varying tilt of the strip including its vertical vibration and the location of heat sources. These factors may differ from furnace to furnace and, therefore, a contact thermometer was developed for *in situ* calibration of the method and for intermittent checking to ensure continued reliable performance. The correction method involves simultaneous measurement of the strip temperature with both the pyrometer and the contact thermometer. The contact thermometer is provided with a thin stainless steel "guard" sheet (40 mm long, 16 mm wide and 0.2 mm thick) onto which a type-K thermocouple is welded. The guard sheet is used to protect the thermocouple from wear and to maintain good heat contact with the steel strip measured.

Figure 19 shows the cross-section of the contact thermometer used in the actual process. To improve the accuracy of measurement, the guard sheet is insulated and has a spring mechanism, which follows the undulating movements of the steel strip. However, the system has certain shortcomings. First, a slight scratch on the steel strip surface is unavoidable due to the contact. Second, the thermometer needs water-cooling to be effective in a furnace. Third, the contact thermometer has high maintenance costs, as the thermocouple guard sheet combination needs to be replaced every 24 h to avoid the growth in measurement errors due to the degradation of thermocouple performance in the high-temperature furnace. After extensive investigation, the resulting repeatability of the measurement with the contact thermometer was evaluated to be better than $\pm 5°C$ at a strip temperature of $800°C$, similar to the ΔT values calculated above.

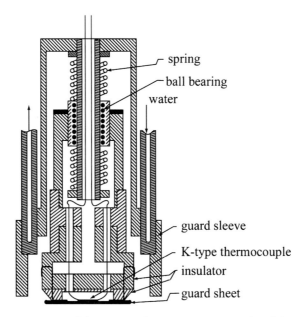

Figure 19 Cross-section of the contact thermometer. (Reproduced from Ref. [28] with permission.)

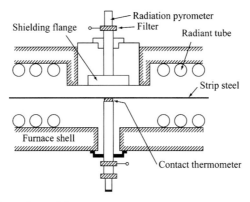

Figure 20 Configuration of the temperature measurement system. (Reproduced from Ref. [28] with permission.)

4.3.1.4. Online measurements in production line. Figure 20 shows the configuration of the online temperature measurement system composed of the water-cooled shielding flange and the contact thermometer for intermittent calibration. Seven systems were installed in 1972 in the CAF for cold-rolled steel and silicon steel, respectively, in Nippon Steel Corp. The emissivity readings measured in the laboratory and in the production line were compared. Results in the production line using the contact-type thermometer coincided with laboratory results within $\pm 5°C$ at $700°C$. The temperature measurement systems greatly contributed to production of steels, until they were finally replaced 15 years later by the Thermometry Reestablished by Automatic Compensation of Emissivity (TRACE) systems described in Section 4.3.4, which took place with the innovation of an efficient CAF. The water-cooled shielding flange was utilized in the TRACE system as a standard apparatus to mitigate background radiation. The contact thermometer for *in situ* calibration was transformed into a simplified system and utilized in the coating line thermometer described in Section 4.4.1.

Other examples of radiation thermometry in steel plants applied to furnaces or to other situations with large background radiation contamination can be found in Refs. [65,66,95,96].

4.3.2. Thermometry utilizing multiple reflections at a roll-strip wedge [29,38]

Yamada et al. devised a simple passive radiation temperature measurement method which is based on multiple reflections of radiation at the wedge formed just after the steel sheet breaks contact with the roll in a furnace [38,97]. The feature of this method is that it is effectively an emissivity-free measurement. By making use of a roll which is already installed for supporting the steel sheet, no auxiliary equipment needs to be introduced for the measurement. This method has the additional advantage that spurious radiation noise ever present in a furnace is rejected because the effective reflectance of the area being measured is close to zero. In the situation considered here, the reflectance of the roll is high enough, its temperature is close enough to that of the bulk steel roll and the number of reflections is large enough that the effective emissivity in the wedge can be treated as unity (cf. Section 3.1).

4.3.2.1. Principle of the multiple reflection method. The effective emissivity at the wedge, formed where the sheet comes out of contact from the roll, was calculated assuming totally specular reflection. Figure 21 shows the relation of the multiple reflections between sheet and roll. Figure 21(a) is an example of the locus of the ray incident upon the wedge, while Figure 21(b) shows the relation between θ_i, x_i and β_i, θ_{i+1}, x_{i+1}, where

θ_i, x_i: $\pi/2$ − incidence angle and position on ith incidence on the steel sheet, respectively;
β_i: incidence angle on roll after ith reflection on the steel sheet;
r: radius of the roll.

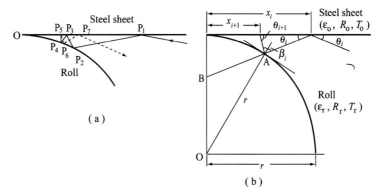

Figure 21 (a and b) Multiple reflections between strip and roll. (Reproduced from Ref. [38] with permission.)

β_i, θ_{i+1} and x_{i+1} are derived successively by considering the two-dimensional geometry and are given by the following equations:

$$\beta_i = \sin^{-1}\left\{\left(1 - \tan\theta_i\frac{x_i}{r}\right)\cos\theta_i\right\}$$

$$\theta_{i+1} = \pi - (\theta_i + 2\beta_i) \tag{19}$$

$$x_{i+1} = r\left\{\cos(\theta_i + \beta_i) + \frac{\sin(\theta_i + \beta_i) - 1}{\tan\theta_{i+1}}\right\}$$

The number of reflections before the ray exits the wedge, with θ_1 and x_1 as parameters, is obtained by applying the above equation iteratively. The number of reflections increases rapidly as the angle of first entry θ_1 decreases and the position of entry x_1 approaches the wedge corner, O (Figure 21). If the roll is such that the direction of the running steel sheet is changed by 90°, sufficient heat transfer occurs between the roll bulk and the steel sheet that the surface temperatures of the roll and the steel sheet can be assumed to be equal. The radiance L_o detected by the radiometer is given by

$$L_o = \varepsilon_{\text{eff}} L_b(T) \tag{20}$$

Considering multiple reflections between the sheet and the roll, similarly as in Equation (1), but in this case with finite number of reflections, the effective emissivity ε_{eff} is represented by $\varepsilon_{\text{eff}} = \varepsilon_{\text{eff1}} + \varepsilon_{\text{eff2}}$, where

$$\varepsilon_{\text{eff1}} = \varepsilon_o\left\{\frac{1 - (R_o R_r)^{n+1}}{1 - R_o R_r}\right\} \tag{21}$$

$$\varepsilon_{\text{eff2}} = \varepsilon_r R_o\left\{\frac{1 - (R_o R_r)^{m+1}}{1 - R_o R_r}\right\} \tag{22}$$

here, R_o and R_r are the reflectances of the steel sheet and the roll, respectively, ε_o and ε_r the emissivities of the steel sheet and the roll, respectively, and n and m the number of reflections on the strip and roll, respectively ($m = n$ or $n-1$).

From this, for instance, when $\varepsilon_r = 0.3$, ε_{eff} can be shown to be effectively equal to 1 for $\theta_1 = 7.5°$ and $0 < x_1 < 0.6r$.

4.3.2.2. Experimental realization. An online experiment into the "wedge" method was carried out before the soaking zone and the heating zone of a CAF as shown in Figure 22 in Nippon Kokan Corp. (currently, JFE Steel Corp.). Measurement windows were installed in the side wall of the tunnel (the "throat") that connected each heat-treatment furnace. The linear scanning pyrometers were angled so that the roll-strip wedge where the steel sheet changes its running direction by 90° was viewed. Figure 22 shows the schematic view of the wedge through the measuring window. The arrows indicate the measuring places. The measuring angle was about 7° and the field of view at which effective emissivity was higher than 0.95 was about 18 mm. The peak value of the temperature readings between the steel surface and the roll was assumed to be the true temperature of the steel surface. However, the system does not possess an *in situ* calibration means, and no attempt was made to verify the accuracy of the online measurement.

The method using multiple reflections in a roll-strip wedge was later demonstrated by Ridley and co-workers on a CAF at Swedish Steel in 1986 [29,70]. The measurement set-up is shown schematically in Figure 23 with the measurement location at the exit of the heating zone of a CAF. The "wedge" thermometer is mounted in a jacket and was inserted from the side so that the line of sight of the thermometer intersected the strip surface 85 mm in front of the roller-strip contact line. The line of sight made approximately an 8° vertical angle with the strip surface, and approximately a 30° horizontal angle with the strip center line. In addition to the "wedge"

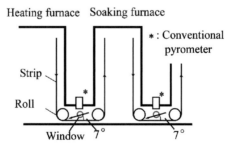

Figure 22   Schematic view of the wedge through the measurement window. (Reproduced from Ref. [97] with permission.)

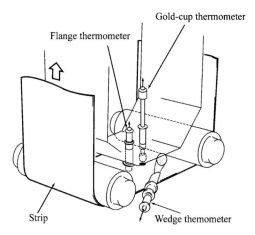

Figure 23 Instrument locations at the exit of the heating zone in a CAF. (Reproduced from Ref. [29] with permission.)

thermometer, a modified gold–cup surface thermometer (similar to the one described in Section 4.1.2 [33]) and a traditional "flange" thermometer using a gas-cooled purge on a sight tube were installed to obtain reference strip temperatures. The advantage of this system over the original work by Yamada et al. [38] is the additional gold–cup surface thermometer, which is a powerful means for *in situ* calibration of this system. The measurement reproducibility of the wedge thermometer readings relative to the reference readings was better than 10°C. The measurement was found to be insensitive to the thermometer type with respect to the operating wavelength or the field of view, which supports the view that the wedge acts as a quasi-blackbody. The conclusion of these trials is that the method has been shown to be practically useful, offering very simple and low cost installation together with improved furnace control.

The technique of utilizing multiple reflections at a roll-strip wedge is now widely applied in steel plants around the world. It is recognized that further work is needed to develop means to verify its accuracy because the gold–cup surface thermometer does not necessarily provide accurate enough temperature of the strip sheet. It is clear that the roll-strip wedge method is quite successful when the roll surface temperature is very close to the sheet temperature. However, when the roll temperature is different from the sheet temperature, *and particularly if it is higher*, then the method is not applicable [98].

A similar method that utilizes multiple reflections between two parallel steel surfaces is reported in Ref. [99].

4.3.3. Conical-cavity fiber optic radiation thermometer [41]

This method, developed by Cielo et al., is a passive radiometric temperature measurement utilizing a sensor consisting of a conical cavity to enhance emissivity by multiple reflections and to eliminate spurious radiation [41]. This sensor was installed at the soaking zone exit in a CAF in Stelco Steel in Canada in 1990. The unique feature of this method is the geometric shape of the conical-cavity reflector with a decentered and tilted fiber optic bundle, which contributes to reduced sensitivity in the fluctuation in the sheet distance and also eliminates spurious reflected radiation. The design is certainly an improvement over the conventional hemispherical reflector (gold-cup surface thermometer [33]) whose performance drops rapidly as the cavity-sheet distance increases. The ambient temperature sensitivity of the optical fiber transmittance is negated through taking the ratio of detector outputs at two wavelengths.

4.3.3.1. Measurement principle of the method and its configuration.

Figure 24 shows a schematic diagram of the conical-cavity sensor, where (a) is the sensor head composed of a conical-cavity with decentered, tilted fiber optics and (b) is the double-wavelength detection module composed of two germanium detectors. The surface of the conical-cavity is a gold-plated copper reflector; thus, the emitted radiation from the steel sheet undergoes multiple reflections between the surfaces of the conical surface and the steel sheet. This results in nearly 100% effective emissivity of the steel-sheet portion that is facing the cavity *if* the top surface is highly reflecting. The radiation transmitted through the fiber optic bundle is detected by two germanium detectors sensitive over the 1.2- and 1.6-µm ($\pm 0.15\,\mu$m) spectral bands, respectively. The temperature of the steel sheet is determined from the ratio of the detector output signals. The reflective conical-cavity has a diameter of 20 cm and a cone half angle of 82.5°.

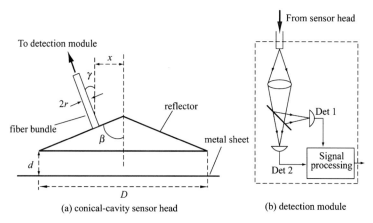

Figure 24 (a and b) Schematic diagram of the conical-cavity sensor. (Reproduced from Ref. [41] with permission.)

The authors evaluated the performance of the conical-cavity sensor thermometer for different values of relevant parameters by a Monte Carlo method. In all computations, the following properties were assumed for the steel sheet, although in real situations conditions may vary: (1) emissivity, $\varepsilon_o = 0.28$, independent of the observation angle, (2) the reflectance of the sheet takes into account its anisotropic topography due to directional rolling, (3) the reflector is perfectly specular with a reflectance of 0.95 and (4) the acceptance angle of the optical fibers collecting the radiation corresponds to a 0.2 numerical aperture [41,98]. Taking into account the four main radiance contributions, the expression for the radiance signal L transmitted by the fibers and collected by the detecting system can be described as follows:

$$L = \varepsilon_{\text{eff}\,o}L_b(T_o) + \varepsilon_{\text{eff}\,r}L_b(T_r) + \varepsilon_{\text{eff}\,e}L_b(T_e) + \varepsilon_{\text{eff}\,f}L_b(T_f) \qquad (23)$$

where the subscripts o, r, e and f refer, respectively, to the metal sheet, the reflector, the isothermal environment and the fiber bundle, and the weighting coefficients ε_{eff} are defined as the corresponding effective emissivities. $L_b(T)$ is the blackbody spectral radiance at the temperature T. The effective emissivities, $\varepsilon_{\text{eff}o}$ and $\varepsilon_{\text{eff}e}$, of the steel sheet and of the environment are about 0.87 and mostly less than 0.02, respectively. The second and the fourth terms of the right-hand side of Equation (23) are generally negligible because these temperatures, T_r and T_f, are low enough when compared with T_o and T_e.

The simulations are described in detail in Refs. [41,98,100]. From these results, a fiber displacement $x = 0.075D$, cone half angle $\beta = 82.5°$ and fiber bundle angle $\gamma = 17.5°$ condition was found to give the resultant effective emissivity close to the peak value. These parameters gave the best results under the following range of in-furnace sheet geometry:

- the variation of the sheet fluttering parameter $2d/D$ between 0.1 and 0.4, where d is the distance between the reflector and the steel sheet and D the diameter of the reflector;
- a tilting of the steel sheet with respect to the reflector axis by $\pm 2°$.

4.3.3.2. Experimental results.

Figure 25 shows a comparison between the conical-cavity pyrometer (indicated as RC pyrometer in the figure) reading of the temperature of steel sheets and the reading from type-K thermocouples attached to the sheets. These tests were made on both oxidized and bright surfaces to assess the emissivity independence of the measurements. The agreement between the pyrometer and the thermocouples is very good over the range $520-800°C$, and in most cases within the thermocouple measurement uncertainty.

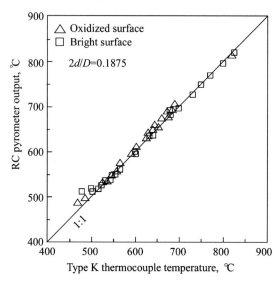

Figure 25 Comparison of temperature readings between the conical-cavity pyrometer and the thermocouples. (Reproduced from Ref. [41] with permission.)

4.3.4. Dual wavelength methods [35,36]

Two-color or ratio pyrometry is based on the assumption that the ratio of the emissivities at the two measurement wavelengths is a constant. However, this assumption is hardly if ever strictly satisfied in real situations and it is necessary to compensate for deviations from this assumption. Many multi-wavelength methods have been devised that purport overcoming this shortcoming, but it has been only the dual wavelength methods that have successfully been applied in the steel industry, particularly to the continuous annealing and the galvanneal process. Generally speaking, great care should be taken when applying multi-wavelength thermometer methods using three or more wavelengths because the measurement often results in unstable and large temperature errors [69,101].

Successful applications of the dual wavelength method have been developed, for instance, by Tanaka et al. [35,102,103] and Metcalfe et al. [36,68,70]. The former method named "TRACE" by its inventors utilizes an *a priori* knowledge of the relationship between two spectral emissivities of the metal surface, and the latter uses a simple relationship of radiance temperatures based on a relationship between two spectral emissivities.

4.3.4.1. The measurement principle of the dual wavelength method 1 [35].
This is the method elaborated by Tanaka and Ohira and known as TRACE [35]. The effective emissivities, ε_x and ε_y, at the two corresponding spectral radiances, $L_x(T)$ and $L_y(T)$, for an assumed temperature, T, are

defined, respectively, as follows:

$$\varepsilon_x = \frac{L_x(T)}{L_{b,\lambda_x}(T)} \tag{24}$$

$$\varepsilon_y = \frac{L_y(T)}{L_{b,\lambda_y}(T)} \tag{25}$$

where $L_{b,\lambda}(T)$ is the blackbody spectral radiance at wavelength λ and temperature T, and λ_x and λ_y the wavelengths. Using Wien's law, the blackbody spectral radiances are expressed as follows:

$$L_{b,\lambda_x}(T) = \frac{2c_{1L}}{\lambda_x^5} \exp\left(-\frac{c_2}{\lambda_x T}\right) = K_x \exp\left(-\frac{c_2}{\lambda_x T}\right) \tag{26}$$

$$L_{b,\lambda_y}(T) = \frac{2c_{1L}}{\lambda_y^5} \exp\left(-\frac{c_2}{\lambda_y T}\right) = K_y \exp\left(-\frac{c_2}{\lambda_y T}\right) \tag{27}$$

where $K_x = 2c_{1L}/\lambda_x^5$ and $K_y = 2c_{1L}/\lambda_y^5$.

By substituting Equation (26) into Equation (24), and Equation (27) into Equation (25), the following expressions are obtained:

$$\varepsilon_x = L_x(T)K_x^{-1} \exp\left(\frac{c_2}{\lambda_x T}\right) \tag{28}$$

$$\varepsilon_y = L_y(T)K_y^{-1} \exp\left(\frac{c_2}{\lambda_y T}\right) \tag{29}$$

Similar equations are obtained for the true temperature T_0 and corresponding emissivities, ε_{x0} and ε_{y0}.

$$\varepsilon_{x0} = L_x(T)K_x^{-1} \exp\left(\frac{c_2}{\lambda_x T_0}\right) \tag{30}$$

$$\varepsilon_{y0} = L_y(T)K_y^{-1} \exp\left(\frac{c_2}{\lambda_y T_0}\right) \tag{31}$$

Dividing Equation (28) by Equation (30), and Equation (29) by Equation (31), respectively, the following equations are derived:

$$\left(\frac{\varepsilon_x}{\varepsilon_{x0}}\right)^{\lambda_x} = \left(\frac{\varepsilon_y}{\varepsilon_{y0}}\right)^{\lambda_y} \tag{32}$$

or

$$\varepsilon_y = g(\varepsilon_x) = \varepsilon_{y0}\left(\frac{\varepsilon_x}{\varepsilon_{x0}}\right)^{\lambda_x/\lambda_y} \tag{33}$$

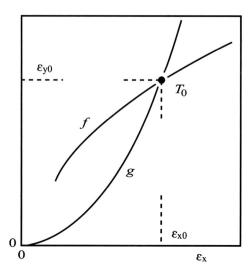

Figure 26 Cross-point coordinates of f-function and g-curve to obtain spectral emissivities. (Reproduced from Ref. [35] with permission.)

Equation (33) gives the relationship between the spectral emissivities, ε_x and ε_y, which is illustrated as the g-curve in Figure 26. When a pair of spectral radiances, $L_x(T)$ and $L_y(T)$, are detected from an object, the g-curve can be generated by varying T in Equations (26) and (27). The true emissivities, ε_{x0} and ε_{y0}, exist somewhere on the g-curve.

If a reciprocal relationship between two emissivities is known *a priori* as

$$\varepsilon_y = f(\varepsilon_x) \tag{34}$$

which is also shown as the f-function in Figure 26, the true emissivities can be obtained as the coordinates of the intersection with the g-curve. In principle, the parametric differences between these two emissivities are not limited to their wavelengths; these could also be related to the direction of observation and/or the mode of polarization.

4.3.4.2. Applications of the TRACE method to cold-rolled and galvannealed steels.
The TRACE method using the polarized emissivities, ε_s and ε_p, at a 40° angle of incidence was applied to a CAF to replace the measurement system described in Section 4.3.1, and another using the dual spectral emissivities was installed in a CGL, at a steel plant of Nippon Steel Corp.

Figure 27 shows the design of the optical system of a dual wavelength thermometer to implement this method. The average f-function for galvann-ealed steel was pre-determined to be a first-order polynomial function as

$$\varepsilon_y = c_0 + c_1\varepsilon_x \tag{35}$$

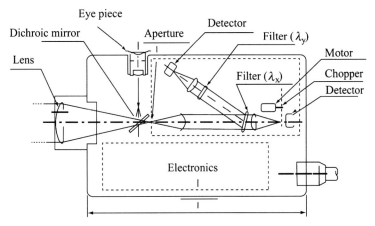

Figure 27 Dual spectral TRACE radiometer. (Reproduced from Ref. [35] with permission.)

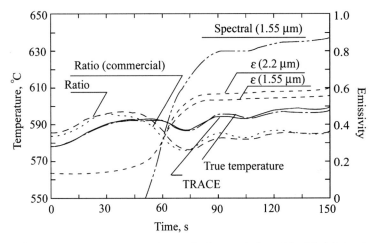

Figure 28 Offline temperature readings of galvannealed steel by TRACE and conventional methods. (Reproduced from Ref. [35] with permission.)

where $c_0 = 0.013$ and $c_1 = 0.926$. The zinc film on the steel plate starts to melt at 419°C, but the spectral emissivities remain constant until the alloy layer forms the most iron-enriched alloy phase. Then the surface abruptly roughens, which results in a sudden increase in emissivity.

Figure 28 shows an offline experimental measurement of the temperature readings of the dual wavelength TRACE method when compared to conventional spectral and ratio methods. The temperatures obtained from the TRACE thermometer showed much better agreement with the true temperatures (as measured by a thermocouple) than those by conventional spectral and ratio methods. Based on these results, the TRACE systems were installed for use in production for both CAF and CGL.

4.3.4.3. The measurement principle of the dual wavelength method 2 [36].
Metcalfe et al. of Land Infrared developed a dual wavelength thermometry which was successfully applied to galvannealing process of steel and hot rolling of aluminum [36,68]. This method utilizes a simple relationship of radiance temperatures linked through a relationship between the two spectral emissivities, but does not use the g-function of the previous method. The online system of this method is equipped with a reference contact probe [33], which collects data *in situ* at the actual intended measurement locations.

The method is based on the observation that sample measurements carried out in laboratory suggested the following generalized relation holds:

$$\varepsilon_1 = a\varepsilon_2^b \tag{36}$$

where a and b are constants, and ε_1 and ε_2 the spectral emissivities at wavelengths λ_1 and λ_2, respectively. For a target amenable to the conventional ratio thermometry, the emissivities obey this relationship when b takes the value of unity.

From Wien's law:

$$\frac{1}{T} - \frac{1}{T_1} = \frac{\lambda_1}{c_2}\ln(\varepsilon_1) \tag{37}$$

$$\frac{1}{T} - \frac{1}{T_2} = \frac{\lambda_2}{c_2}\ln(\varepsilon_2) \tag{38}$$

where c_2 is the second radiation constant. From Equations (36)–(38), the following equation is obtained:

$$\frac{1}{T} - \frac{1}{T_1} = A\left(\frac{1}{T_1} - \frac{1}{T_2}\right) + B \tag{39}$$

where $A = b\lambda_1/(\lambda_2 - b\lambda_1)$ and $B = A\lambda_2 \ln(a)/(bc_2)$. One can obtain the temperature T of the target from two radiance temperatures, T_1 and T_2, through Equation (39). The parameters A and B are determined by a linear fit to the empirical "1/T plot" with $1/T - 1/T_1$ ordinate and $1/T_1 - 1/T_2$ abscissa.

4.3.4.4. In situ measurements.
A gold-cup pyrometer described in Section 4.1.2 was especially developed as an easy to use reference "contact" thermometer probe for galvannealed steel. This was used to determine *in situ* the wavelength relationship between the two emissivities with minimum uncertainty [36].

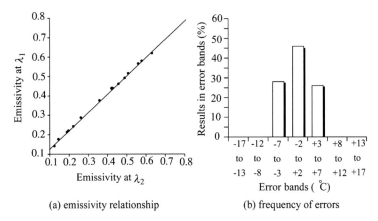

(a) emissivity relationship   (b) frequency of errors

Figure 29   (a) Functional emissivity relationship at two wavelengths and (b) frequency of temperature errors in readings of galvannealed steel. (Reproduced from Ref. [36] with permission of IMEKO Secretariat.)

Figure 29 shows (a) the functional relationship between emissivities at two wavelengths ($\lambda = 2.1$ and $2.4\,\mu m$) measured on site at a galvannealing plant and (b) the frequency plot of the difference between the ratio thermometer measurement and the reference value given by the gold–cup "contact" probe at the same location. It should be emphasized that the gold–cup thermometer gives the *in situ* reference. Delicaat and Leek also carried out the online application at a galvannealing process [37] that was essentially the same method as Metcalfe et al. described above.

Other similar applications of dual- and multi-wavelength methods to galvannealed steel can be found in literature [104–106], though the caution given concerning multi-wavelength methods given above should be observed.

4.4. Coating lines

In this section, two radiometric temperature measurements for steel coating processes are described. One application is to insulation film coating on silicon steels and the other to the coil coloring process of galvanized steels. The former is a passive method which utilizes multiple reflections of thermal radiation inside a cylindrical cavity; the latter is an active one which uses the irradiation by an infrared lamp onto the strip surface. Both are examples of emissivity-compensated radiation thermometry which were successfully implemented in production.

4.4.1. Simultaneous measurement system of temperature and emissivity using a cylindrical cavity [40]

This section describes a method which was applied successfully to the drying and coating lines of silicon steels in Nippon Steel Corp. in 1975

[40,107]. Silicon steel is the most important electromagnetic steel which features suppressed core loss due to the reduced eddy current component owing to its relatively high resistivity. The material is widely used in electrical products such as generators, motors and transformers. The described instrument is believed to be the very first online system of simultaneous measurement of temperature and emissivity put to practical use in a steel process.

The principle of the method is based on the observation that the two unknown quantities, the temperature and the emissivity, can both be uniquely determined if two simultaneous equations, each including these two quantities, are available. In order to obtain such equations, two measurements are made: one utilizing multiple reflections of radiation in a cylindrical cavity which has a specularly reflecting gold–plated interior surface and one without the cavity. For the measurement with the cavity, the blackbody condition under multiple reflections discussed in Section 3 is not satisfied, and the emissivity is enhanced only by a certain factor. However, this factor can be obtained as a function of the object emissivity, so that no additional unknown quantity is introduced by the second measurement. The measurement principle is illustrated in Figure 30.

In Figure 30(a), an open cylinder is placed above the object to be measured. There is a gap, d, between the object and the bottom opening of the cylinder. In Figure 30(b), a cylindrical cavity with a base surface having a small aperture of diameter, a, is placed above the object. The aperture is much smaller than the diameter of the cavity opening. Identical radiometers located at the upper end detect the radiances through the opening in Figure 30(a) and the aperture in Figure 30(b), respectively. In Figure 30(a), assuming negligible thermal radiation contribution from the cylinder, the

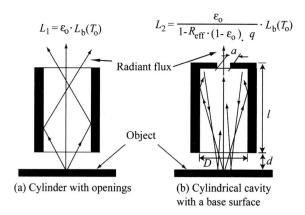

$$L_1 = \varepsilon_0 \cdot L_b(T_0) \qquad L_2 = \frac{\varepsilon_0}{1 - R_{\text{eff}} \cdot (1 - \varepsilon_0) \cdot q} \cdot L_b(T_0)$$

(a) Cylinder with openings

(b) Cylindrical cavity with a base surface

Figure 30   (a and b) Measurement principle. (Reproduced from Ref. [40] with permission.)

radiance, L_o, is proportional to the emissivity, ε_o, of the object as given in the following equation:

$$L_1 = \varepsilon_o L_b(T_o) \tag{40}$$

where T_o is the surface temperature of the object and $L_b(T_o)$ the blackbody radiance at temperature T_o. On the other hand, in Figure 30(b), the radiation from the object is multiply reflected between the object and the cavity so that the radiance increases. Assuming negligible thermal radiation from the cavity itself, constant cavity effective reflectance, R_{eff}, for a constant cavity surface reflectance, and direction independent object emissivity, ε_o, L_2 is derived as

$$L_2 = \varepsilon_o L_b(T_o) \sum_{n=0}^{\infty} (R_{eff}(1 - \varepsilon_o)q)^n \tag{41}$$

where the zeroth term in the series corresponds to the radiance of the object itself and successive nth term corresponds to the increased radiance due to nth interreflection between the cavity and the object. $(1 - \varepsilon_o)q$ is the fraction of radiation reflected back into the cavity where the efficiency coefficient, q $(0 < q < 1)$, depends on the cavity dimensions and the gap d. As the product $R_{eff}(1 - \varepsilon_o)q$ is less than one, Equation (41) can be written as

$$L_2 = \frac{\varepsilon_o}{1 - R_{eff}(1 - \varepsilon_o)q} L_b(T_o) \tag{42}$$

If the value $R_{eff}q$ is known, both the temperature and the emissivity of the object can be obtained by solving Equations (40) and (42) simultaneously as follows. By taking the ratio L_1/L_2, the following equation is derived:

$$\frac{L_1}{L_2} = 1 - R_{eff}(1 - \varepsilon_o)q \tag{43}$$

The above equation can be rewritten as

$$\varepsilon_o = 1 - \frac{1 - L_1/L_2}{R_{eff}q} \tag{44}$$

The emissivity is obtained from the above equation. The temperature can then be obtained from Equation (40) on substituting the value of ε_o obtained above.

The principle of the measurement is realized by the apparatus illustrated in Figure 31. A cylinder is placed above the object to be measured. A fan-shaped rotating chopper is placed close to the open top of the cylinder. There is a slit of width a in the chopper blades. The radiometer detects the radiance from the object through this slit. The specular surfaces of the cylinder interior and the chopper blade are gold plated. When the upper

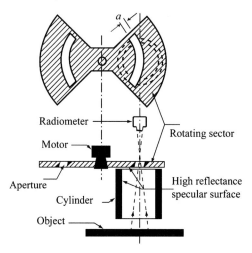

Figure 31 Schematic of the measurement apparatus. (Reproduced from Ref. [40] with permission.)

opening of the cylinder is not covered by the chopper, the situation is equivalent to Figure 30(a), and when it is covered it corresponds to Figure 30(b). The same radiometer is applied for the two situations, and therefore the radiometer response coefficient, which contains the constant of calibration as well as other measurement constants including the solid angle of the target and the effective wavelength and bandwidth of the instrument, is canceled when calculating L_1/L_2, the ratio of the radiance signals, in Equation (44). The value for $R_{\text{eff}}q$ is determined through Equation (43) by plotting the relationship between L_1/L_2 and $(1-\varepsilon_0)$, the latter obtained experimentally *a priori* for various types of insulating film-coated steels with both organic and inorganic substances, and applying a linear fit. Figure 32 shows that L_2/L_1 and ε are well correlated in spite of a large change in the emissivity. The scatter corresponds to relative uncertainties for ε_0 and T_0 of $|\Delta\varepsilon/\varepsilon_0| \leq 0.08$ and $|\Delta T/T_0| \leq 0.01$, respectively, in the temperature range from 200 to 500°C.

In Figure 33, the results of typical online experiments are displayed, where part (a) indicates the readings of the strip temperature, T_r, by the contact thermometer which is a simplified version of the one described in Ref. [28], part (b) is the emissivity readings determined with the online measurement system and part (c) is the difference, ΔT, between T and T_r, where T is the temperature indicated by the online measurement system. The horizontal axis of the figure is the coil number of the strips, where S1−S5 and Th1−Th6 identify the coated steel strips and the threading strips, respectively. The total estimated measurement error for the contact thermometer, at a line speed of 1.7 m/s, was ±5°C. Figure 33 clearly

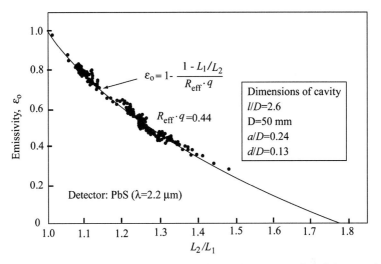

Figure 32 Experimental relation between L_2/L_1 and ε_0. (Reproduced from Ref. [40] with permission.)

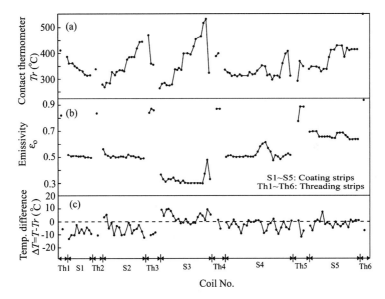

Figure 33 (a–c) Results of *in situ* measurements. (Reproduced from Ref. [40] with permission.)

shows that the strip temperature ranged between 300 and over 500°C (see (a)), and the emissivity indicated large changes from 0.3 to 0.9 for the different coated coils (see (b)). The emissivities of the threading coils (Th1−Th6) were all about 0.8 because their surfaces were blackened from repeated use. Despite these very different measurement conditions, the temperature difference, ΔT, was mostly within $\pm 10°C$ (see (c)).

An auxiliary apparatus for the maintenance of the online temperature measurement system and the considerations for the most suitable cylindrical cavity dimensions are described in detail in Refs. [40,107], respectively.

4.4.2. Emissivity compensation utilizing the reflected image of a linear light source [44,108]

An emissivity-compensating radiation thermometer has been applied successfully to the coil coloring line and is contributing to the improved control of the coating quality in production. In the coil coloring line, galvanized steel sheet is coated continuously by a coating roll, and then is traversed through a baking oven. The baking condition determines the coating quality and the peak temperature needs to be controlled.

The emissivity of coated steel strip varies with the type of coating as well as with the degree of baking and the roughness of the substrate galvanized steel sheet. The thermometry method described here utilizes a reference light source to measure the reflectance and obtain in situ the emissivity through Kirchhoff's law and thereby to determine the strip temperature at the exit of the baking oven. In the continuous coloring line, the steel strip traverses the length of the baking oven and then some extra distance after emerging from the oven without touching a supporting roll until the coating is dry. Therefore, at the exit of the baking oven, the position of the steel sheet may vary drastically, including occasional tilting to the side, which makes the reflectance measurement a complex task. To facilitate the reflectance measurement from a remote position in the presence of such large and irregular movements of the target, a system consisting of a linear light source and a scanning radiation thermometer was devised.

The system is shown schematically in Figure 34. The linear light source is aligned parallel to the strip in the direction of its width and at a distance of 300 mm. A one-dimensional scanning radiation thermometer measures the

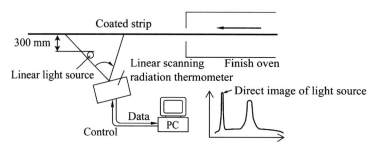

Figure 34 Schematic diagram of the coil coloring line temperature measurement system. (Reproduced from Ref. [44] with permission.)

radiance distribution on the strip surface in the direction perpendicular to the orientation of the light source. The orthogonal arrangement of the linear source and the scanning detector ensures that the reflection of the source on the strip surface is always detected. The one-dimensional scanned image consists of the reflected image of the source superimposed on the thermal radiation of the strip. By obtaining the scan when turning on and off the light source, the reflected image can be obtained as their difference. The reflected image is blurred, and the degree of the blur can be related to the surface roughness of the strip: when the surface reflection is specular the image is sharp, whereas when it is diffuse the image is spread along the length of the strip.

A typical coated surface shows a reflection which is a combined reflection of two distinct features: a near specular and a totally diffuse reflection. The former can be interpreted as the reflection at the coating surface, and the specularity is related to the substrate roughness. The latter, which actually consists 80–90% of the reflection, can be interpreted as the reflected light from the pigment within the color coating.

The procedure to obtain the true temperature from the linear scanned images is shown in Figure 35. From the reflected image of the light source, the nearly specularly reflected term and totally diffusely reflected term are separated; see Figure 35(b). For each term, the two-dimensional angular reflected radiance distribution is reconstructed from the image. Both distributions are individually hemispherically integrated, whose sum, divided by the source radiance monitored simultaneously by the scanning thermometer, gives the hemispherical reflectance of the coated strip. Kirchhoff's law is subsequently applied to give the emissivity enabling the true temperature of the coated strip to be obtained.

In the above process, several empirical assumptions are made. (1) The near specular reflection is assumed to have angular distribution with rotational symmetry with the direction of the reflection peak as the axis of symmetry, and the shape of the angular distribution for a certain surface stays the same for all incidence angles. (2) For the near specular reflection, the contribution of the length of the light source is ignored: the source is treated as a point source. (3) For the totally diffuse reflection, the detected reflected signal is a contribution of the Lambertian reflection integrated along the length of the source. A system for *in process* measurement was constructed, which consisted of a linear light source of a rod waveguide with a roughened surface illuminated by a halogen lamp, and a scanning radiation thermometer with an InGaAs detector of central wavelength 1.6 μm and a rotating mirror of scanning angle of $50°$.

The results of the measurements are shown in Figure 36. The emissivity measured *in situ* is compared with the measurement of a sample of the same material measured offline by a spectrometer, and the two agree to within $\pm 2°C$ when converted to measurement error in temperature. A direct

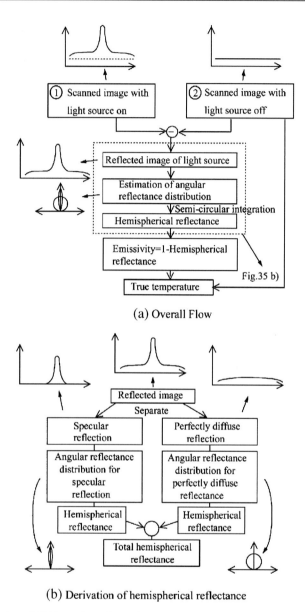

(a) Overall Flow

(b) Derivation of hemispherical reflectance

Figure 35 (a and b) Procedure for emissivity compensation for coated steel strips. (Reproduced from Ref. [44] with permission.)

comparison of the measurement of the temperature was made by welding a thermocouple onto the surface of a test coil. Two measurement runs were performed. The results are shown in Figure 37. At the exit of the baking oven, the two measurements are seen to agree within $\pm 5°C$. These results

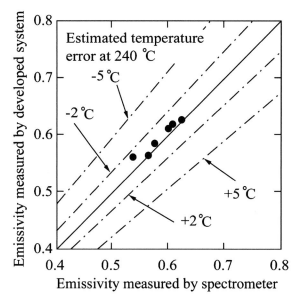

Figure 36 Comparison of emissivity obtained online by the system, and offline by a spectrometer. (Reproduced from Ref. [44] with permission.)

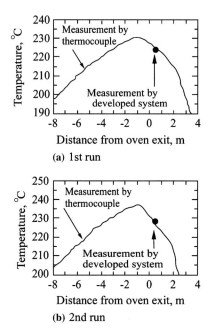

Figure 37 (a and b) Comparison of measurement by the system and coil temperature profile measured by a thermocouple. (Reproduced from Ref. [44] with permission.)

show the practicality of the method. The system has since been installed in the Coil Coloring Line at NKK Corporation (currently, JFE Steel).

An example of application of linear scanning radiation thermometer can be found in Ref. [109]. Another example of an "active" emissivity-compensating radiation thermometry can be found in Ref. [110].

5. SUMMARY

Steel remains the most widely used and the most indispensable material for industrial and commercial products, and is the core material for building and transportation infrastructures worldwide. Decades after the steel industry was considered mature, continued research and development for new and high-quality steel products is still being enthusiastically pursued in aspects of both metallurgical investigation and process innovation. In accordance with these trends, the importance of radiometric temperature measurement in future steel processes is anticipated to increase. These trends are summarized in Section 1. Section 2 describes how radiometric temperature measurement methods have been applied to molten iron and steel. As described in Section 3, radiation thermometry in the steel industry has to overcome problems of unknown emissivity, background radiation and adverse environment, resulting in an expensive and rugged apparatus. This very often becomes an obstacle for implementation and commercialization, not only technically but also financially. This issue is not unique to the steel industry, and especially is a problem for thermometry in semiconductor processing, as discussed in Chapter 3 (this volume). We anticipate mutual benefit of collaborative developments in this field between these two industrial sectors. In Section 4, concrete examples of unique radiometric temperature measurement methods and systems are summarized. Emissivity compensation, auxiliary equipments against harsh environment and *in situ* calibration are described. Advances in radiometric temperature measurement systems have been facilitated by the recent rapid progress in electronic and optical devices including computers, optical sensors and signal processing technology. Two-dimensional temperature measurement of molten iron at the BF described in Section 4.2.2 is a typical example.

For more precise control of the steel quality, it is necessary to analyze and control the metallurgical phenomena occurring in the steel while it is being produced, and to effect this data fusion is required in which temperature measurement will play a central role. Even a superb and excellent measurement method will be ultimately replaced with a new and better one in the long run. We believe that the examples given in this chapter are not only informative, but will also contribute to forming new innovative ideas of radiation thermometry.

REFERENCES

[1]   World Steel Association, "Crude Steel Production," http://www.worldsteel.org/ (2008).

[2]   JFE 21st Century Foundation; "An Introduction to Iron and Steel Processing," http://www.jfe-21st-cf.or.jp/index2.html (2008).

[3]   UK Steel, "Making Steel," http://www.eef.org.uk/uksteel/steel/about+the+ industry/making+steel/Making_steel.htm (2008).

[4]   American Iron and Steel Institute, "How a Blast Furnace Works?" http://www. steel.org/AM/Template.cfm?Section = Home&template = /CM/HTMLDisplay. cfm&ContentID = 5433 (2008).

[5]   Wikipedia, "Basic Oxygen Steelmaking," http://en.wikipedia.org/wiki/Basic_ oxygen_steelmaking (2008).

[6]   Wikipedia, "Continuous Casting," http://en.wikipedia.org/wiki/Continuous_ Casting (2008).

[7]   Matter, "Hot Rolling Strip Steels," http://www.matter.org.uk/steelmatter/forming/ 4_5.html (2008).

[8]   Steeluniversity, "Seamless Tube Rolling," http://www.steeluniversity.org/content/ html/eng/default.asp?catid = 199&pageid = 2081272066 (2008).

[9]   T. Imae, N. Nomura, and S. Miyoshi, *High Quality Production Technology at the Chiba Works No. 3 Hot Strip Mill*, Kawasaki Steel Technical Report No. 37, pp. 59-64 (1997).

[10] N. Mihara, N. Ide, and M. Yamamoto, "Development of Temperature Control System of Hot Strip Mill Run Out Table," in *Proc. TEMPBEIJING 1997, Int. Conf. of the Temperature and Thermal Measurements*, Beijing, pp. 232−239 (1997).

[11] Land Instruments International, "Hot Rolling Mill," http://landinst.com/infrared/ application/iron_steel/hotrolling.htm (2008).

[12] H. Nikaido, S. Isoyama, N. Nomura, K. Hayashi, K. Morimoto, and H. Sakamoto, *Endless Hot Strip Rolling in the No. 3 Hot Strip Mill at the Chiba Works*, Kawasaki Steel Technical Report No.37, pp. 65−72 (1997).

[13] M. Kido, A. Sugihashi, H. Yamamoto, K. Maeda, N. Hamada, K. Minamida, and T. Kikuma, *Development of 45-kW Laser Welding System for Continuous Finish Rolling*, Nippon Steel Technical Report No. 89, pp. 91−95 (2004).

[14] R. Pradhan, "Continuous Annealing of Steel," in *Volume 4 Heat Treating, ASM Handbook*, edited by ASM, ASM International, Ohio, pp. 56−66 (1991).

[15] F. Yanagishima, Y. Shimoyama, M. Suzuki, H. Sunami, T. Haga, Y. Ida, and T. Irie, *Characteristics and Operation of Multipurpose Continuous Annealing Line at Chiba Works*, Kawasaki Steel Technical Report No. 2, pp. 1-13 (1981).

[16] K.Yahiro, H. Shigemori, K. Hirohata, T. Ooi, M. Haruna, and K. Nakanishi, "Development of Strip Temperature Control System for a Continuous Annealing Line," in *Proc. IECON 1993, 19th Annu. Conf. of the IEEE Industrial Electronics Society*, Maui, pp. 481−486 (1993).

[17] T. Kawasaki, *Stainless Steel Production Technologies at Kawasaki Steel − Features of Production Facilities and Material Developments*, Kawasaki Steel Technical Report No. 40, pp. 5−15 (1999).

[18] T. Kubota, M. Fujikura, and Y. Ushigami, "Recent Progress and Future Trend on Grain-Oriented Silicon Steel," J. Magnetism Magn. Mater. **215/216**, 69−73 (2000).

[19] N. Morito, M. Komatsubara, and Y. Shimizu, *History and Recent Development of Grain Oriented Electrical Steel at Kawasaki Steel*, Kawasaki Steel Technical Report No. 39, pp. 3−12 (1998).

[20] K. Matsuda and Y. Mashino, *Progress of Mechanical Equipment, Process Control and Instrumentation Technologies for Innovation in Steel Production Processes*, Kawasaki Steel Technical Report No. 44, pp. 52−61 (2001).

[21] A. Torao, K. Asano, and H. Takada, *Recent Activities in Research of Measurement and Control*, Kawasaki Steel Technical Report No. 41, pp. 77–82 (1999).

[22] G. R. Peacock, "Verifying Strip and Slab Surface Temperatures On-line," Iron Steel Eng. **72**, 34–38 (1995).

[23] G. R. Peacock, "A Review of Non-Contact Process Temperature Measurements in Steel," in *Proc. SPIE*, edited by D. H. LeMieux and J. R. Snell, Jr., Vol. 3700, SPIE, Orlando, pp. 171–189 (1999).

[24] G. R. Peacock, "Ratio Radiation Thermometers in Hot Rolling and Galvannealing of Steel Strips," in *Temperature, Its Measurement and Control in Science and Industry*, edited by D. C. Ripple, AIP, New York, pp. 789–793 (2002).

[25] G. R. Peacock, "Radiation Thermometers in Steel and Metals Processing," in *Temperature, Its Measurement and Control in Science and Industry*, edited by D. C. Ripple, AIP, New York, pp. 813–817 (2002).

[26] G. R. Peacock, "A Type-S Radiation Thermometer for Steel Mills," in *Proc. SPIE*, edited by K. E. Cramer and X. P. Maldague, Vol. 5073, SPIE, Orlando, pp. 85–91 (2003).

[27] Key to Metals AG, "Welcome to KEY to METALS · STEEL," http://steel.keytometals.com/default.aspx?ID = Home&LN = EN (2008).

[28] T. Iuchi, J. Ohno, and R. Kusaka, "Temperature Measurement System of Steel Strips in a Continuous Annealing Furnace," Trans. Iron Steel Inst. Jpn. **16**, 195–203 (1976).

[29] I. Ridley and T. G. R. Beynon, "Infra-red Temperature Measurement of Bright Metal Strip using Multiple Reflection in a Roll-Strip Wedge to enhance Emissivity," Measurement **7**, 171–176 (1989).

[30] K. Tamura, T. Iwamura, and K. Kurita, "Improvement of Traceability for Radiation Pyrometers in the Steel Industry," in *Temperature, Its Measurement and Control in Science and Industry*, edited by J. F. Schooley, AIP, New York, pp. 479–483 (1982).

[31] R. Barber, "Furnace Load Temperature during the Heating Process," Ind. Process Heating (February) 24–28 (1967).

[32] T. P. Murray, "Use of Silicon-Cell Pyrometers on Continuous Annealing Lines," Iron Steel Eng. (February) 33–39 (1971).

[33] M. D. Drury, K. P. Perry, and T. Land, "Pyrometers for Surface Temperature Measurement," J. Iron Steel Inst. **169**, 245–250 (1951).

[34] M. Sugiura, T. Yamazaki, R. Nakao, et al., *Development of New Technique for Continuous Molten Steel Temperature Measurement*, Nippon Steel Technical Report No. 89, pp. 23–27 (2004).

[35] F. Tanaka and H. Ohira, "Thermometry Reestablished by Automatic Compensation of Emissivity; The TRACE Method," in *Temperature, Its Measurement and Control in Science and Industry*, edited by J. F. Schooley, AIP, New York, pp. 895–900 (1992).

[36] S. F. Metcalfe, A. D. Tune, and T. Beynon, "Dual Wavelength Thermometry applied to Hot Rolling of Aluminium and Galvannealing of Steel," in *Proc. XIII IMEKO World Congress*, Torino, pp. 1445–1450 (1994).

[37] C. L. A. Delicaat and T. L. M. Leek, "On-Line Application of a Dual-Wave Pyrometeter during Galvannealing," in *Proc. TEMPMEKO 1996, 6th Int. Symp. on Temperature and Thermal Measurement in Industry and Science*, edited by P. Marcarino, Levrotta & Bella, Torino, pp. 353–358 (1996).

[38] T. Yamada, E. Makabe, N. Harada, and K. Imai, *Development of Radiation Thermometry using Multiple Reflection*, Nippon Kokan Technical Report Overseas No. 41, pp. 126–134 (1984).

[39] Y. Yamada, A. Ohsumi, Z. Yamanaka, and T. Yamada, "Temperature Measurement of Molten Metal by Immersion-Type Optical Fiber Radiation Thermometer,"

in *Proc. TEMPMEKO 1996, 6th Int. Symp. on Temperature and Thermal Measurement in Industry and Science*, edited by P. Marcarino, Levrotta & Bella, Torino, pp. 347–352 (1996).

[40]   T. Iuchi, "A Method for Simultaneous Measurement of Both Temperature and Emissivity, and Its Application to Steel Processing," Trans. Iron Steel Inst. Jpn. **19**, 474–483 (1979).

[41]   P. Cielo, J.-C. Krapez, M. Lamontegne, J. G. Thomson, and M. G. Lamb, "Conical-Cavity Fiber Optic Sensor for Temperature Measurement in a Steel Furnace," Opt. Eng. **32**, 486–493 (1993).

[42]   D. Kelsall, "An Automated Emissivity-Compensated Radiation Pyrometer," J. Sci. Instrum. **40**, 1–4 (1963).

[43]   T. P. Murray, "Polaradiometer – A New Instrument for Temperature Measurement," Rev. Sci. Instrum. **38**, 791–798 (1967).

[44]   Y. Yamada, D. Yuasa, T. Manabe, H. Suzuki, and N. Inoue, *Emissivity Compensating Thermometer for Coil Coating Line*, NKK Technical Report No. 161, pp. 100–104 (1998). (in Japanese).

[45]   E. Schreiber and G. Neuer, "The Laser Absorption Pyrometer for Simultaneous Measurement of Surface Temperature and Emissivity," in *Proc. TEMPMEKO 1996, 6th Int. Symp. on Temperature and Thermal Measurement in Industry and Science*, edited by P. Marcarino, Levrotta & Bella, Torino, pp. 365–370 (1996).

[46]   G. J. Edwards, A. P. Levick, and Z. Xie, "Laser Emissivity Free Thermometry (LEFT)," *Proc. TEMPMEKO 1996, 6th Int. Symp. on Temperature and Thermal Measurement in Industry and Science*, edited by P. Marcarino, Levrotta & Bella, Torino, pp. 383–388 (1996).

[47]   E. Hagen and H. Rubens, "Über Beziehungen des Reflexions und Emissionsvermögens der Metalle zu ihren elektrischen Leitvermögen," Ann. Physik **11**, 873–901 (1903).

[48]   P. Drude, "Zur Elektronentheorie der Metalle," Ann. Physik **1**, 566–613 (1900).

[49]   P. Drude, "Zur Elektronentheorie der Metalle," Ann. Physik **3**, 369–402 (1900).

[50]   N. W. Snyder, "Radiation in Metals," Trans. ASME (May) 541–548 (1954).

[51]   D. P. DeWitt and R. S. Hernicz, "Theory and Measurement of Emittance Properties for Radiation Thermometry Applications," in *Temperature, Its Measurement and Control in Science and Industry*, edited by H. H. Plumb, ISA, Pittsburgh, pp. 459-481 (1972).

[52]   R. Siegel and J. R. Howell, *Thermal Radiation Heat Transfer*, Taylor & Francis, New York, 3rd ed. (1992).

[53]   E. M. Sparrow and R. D. Cess, *Radiation Heat Transfer*, Hemisphere Publishing Corp., Washington, DC, Augmented ed. (Chapter 2) (1978).

[54]   M. Born and E. Wolf, *Principle of Optics*, Cambridge University Press, Cambridge, 7th ed. (1999).

[55]   R. R. Brannon and R. J. Goldstein, "Emittance of Oxide layered on a Metal Substrate," J. Heat Transfer **92**, 257–263 (1970).

[56]   D. P. DeWitt and H. Kunz, "Theory and Technique for Surface Temperature Determinations by measuring the Radiance Temperature and the Absorptance Ratio for Two Wavelengths," in *Temperature, Its Measurement and Control in Science and Industry*, edited by H. H. Plumb, ISA, Pittsburgh, pp. 599–610 (1972).

[57]   G. Neuer, F. Güntert, and R. Wörner, "Radiation Thermometry as an Appropriate Method to observe the Oxide Growth on Metal-Surfaces," Thermochim. Acta **85**, 295–298 (1985).

[58]   G. Neuer, "On the Influence of Radiation Properties when Measuring Temperatures of Oxidized Metals and Ceramics," in *Practical surface pyrometry workshop 1993*, edited by R. Kaarls, NMi, Delft, pp. 1–12 (1993).

[59] Y. Tamura and K. Hiramoto, "The Behavior of Spectral Emissivity of Metal in
      Oxidation Process," in *Proc. TEMPBEIJING 1997, Int. Symp. on Temperature
      Measurement in Industry and Science*, Beijing, pp. 45–50 (1997).
[60] W. Bauer, W. Gräfen, and M. Rink, "Spectral Emissivities of Heat-Treated Steel
      Surfaces," in *Temperature, Its Measurement and Control in Science and Industry*, edited by
      D. C. Ripple, AIP, New York, pp. 807–811 (2002).
[61] T. Furukawa and T. Iuchi, "Experimental Apparatus for Radiometric Emissivity
      Measurements of Metals," Rev. Sci. Instrum. **71**, 2843–2847 (2000).
[62] P.-J. Krauth, "Development of an Industrial IR Sensor to continuously measure
      Total Absorptivity of Steel Sheet On-line: Application to the Optimization of the
      Thermal Treatment within an Annealing Furnace," in *Temperature, Its Measurement
      and Control in Science and Industry*, edited by D. C. Ripple, AIP, New York,
      pp. 795–800 (2002).
[63] T. Iuchi, T. Furukawa, and S. Wada, "Emissivity Modeling of Metals during the
      Growth of Oxide Film and Comparison of the Model with Experimental Results,"
      Appl. Opt. **42**, 2317–2326 (2003).
[64] L. del Campo, R. B. Pérez-Sáez, X. Esquisabel, I. Fernández, and M. J. Tello,
      "New Experimental Device for Infrared Spectral Directional Emissivity Measure-
      ments in a Controlled Environment," Rev. Sci. Instrum. **77**, 113111 (2006).
[65] J. E. Roney, "Steel Surface Temperature Measurement in Industrial Furnaces by
      Compensation for Reflected Radiation Errors," in *Temperature, Its Measurement and
      Control in Science and Industry*, edited by J. F. Schooley, AIP, New York,
      pp. 491–503 (1982).
[66] T. Iuchi, "A New Radiation Thermometry of a Material in a High Temperature
      Furnace," in *Proc. TEMPBEIJING 1986, Int. Symp. on Temperature Measurement in
      Industry and Science*, Beijing, pp. 291–298 (1986).
[67] R. Barber and T. Land, "The Place of Photovoltaic Detectors in Industrial
      Pyrometry," in *Temperature, Its Measurement and Control in Science and Industry*, edited
      by C. M. Herzfeld and A. I. Dahl, Reinhold Publishing Corp., New York,
      pp. 391–403 (1962).
[68] S. Metcalfe, "Practical Temperature Measurement of Aluminium Bright Strip using
      Dual Wavelength Infrared Thermometry," in *Proc. TEMPMEKO 1993, 5th Int.
      Symp. on Temperature and Thermal Measurement in Industry and Science*, Prague,
      pp. 133–144 (1993).
[69] P. Saunders, *Radiation Thermometry, Fundamentals and Applications in the Petrochemical
      Industry*, SPIE Press, Bellingham, Washington (2007).
[70] G. R. Beynon and I. Ridley, *Continuous Annealing Furnace Trial at SSAB Domnarvet*,
      Land Infrared Ltd., New Development and Applications, Report No. NDA 142
      (1986).
[71] T. P. Murray, "The Polarized Radiation Method of Radiation Thermometry,"
      in *Temperature, Its Measurement and Control in Science and Industry*, edited by H. H.
      Plumb, ISA, Pittsburgh, pp. 619–623 (1972).
[72] V. C. Tingwaldt, "Ein einfaches optisch-pyrometrisches Verfahren für direkten
      Ermittlung wahrer Temperaturen glühender Metalle," Z. Metallkunde **51**,
      116–119 (1960).
[73] G. J. Edwards and A. P. Levick, "Recent Developments in Laser Absorption
      Radiation Thermometry at the NPL," in *Proc. TEMPMEKO 1999, 7th Int. Symp.
      on Temperature and Thermal Measurement in Industry and Science*, edited by J. F.
      Dubbeldam and M. J. de Groot, IMEKO/Nmi VSL, Delft, pp. 619–624 (1999).
[74] T. Loarer, J. J. Greffet, and M. Huetz-Aubert, "Noncontact Surface Temperature
      Measurement by Means of a Modulated Photothermal Effect," Appl. Opt. **29**,
      979–987 (1990).

[75] T. Loarer and J. J. Greffet, "Application of the Pulsed Photothermal Effect to Fast Surface Temperature Measurements," Appl. Opt. **31**, 5350−5358 (1992).

[76] G. J. Edwards, A. P. Levick, G. Neuer, et al., "Laser Absorption Radiation Thermometry and Industrial Temperature Measurement − The Result of an EC Collaborative Project (SMT4-CT95-2003)," in *Proc. TEMPMEKO 1999, 7th Int. Sym. On Temperature and Thermal Measurements in Industry and Science*, edited by J. F. Dubbeldam & M. J. de Groot, IMEKO/NMi VSL, Delft, pp. 613−618 (1999).

[77] S. N. Park, B. H. Kim, H. S. Kim, and B. H. Son, "Recent Development of a Laser Emissivity Free Thermometer at the KRISS," in *Proc. TEMPMEKO 2001, 8th Int. Symp. On Temperature and Thermal Measurements in Industry and Science*, edited by B. Fellmuth, J. Seidel, and G. Scholze, VDE VERLAG, Berlin, pp. 199−203 (2002).

[78] E. W. M. van der Ham, R. Bosma, and P. R. Dekker, "Nonconventional Two-Color Pyrometry at NMI-VSL," in *Proc. TEMPMEKO 2004, 9th Int. Sym. on Temperature and Thermal Measurements in Industry and Science*, edited by D. Zvizdic, LPM/FSB, Dubrovnik, pp. 557−562 (2005).

[79] H. J. Hoge, "Temperature Measurement in Engineering," in *Temperature, Its Measurement and Control in Science and Industry*, edited by H. C. Wolfe, Reinhold Publishing Corp., New York, pp. 287−325 (1955).

[80] L. O. Sordahl and R. B. Sosman, "The Measurement of Open-Hearth Bath Temperature," in *Temperature, Its Measurement and Control in Science and Industry*, Reinhold Publishing Corp., New York, pp. 927−936 (1941).

[81] H. D. Baker, E. A. Ryder, and N. H. Baker, "Chapter 9 Liquids," in *Temperature Measurement in Engineering*, Wiley, New York, Vol. 2, pp. 279−305 (1961).

[82] J. A. Hall, "Fifty Years of Temperature Measurement," J. Sci. Instrum. **43**, 541−547 (1966).

[83] F. H. Schofield and A. Grace, "A 'Quick-Immersion' Technique for High-Temperature Measurement on Fluids," in *Temperature, Its Measurement and Control in Science and Industry*, Reinhold Publishing Corp., New York, pp. 937−945 (1941).

[84] W. C. Heselwood and D. Manterfield, "Liquid steel temperature measurement," Platinum Met. Rev. **1**, 110−118 (1957).

[85] J. A. Stevenson, "Temperature Measurement with the Expendable Immersion Thermocouple-A Survey of Works Experience," Platinum Met. Rev. **7**, 2−6 (1963).

[86] Z. Kunlun and L. Yinwen, "A Research on Touching Radiation Pyrometer," in *Proc. TEMPMEKO 1993, 5th Int. Symp. on Temperature and Thermal Measurement in Industry and Science*, Prague, pp. 294−298 (1993).

[87] M. Sugiura, S. Matsuzaki, H. Yamamoto, and Y. Ootani, "Continuous Tapped Iron Temperature Measurement using Radiation Thermometry," CAMP-ISIJ **20**, 294 (in Japanese) (2007).

[88] Y. Yamada, T. Yamada, A. Ohsumi, T. Itakura, H. Wakai, and Z. Yamanaka, "Temperature Measurement of Molten Metal by Immersion-Type Optical Fiber Radiation Thermometer," in *Proc. 13th Sensing Forum, Society of Instrum. and Control Eng.*, Tokyo, pp. 57−62 (1996) (in Japanese).

[89] Y. Hayasaka, A. Sakai, M. Sakurai, H. Wakai, K. Mori, and T. Maehara, *New Measurement Method of Smelting Temperature and Application to Blast Furnace*, NKK Technical Report No. 178, pp. 32−36 (2002) (in Japanese).

[90] T. Isawa, R. Kawabata, M. Komatani, and I. Kikuchi, "Continuous Temperature Measurement by an Optical Fiber," CAMP-ISIJ **10**, 776 (in Japanese) (1997).

[91] Y. Ueno, "High Accuracy Measurement of Welding Point Temperature by Optical Fiber Thermometer," in *JFE-TEC News*, No. 7, p. 1 (2006) (text and original title in Japanese).

[92]   K. M. Shvarev, V. S. Gushchin, B. A. Baum, and V. P. Gel'd, "Optical Constants of
        Iron Alloys with Carbon in the Temperature Interval 20−1600°C," High
        Temperature **17**, 57−61 (1979).
[93]   M. Inui, A. Mori, T. Okada, T. Yamamoto, and A. Koyama, "Development of
        Continuous Measurement of Molten Steel Temperature on STB," CAMP-ISIJ **2**,
        216 (in Japanese) (1989).
[94]   N. Ramaseder and J. Heiss, "Continuous Temperature Monitoring in Metallurgical
        Vessels," Metallurgical Plant Technol. Int. **2**, 48−53 (2001).
[95]   Y. Tamura, M. Tatsuwaki, T. Sugiyama, T. Yokoi, M. Sano, and M. Koriki,
        "Temperature Measurement of Steel in the Furnace," in *Temperature, Its
        Measurement and Control in Science and Industry*, edited by J. F. Schooley, AIP, New
        York, pp. 505−512 (1982).
[96]   T. Iuchi, "Radiation Thermometry of Low Emissivity Metals near Room
        Temperature," in *Temperature, Its Measurement and Control in Science and Industry*,
        edited by J. F. Schooley, AIP, New York, pp. 865−869 (1992).
[97]   A. Honda, A. Takekoshi, T. Yamada, and N. Harada, "New Radiation
        Thermometry using Multiple Reflection for Temperature Measurement of Steel
        Sheets," in *Temperature, Its Measurement and Control in Science and Industry*, edited by
        J. F. Schooley, AIP, New York, pp. 923−927 (1992).
[98]   J.-C. Krapez, P. Cielo, and M. Lamontagne, "Reflective-Cavity Temperature
        Sensing for Process Control," in *Temperature, Its Measurement and Control in Science
        and Industry*, edited by J. F. Schooley, AIP, New York, pp. 877−882 (1992).
[99]   T. Yamamoto, Y. Adachi, Y. Tamura, I. Sakaguchi, T. Uemura, and S. Inoue,
        "Radiation Thermometry Method using Multi-Reflection between Two Parallel
        Steel Sheet Surfaces," in *Temperature, Its Measurement and Control in Science and
        Industry*, edited by J. F. Schooley, AIP, New York, pp. 933−938 (1992).
[100]   J.-C. Krapez, P. Cielo, and M. Lamontagne, "Reflecting-Cavity IR Temperature
        Sensors: An Analysis of Spherical, Conical and Double-Wedge Geometries," in *Proc.
        SPIE*, edited by A. H. Lettington, SPIE, London, Vol. 1320, pp. 186−201 (1990).
[101]   P. B. Coates, "Multi-Wavelength Pyrometry," Metrologia **17**, 103−109 (1981).
[102]   F. Tanaka and D. P. DeWitt, "Theory of a New Radiation Thermometry Method
        and an Experimental Study using Galvannealed Steel Specimens," Trans. Soc.
        Instrum. Control Eng. **25**, 1031−1037 (1989).
[103]   F. Tanaka, H. Ohira, and M. Masuda, *Application of TRACE Method: New Method for
        Simultaneous Determination of Temperature and Two Spectral Emissivities*, Nippon Steel
        Technical Report No. 49, pp. 70−75 (1991).
[104]   L. K. Zentner, D. P. DeWitt, D. A. White, and D. R. Gaskell, "Dual-Wavelength
        Emissivity Compensation Algorithms for Galvannealed Steel," in *Temperature, Its
        Measurement and Control in Science and Industry*, edited by J. F. Schooley, AIP, New
        York, pp. 861−864 (1992).
[105]   K. Hiramoto, T. Yamamoto, and C. Uematsu, "Development of Multiwave-
        length Pyrometer," in *Proc. XIII IMEKO World Congress*, Torino, pp. 1456−1461
        (1994).
[106]   D. P. Hill, R. L. Shoemaker, D. P. DeWitt, et al., "Relating Surface Scattering
        Characteristics to Emissivity Changes during the Galvanneal Process," in *Proc. SPIE*,
        SPIE, San Diego, Vol. 1165, pp. 62−71 (1989).
[107]   T. Iuchi and R. Kusaka, "Two Methods for Simultaneous Measurement of
        Temperature and Emittance using Multiple Reflection and Specular Reflection, and
        Their Applications to Industrial Processes," in *Temperature, Its Measurement and
        Control in Science and Industry*, edited by J. F. Schooley, AIP, New York,
        pp. 491−503 (1982).

[108] Y. Yamada, D. Yuasa, M. Uesugi, and T. Yamada, "Emissivity Compensating Thermometer by a Scanned Image of a Linear Light Source," in *Proc. SICE Annual Conference 1996*, pp. 321–322 (1996).

[109] T. Yamada, N. Harada, and M. Koyanagi, "Temperature Distribution Measurement with a Silicon Photodiode Array," in *Temperature, Its Measurement and Control in Science and Industry*, edited by J. F. Schooley, AIP, New York, pp. 395–400 (1982).

[110] Y. Fukutake, N. Iida, T. Iwamura, I. Hishikari, T. Ide, and T. Suzuki, "Auto-Emissivity-Compensated Radiation Thermometry with Variable-Area Reference Radiator," in *Temperature, Its Measurement and Control in Science and Industry*, edited by J. F. Schooley, AIP, New York, pp. 849–853 (1992).

CHAPTER 5

THERMAL IMAGING IN FIREFIGHTING AND THERMOGRAPHY APPLICATIONS

Francine Amon[1] *and* Colin Pearson[2]

Contents

1. Introduction   279
   1.1. Basic TIC components   280
   1.2. Figures of merit   287
2. Thermography Engineering Applications   287
   2.1. Comparative thermography   288
   2.2. Electrical thermography   293
   2.3. Building thermography   299
   2.4. Mechanical plant thermography   310
3. Thermal Imaging in Firefighting   316
   3.1. Thermal environment   316
   3.2. Thermal imaging operations in the fire service   319
   3.3. Design guidelines   321
4. Standards   327
5. Summary   328
Acknowledgments   328
References   328

1. INTRODUCTION

Thermal imaging cameras (TIC)[1] are noncontact temperature measurement devices that determine apparent temperature from the

---

[1] The acronym TIC shall be used throughout this chapter in reference to both the singular and plural forms, that is thermal imaging camera(s), as befits the context of the sentence in which it is used.

---

[1] Building and Fire Research Laboratory, National Institute of Standards and Technology, Gaithersburg, MD 20899, USA
[2] BSRIA Design and Facilities Innovation, Old Bracknell Lane West, Brackness, Berkshire, RG12 7AH, UK

Experimental Methods in the Physical Sciences, Volume 43
ISSN 1079-4042, DOI 10.1016/S1079-4042(09)04305-7

infrared (IR) radiation emitted by objects. TIC produce a thermal image of a scene that provides information about both its temperature and radiative properties. Thermal imaging is extremely valuable for tasks for which contact temperature measurements are not possible or cannot provide adequate information, for example, location of leaks, electronic defects, moving objects, or assessment of safety. The spectral responsivity of the detectors used in commercial uncooled TIC can be selected to avoid atmospheric absorption and thus be useful in fire control circumstances [1]. TIC can provide first responders with critical information to assess a fire incident, track fire growth, locate victims and other first responders, and establish egress routes. During the past 15–20 years, IR technology for firefighting applications has matured to the point that most first-responder organizations, in many countries, now possess and routinely use TIC.

Besides these emergency service applications, TIC are now commonly used in industry in, for example, preventative maintenance, building surveys and nondestructive evaluation and testing. This chapter begins with a review of TIC, their construction and performance, followed by applications in industry and firefighting.

1.1. Basic TIC components

A TIC is a complex system of assembled components. Each component serves a purpose in the goal of transforming invisible IR radiation into an informative image that can be viewed by a human. A basic schematic of the components of a TIC system is shown in Figure 1. The incident thermal radiation on the TIC lens is focused on the detector. The signals generated by the detector are then manipulated into a video signal that is sent to the display. Some TIC also have a separate video output port that can be used to record a similar image on videotape. The following sections will focus upon the design, function, and operation of the components of the imaging

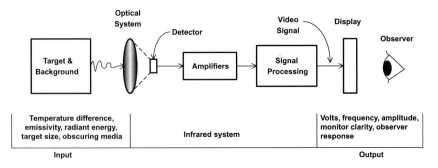

Figure 1 Basic schematic of TIC components.

system. This discussion will include the transmissive properties of the optics, the electrical response of the sensor array to incident radiative flux to the internal amplification, multiplexing, interpolation, and filtering devices, and the conversion of the data into useful output signals through image recording and liquid crystal display (LCD) luminance.

For most thermographic applications and all firefighting applications, the observer's interface with the TIC is the display screen. The overall usefulness of the TIC is therefore very much dependent on the ability of the observer to interpret the image displayed on the screen well enough to perform a task or operation successfully.

1.1.1. Target and background

The optimum characteristics of the TIC chosen or designed for use for a specific application depend in large part on the target-background temperature difference, emissivity, radiant energy, target size, and obscuring media present during use. The TIC must be capable of differentiating between the target and the background, particularly when they are near the same apparent temperature. Emissivity is a property of surfaces that describes their ability to emit radiant energy, and can affect the measured temperature of a surface when TIC are used. The ratio of the emitted radiant energy of a surface to that of a blackbody at a particular temperature defines the emissivity of the surface. If a surface behaves like a blackbody (it absorbs and re-emits all energy incident upon it), it will have an emissivity of 1. However, a surface that reflects all incident radiant energy has an emissivity of 0 and will appear to be the temperature of the reflected radiant energy rather than the temperature of the surface itself. The apparent temperature is the temperature of a surface that is measured without considering its emissivity. It is possible that a target and background may have counteracting temperatures and emissivities such that their apparent temperatures appear to be similar when viewed with a TIC.

In addition to having sufficient thermal sensitivity to differentiate between thermally similar targets and backgrounds, a TIC must be capable of detecting high-radiance targets and backgrounds without saturating the detector *if* the normal use of the TIC requires viewing hot surfaces or scenes having a wide range of surface temperatures. The size and amount of detail of the target determine the necessary spatial resolution for the TIC. Lastly, consideration must be given to the environment between the target/background and the optical system of the TIC. For example, if the TIC is to be used in an industrial setting to view a reaction vessel located behind a window, that window must transmit thermal radiation in the sensitive wavelength region of the TIC. As another example, firefighters must be able to "see" through heavy smoke.

1.1.2. Optical system

IR optics are designed to focus the image of the thermal scene onto the detector array. They are simple lenses similar to those found in any optical device, with the exception that the material used must be transmissive in the wavelengths of radiation being detected. Germanium (Ge) and zinc selenide (ZnSe) are the two most common materials used to construct optics for TIC because they transmit radiative energy in the $8-14\,\mu m$ range (Figure 2.) The lens design is important for reducing aberrations and internal reflections. The lens is also designed to provide a field of view (FOV) appropriate to the intended use of the TIC. TIC designed for first responders often have an FOV from $40°$ to $60°$, while the FOV for thermographic applications may be much narrower, depending on the specific needs of the user.

1.1.3. Detector

The oldest and most traditional method of detecting thermal radiation is a cryogenically cooled quantum or photon detector. In recent years, a great deal of research has moved away from these sensors and focused upon thermal detectors, which are constructed of a material with an electrical property that changes dramatically with temperature. This electrical property can then be measured to determine the amount of irradiation

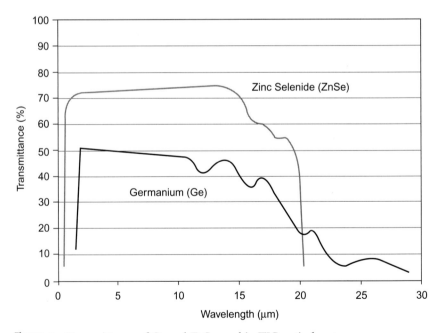

Figure 2   Transmittance of Ge and ZnSe used in TIC optical systems.

incident upon the detector from changes in the material temperature. Consequently, most handheld TIC used by firefighters use thermal detectors because they do not require advanced cooling systems or large power requirements needed by quantum detectors. However, other thermographic applications may require the additional thermal sensitivity offered by quantum detectors. The detector material should have high absorption coefficients in the desired spectral region or be coupled with another appropriate material. These pyroelectric and microbolometer sensors are then fabricated into large arrays and placed in the focal plane of the optical system. These devices, often referred to as focal plane arrays (FPAs), are the primary types of thermal detectors used for handheld TIC [1].

Pyroelectric detectors measure radiation by sensing changes in polarity of the detector material due to exposure to incoming radiation. The detector is constructed of a polar, dielectric material containing a surface charge that is proportional to changing temperature. The detector material directly generates an electric signal by acting as a capacitor whose charge, ΔQ, is described by

$$\Delta Q = Ap\Delta T \tag{1}$$

where A is the area of the capacitor, p the pyroelectric coefficient of the material, and ΔT the measured temperature difference corresponding to ΔQ. Pyroelectric detectors can be constructed of ferroelectric metals, polymers, or pyroelectric crystals [2]. The faces of pyroelectric materials take on opposite electric charges. When the temperature is steady, these charges are kept balanced by internal charges; however, the magnitude of the polarization changes as the temperature of the sensing material changes during the transient temperature condition. This effect can also be enhanced in ferroelectric materials by applying an additional, external electric field. This field acts to change the dielectric permittivity of the sensing material and thus increases the apparent pyroelectric coefficient and is known as the field-enhanced pyroelectric effect. TIC sensors constructed from barium strontium titanate (BST) take advantage of this effect and are commonly used for handheld imaging devices. Further information on this topic can be found in Refs. [3,4].

Because pyroelectric detectors measure transient changes in temperature, they require the use of a rotating chopper to continuously reference the temperature of each pixel to ambient and to keep the temperature of the sensing material dynamic. For this reason these detectors are very low in thermal mass so that the temperature can change very rapidly [5]. An example of a pyroelectric straight blade chopper and the resulting temperature of a pixel are shown in Figure 3. The detected signal of the pyroelectric detector will be proportional to both the absolute temperature and the slope of temperature with respect to time.

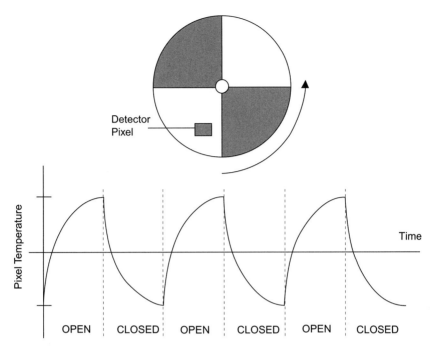

Figure 3   Temperature response of a BST pixel to a chopper.

A microbolometer functions on the principle of varying electric resistance with changes in temperature. A bias voltage is applied to the detector and the temperature is determined from the amount of current passing through the circuit [1]. Microbolometers are typically made from vanadium oxide (VOx) or amorphous silicon (ASi). These materials are chosen for their strong resistive response to temperature and the ability to construct them into detector arrays. A typical value of the temperature coefficient of resistance (TCR) for these materials is about $-0.003\,\mathrm{K}^{-1}$, and they are capable of producing resistances in microarrays of about $10\,\mathrm{k\Omega}$. The voltage response of a pixel, V_{pixel}, is then proportional to

$$V_{\mathrm{pixel}} \propto \kappa R I_b [T(t) - T_0] \tag{2}$$

where κ is the TCR, R the resistance, I_b the bias current, T_0 the average temperature, and $T(t)$ the temperature as a function of time [2].

Each detector pixel contains a small layer of thermally resistive material coupled with an IR-absorbing material. The detector is supported over the circuitry by two small legs in order to reduce the thermal conductivity while still maintaining electric contact. The detector itself contains VO_x or ASi bonded to another layer, in this case silicon nitride, which acts to increase the absorption of the thermal radiation in the spectral band of

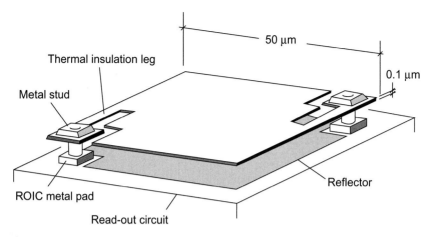

Figure 4   Schematic of a microbolometer pixel (not to scale).

interest. An example of a microbolometer pixel is shown in Figure 4. The individual pixels are constructed as part of an FPA, often 640×480, 320×240, or 160×120 for commercial TIC. The FPA is stored in a near vacuum to eliminate convection of heat between pixels. The individual pixel can only lose heat by radiating back into the environment or by conduction through the electrical contacts. This allows the detector to respond quickly to changes in the temperature of the scene. One advantage of bolometers over pyroelectric devices is that they do not require the use of a chopper for temperature reference; instead they measure the absolute temperature of the pixel, rather than a variation in that temperature [6].

1.1.4. Amplifiers and signal processing

The signal from a microbolometer is detected by measuring the voltage response of the pixel as described by Equation (2). The charge signal on a pyroelectric detector can be measured in two ways. The voltage or the current can be measured and both should be roughly proportional to $T_{target}^4 - T_{chopper}^4$. Once a signal has been generated by the detector, it must be delivered into a circuit and processed into a coherent video image. Each detector pixel is connected to an electronic readout system. There are two options for integrating a readout circuit to the detector pixels: monolithic or hybrid design.

The monolithic design incorporates the readout integration circuit (ROIC) on the same layer as the sensing material. While this permits simpler and cheaper manufacturing, it also reduces the amount of sensing area on each pixel, or the fill factor. The increased space taken up by the output electronics reduces the sensitivity of the detectors to small temperature differences or low input temperatures. In general, the monolithic design sacrifices image quality for reduced manufacturing cost.

The hybrid design separates the detector elements and the signal transmission network onto two separate layers in electrical contact. While this process is more complex and expensive to manufacture, the fill factor can be raised from about 55% for monolithic arrays to around 90% for hybrid designs. In addition to improving thermal sensitivity, the hybrid design also improves cooling efficiency, which leads to increased TIC battery life and reliability. The decision to design an FPA with monolithic or hybrid design is based upon the need for image quality versus cost.

After each pixel has been connected electrically, the signals must be compiled and formatted for signal processing in a multiplexer. The array signals can be multiplexed by a charge-coupled device (CCD) or by a complementary metal-oxide semiconductor (CMOS) device.

A CCD device transfers electrons from each detector down an entire row of detectors to a readout device at the end of each row. This design allows for a less complex pixel design with a high fill factor and uniformity with a relatively slow readout speed but can be subject to errors through charge loss along rows or by charge overflow between adjacent pixels, known as blooming.

A CMOS device directly outputs a voltage signal from each pixel into the readout device. This requires the use of transistors on the pixel, reducing the uniformity and fill factor, but increasing the readout speed.

Once an ROIC has been implemented, the compiled signal is electronically adjusted to produce a crisp, coherent, and contrasted image. These adjustments include amplification and gain based on dynamic range control, interpolation to smooth the image from pixel to pixel, and various other complex algorithms designed to improve image quality. Because processing algorithms are unique to each TIC and are the intellectual property of each manufacturer, individual functions will not be discussed in detail.

In addition to filtering the signal and adjusting the gain to produce a useful image, some TIC have an automatic mode shift operation. When a certain trigger is reached, the thermal sensitivity of the TIC is reduced so that a larger temperature range can be surveyed. The trigger to this function may be different for each TIC. Some TIC shift when radiation from a specific transition temperature is detected; others shift when a certain percentage of the pixels become saturated; and other TIC contain several gradual mode shifts inherent in the gain electronics.

1.1.5. Output signals
Output signals typically transmit video data to a recording device or to a central location. Firefighting TIC sometimes make use of radio transmitters

to relay the video signal to a central controller where the data from all TIC at an emergency event can be observed to define a unified strategy for firefighting or rescue.

The image is further modified and digitized for transmission to a display, where the pixel intensities are converted into a projected luminance visible to the human eye. The display may be the only video output available to the user of the TIC. There are many display technologies available that can be integrated into TIC systems, although LCDs are the most common.

1.2. Figures of merit

The detection limit of thermal sensors is most commonly described by the noise equivalent temperature difference (NETD). This is essentially the temperature difference detected by the sensor that creates a differential signal equivalent to the amount of signal variation caused by noise. There are two basic methods of decreasing NETD. First, the effect of the noise can be reduced by decreasing the total heat capacity (mc_p) of the detector element where m is mass and c_p is the thermal capacity, which will increase the temperature change from incoming radiation and thus increase the signal from an equivalent input. Second, the intensity of noise can also be reduced by decreasing the total heat conduction from the detector element (G) [2]. The temperature of the sensing element is determined from

$$mc_p \frac{dT}{dt} + G(T - T_{\text{base}}) = \alpha(T)\Phi_f(t) \qquad (3)$$

where α is the absorption coefficient for spectral radiation, T_{base} the temperature of the connecting substrate, and $\Phi_f(t)$ the incident radiant flux as a function of time. For microbolometer detectors, the measured electric resistance will be a function of T, the temperature of the detector pixel. For field-enhanced pyroelectric devices, the measured polarization will be a function of both dT/dt and T, and the incoming radiation $\Phi_f(t)$ will be a waveform modulated by the chopper device [3].

2. THERMOGRAPHY ENGINEERING APPLICATIONS

There is a huge range of applications of IR thermal imaging in the fields of engineering: mechanical, electrical, and civil. It is impossible to cover all of them, but this section outlines a sample of applications, the benefits of thermography, and some of the problems encountered in achieving these benefits. There are broadly two demands for thermographic inspection for engineering plant, equipment, and structures:

1. Routine inspection, usually to check that equipment is intact or operating as expected (condition survey, CS), and
2. Regular inspection to compare with previous results (condition monitoring, CM).

In many cases the inspection techniques are identical for both condition monitoring and condition survey; both use nondestructive testing (NDT) methods that test the condition without affecting the operation of a structure, machine, or plant item.

There are several recognized IR thermography techniques in use throughout industry. Comparative thermography is the most common technique, and it is normally used to provide the best available data without the strict rigor of actual temperature measurements.

2.1. Comparative thermography

Comparative thermography can be either quantitative or qualitative. The quantitative technique requires the determination of a temperature value to distinguish the severity of a component's condition. This value is determined by comparing the target's temperature to that of similar objects or baseline data. For high-emissivity surfaces, both absolute temperature and differential temperature (ΔT) values are typically reliable, provided good measurement techniques are followed. Temperature and ΔT values of low-emissivity surfaces are often unreliable due to surface and environmental variations. In addition, many applications also require assigning values to observed thermal patterns for the purposes of analysis, trending, designating severity levels, and assigning priorities. However, there are many applications where quantitative data are not required to monitor the condition of machinery, or to diagnose a problem and recommend the appropriate corrective action. In these cases, qualitative techniques may be more than adequate.

2.1.1. Comparative quantitative thermography

The comparative quantitative thermography method is an accepted and effective method for evaluating the condition of a machine or component by estimating temperatures. It is very difficult to determine precisely the actual temperatures of a component, using infrared thermography (IRT) in the field. This is due to the physics of IRT that must take into consideration multiple parameters such as emittance, reflectance, transmittance, and background temperature. Estimates of these IRT considerations can be readily made to obtain a component's approximate temperature, which, in most cases, is more than sufficient to determine the severity of an adverse condition.

Comparative measurement, unlike qualitative measurement, identifies a thermal deficiency by comparing the temperatures obtained using a consistent emissivity value default, for those surfaces of similar emissivity, that is, across the surface of a single machine or between the surfaces of similar machines. The ΔT between two or more identical or similar surfaces is measured numerically. Assuming that the environmental conditions and surface properties for both components are similar, the ΔT for the given piece of equipment is recorded as being the degrees above the normal operating temperature of the similar equipment.

As an example of comparative quantitative thermography, two machines are operating in the same environment and under the same load conditions, and one is experiencing an elevated temperature. This is usually an indication that a deteriorating condition may exist. However, the determination of the ΔT would then assist in establishing the severity of the condition. In this example, a $5°C$ ΔT would be considered minor, whereas a $100°C$ ΔT may be considered to be critical. Also, knowing the approximate value of the elevated temperature would provide an indication that the temperature of a component may be approaching specified limits. Therefore, while qualitative measurements can also detect deficiencies, it is the quantitative measurements that have the capability of determining severity.

The comparative measurement technique uses quick emissivity estimates, the reflected apparent temperature, and component distance measurements. The emissivity factors of the materials are obtained from tables or through experience. It is possible to check the emissivity of the most commonly encountered materials to assign default values that can then be used when inspecting components with these materials. The method can be found in the Standard ISO 18434-1 [7]. Thermographers produce tables of emissivity for materials they often encounter for their own reference.

The estimated emissivity, distances, and reflected apparent temperatures are entered into the TIC, and a temperature value for each component is shown in the image. This type of measurement is effective when surveying many components. It is quick and provides useful information for determining the severity of a component's condition.

2.1.2. Comparative qualitative thermography

Comparative qualitative measurement compares the thermal pattern or profile of one component to that of an identical or similar component under the same or similar operating conditions. When searching for differing thermal patterns or profiles, an anomaly is identified by the intensity variations between any two or more similar objects, without assigning temperature values to the patterns. This technique is quick and easy to apply, and it does not require any adjustments to the IR instrument to compensate for atmospheric or environmental conditions, or surface emissivity. Although the result of this type of measurement can identify a deficiency, it does not provide a level of severity.

This IR thermography technique is used throughout most industries. It is very effective in identifying hot bearings or other abnormally hot machine components, hot spots in electrical equipment, undesirable hot electrical connections, leaking or blocked fluid heat exchange equipment and components (tubes), and fluid leaks from pressure vessels, pipes, and valves.

2.1.3. Factors to be considered in engineering thermography

The following sections describe aspects of the use of engineering thermographic TIC. It is important to consider these issues when using a thermographic TIC in order to optimize the time spent and to fully utilize the functions available on the TIC.

2.1.3.1. Object size and distance. As with spot radiometers, it is important to ensure the object fills enough of the FOV to get a reliable temperature measurement. For thermographic TIC this problem is significantly reduced. FPA detectors typically have over 76,800 detector elements or cells, in a 320×240 matrix. The FOV is therefore projected onto this array of detector elements. Each detector operates as a spot radiometer with individual FOV angles, usually expressed in milliradians rather than degrees. This value is often referred to as the instantaneous field of view (IFOV), which is a reminder of the history of thermal imaging systems in which a single detector was used to scan the FOV through rotating and oscillating mirrors. The smaller the angle, the greater the spatial resolution, thus enabling objects to be detected and accurately resolved at greater distances. The discernible object size will depend on the lens used. Errors in measurement and failure to resolve faults will occur if the object is too small or the distances too great. To ensure accurate resolution of small objects, the objects should be covered by at least three pixels of the array. For example, if the FOV of a 320×240 pixel array is $1\,\text{m}$ wide, the smallest accurately resolvable object will be $3 \times 1{,}000\,\text{mm}/320 = 9.4\,\text{mm}$. A smaller object less than three pixels wide might not fill any of the pixels, and hence the temperature detected on all pixels would also be affected by the background temperature.

The minimum discernible object size is affected by lens field angle, distance, and TIC resolution, as defined by

$$O = \frac{6d}{n}\tan\left(\frac{A}{2}\right) \tag{4}$$

where O is the minimum discernible object size in millimeters, d the distance between lens and object in millimeters, n the number of pixels in the TIC field width or height, and A the TIC field angular width or height in degrees. The minimum discernible object size can be plotted against distance for the lens and number of pixels used as in the example seen in Figure 5. Experts in geometry will see that the discernible object size is in fact 2.828 × the pixel pitch, but it is safer to assume 3 × . The minimum detectable object size clearly varies with number of pixels in the image and distance from the object. A set of curves can be produced for any detector and lens combination. An example can be seen in Figure 6.

2.1.3.2. Reflected ambient temperature. Sometimes known as "background temperature," a correct estimate of the reflected ambient temperature is crucial to estimating temperatures with thermal imaging. The thermal radiation from an object depends on the temperature of the surfaces facing it. Reflected ambient temperature is a term used to mean the

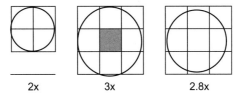

2x   3x   2.8x

Figure 5 Geometric demonstration of minimum detectable object size in terms of numbers of pixels.

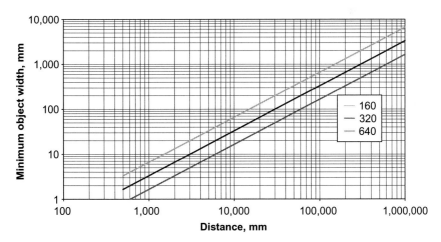

Figure 6 Minimum detectable object size at a distance with different numbers of pixels using a 20° lens.

average effective temperature of those surfaces in the hemisphere facing the object. It is based on the amount of incoming thermal radiation, being the temperature of surfaces with emissivity of 1.0 that would produce the same incoming thermal radiation. It can be estimated by one of two methods that are described in more detail in ISO 18434. The first involves placing a multifaceted reflector, crinkled aluminum foil for example, in the plane of the object and using the TIC to measure its average apparent temperature with unity emissivity. This is often difficult to perform in electrical investigations, and, of course, when the object is at high voltage it should not be attempted. An alternative method involves estimating the average temperature of surfaces in the hemisphere facing the object by sweeping the TIC across the hemisphere and calculating average apparent temperature.

2.1.3.3. Reflection and transmission.

The thermal radiation from an object is made up of radiation reflected, transmitted, and emitted by it. The first two factors are often forgotten. Transmitted radiation is only found with objects that are transparent to IR radiation. Most materials, including most window glass, are only slightly transparent to IR, so in most cases there is no transmitted radiation to contend with. Glasses are optimized to provide excellent transmittance throughout the total visible range from 400 to 800 nm (0.4−0.8 μm). Usually the transmittance range spreads also into the near IR regions up to 1.5 μm, but the TIC used by thermographers have their peak sensitivity in the 2−6 μm (short wave) or 7−14 μm (long wave) ranges. As shown in Figure 7, there is significant transmission through thin glass in the short-wave band and almost none in the long-wave band. But consider a very hot object behind a glass window; this may still be detected by a long-wave TIC. There is another situation in which heat can be apparently transmitted through glass. If there is a heat source close to the window, it will warm the glass by radiation and convection

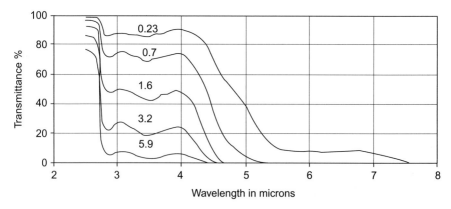

Figure 7   Transmittance of soda–lime glass at different thicknesses (in mm).

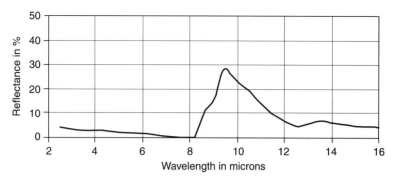

Figure 8   Spectral reflectance of plain soda-lime glass in the long-wave band.

thereby increasing its temperature. This is often seen in building surveys where radiators are placed below or in front of the window, when viewed from inside. Normally a long-wavelength TIC will give a better representation of the glass temperature, but there is still considerable reflection from glass surfaces in the long-wave band as seen in Figure 8.

The situation is further complicated because windows in buildings are increasingly using coated glass to reduce heat loss and solar gain. The purpose of these coatings may be to reduce emittance in the IR bands to reduce heat loss. This has the effect of increasing IR reflectance, although the glass may not appear to be reflective in the visible spectrum.

2.2. Electrical thermography

The distribution and use of electricity produces heat, either intentionally or as a side effect. The movement of this heat is important to designers, users, and maintenance engineers, and temperature anomalies are the key to distinguishing normal or abnormal working conditions. Electrical applications of thermography include inspections of cables and busbars, cable and busbar terminations, transformers, switches, circuit breakers, fuses, control equipment, motors, motor control centers, overhead lines, and uninterruptible power supply (UPS) batteries. Each type of component has a characteristic thermal pattern in normal operation and typical modes of failure. The experienced thermographer recognizes these patterns and can diagnose faults that might indicate incipient failures. This section looks at the generation, detection, and analysis of heat from the flow of electricity.

2.2.1. Principles: Why thermal radiation is important in electrical systems

Electric energy or electricity is a form of energy that can be transformed to other forms of energy such as mechanical energy. Electric energy is measured in joules (J) or kilowatt-hours (kWh). Electric power measured in watts (W) is the rate at which electric energy is being used, stored, or transferred. This dissipated electric power in a resistive circuit with a current (*I*) and resistance (*R*) is given by

$$P = I^2 R \tag{5}$$

The dissipation of electrical power in electrical conductors produces thermal energy that causes them to warm. Most faults in electrical installations and equipment, such as poor electric contact, result from changes in resistance and subsequently changes in heat generation. Since IR detectors are used to measure the radiant heat flux related to the temperature of a surface, they are sensitive to changes in electrical resistance. Most TIC now calculate the temperature of the object from the heat flux, emissivity, background temperature, air temperature, and distance by the equation

$$U_{obj} = \frac{1}{\varepsilon\tau} U_{tot} - \frac{1 - \varepsilon}{\varepsilon} U_{amb} - \frac{1 - \tau}{\varepsilon\tau} U_{amb} \tag{6}$$

where U_{obj} is the calculated TIC output voltage for a blackbody of temperature T_{obj}, the object emittance is ε, the atmospheric transmittance is τ, the (effective) temperature of the object surroundings or the reflected ambient temperature is T_{amb}, and the temperature of the atmosphere is T_{atm}.

Most electrical failures result in heat generation such as increased resistance of cable terminations or contacts due to corrosion or other problems. It is always helpful to have another similar component to compare with the suspect item, as seen in Figures 9 and 10. Sometimes a thermal image recorded in a previous survey may be used. Any changes in apparent temperature should be investigated to identify the anomaly as a fault or an expected change. Benign changes in apparent temperature can be a result of changes in environmental factors such as ambient temperature, electrical load, emissivity due to corrosion or dirt, and reflected ambient temperature. In some circumstances, such as failure of one phase of a three-phase circuit, a lower temperature can indicate a fault.

2.2.2. Detectability of early signs

The speed at which faults develop dictates the frequency that they should be inspected. The interval between inspections should be less than the time for a fault to develop from a perceptible level to a critical level. Electrical installations that suffer from vibration, moisture, corrosive atmosphere, changing load, changing ambient temperature, or frequent intrusive maintenance are likely to develop faults faster than installations under constant load in a constant environment. The history of failure on an installation should be used to determine inspection frequency. Annual inspections are commonly used in benign environments.

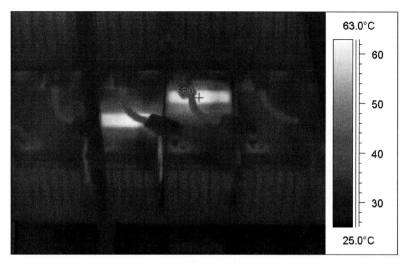

Figure 9 UPS batteries under test load showing two batteries with failed cells among good batteries.

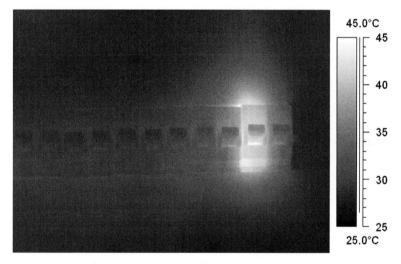

Figure 10 A row of miniature circuit breakers with one poor connection giving a surface temperature of 50°C.

Faults can be detected early if a history of previous thermographic surveys shows a constant apparent temperature. A change as small as 1° can be used as an early fault indicator, if it cannot be explained by environmental factors. Many examples can be found in other books such as the British Institute of Non-Destructive Testing (BINDT) thermography handbooks [8].

Thermographic surveys should be undertaken under normal operating maximum load conditions and in the normal operating environment. According to Equation (5), heat generated is proportional to the square of the current. If the current is halved, heat generation will be reduced by 75%. In thermography of circuits with less than full load, the temperature rise at full load can be estimated by

$$\Delta T_{\text{(full load)}} = \Delta T_{\text{(measured)}} \times \left(\frac{\text{full load current}}{\text{measured current}} \right)^{n} \tag{7}$$

If heat exchange is mainly radiant, as in thin wires in still air, $n = 2$. For thick busbars (where convection is significant) with heat generated at a point of high resistance rather than along the conductor, n tends to 1.6. Where convection is significant, such as on overhead lines in windy conditions, n can range from 0.6 to 1.1. If run and standby equipment are present, the thermographer should note which is running and preferably repeat the survey with the alternate equipment under load.

2.2.3. Quantitative vs. qualitative thermography in electrical applications

Comparative qualitative thermography can reveal differences between the thermal patterns on similar electrical components. Two fuses at different temperatures clearly show up differently in a thermal image. Without actual temperature data it is usually impossible to say with any confidence whether the observed thermal difference is significant. This method is widely used to quickly scan a large number of components. The thermographer can then concentrate on the few thermal patterns and use comparative quantitative thermography to identify whether these are anomalies, faults, or just normal variations between components.

The thermographer needs to have a clear guide to determine what temperatures are acceptable for the equipment surveyed. Temperature limits can be taken from manufacturer's data or knowledge of the type of materials used. Electrical insulating materials have different temperature limits. Commonly used polyvinyl chloride (PVC) insulation should not be operated at greater than 70°C, but other insulation rated at a higher temperature is often used if required by the operational circumstance. Busbar operating temperature may be limited by the rating of the insulated support components, often 110°C. Busbar temperature can be difficult to measure because of their low emissivity. New busbars have emissivity of between 0.01 and 0.10. A useful feature of busbar systems is that there are often parallel bars as shown in Figure 11. Between the bars there is repeated reflection creating an area with similar features to a blackbody cavity. The emissivity here is close to 1 and so the temperature of 73°C could easily be measured.

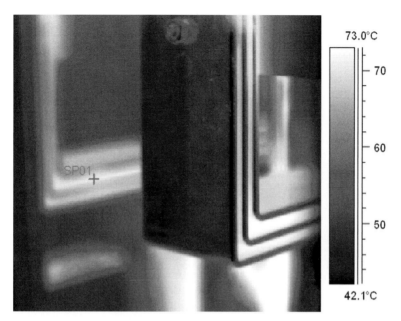

73.0°C

─ 70

─ 60

─ 50

42.1°C

Figure 11 Busbars, showing detection of temperature when emissivity is low.

In any electrical conductor there is resistance and therefore a potential temperature rise above ambient. A perfect connection in this conductor should maintain the same temperature while in operation. Therefore, one of the main criteria used for electrical connections is the differential temperature with respect to the conductor. There is general agreement about the severity of differential temperatures in low-voltage installations. In the USA, the recommendations are published in the guidelines provided in Table 1 by the International Electrical Testing Association (NETA) [9].

There are other condition monitoring techniques that can be used in conjunction with thermal imaging to make it even more effective. In high-voltage installations, ultrasonic detection can be used to check for electrical discharge. Acoustic noise is an early sign of insulation breakdown that can be remotely sensed without line of sight to the source. It is therefore a useful safe precursor to a thermographic survey to check that there is no electrical discharge taking place in the area before opening enclosures.

2.2.4. Material limits

Electrical component temperature may be limited by the physical properties of a material. Some plastics used for insulation melt at about 180°C but soften to an unusable state at temperatures as low as 100°C. There may be other materials in the vicinity that demand low limits. The ignition temperature of wood is approximately 320−350°C, but

Table 1 Guidelines for electrical installations (NETA).

Problem classification	Phase to phase temperature rise	Comments
Minor	1°C to 10°C	Repair in regular maintenance schedule; little probability of physical damage
Intermediate	10°C to 30°C	Repair in the near future (2–4 weeks); watch load and change accordingly; inspect for physical damage; probability of damage in the component, but not in the surrounding components
Serious	30°C to 70°C	Repair in immediate future (1–2 days); replace component and inspect the surrounding components for probable damage
Critical	Above 70°C	Repair immediately (overtime); replace component; inspect surrounding components; repair while TIC is still available to inspect

temperatures do not need to get this high to present a fire hazard. Long exposure to temperatures as low as 120°C causes cellulose materials like wood to degrade into pyrophoric carbon. In this process the character of the material is changed to almost pure carbon that burns much more easily.

Atmospheres containing dust or fumes can present an even greater hazard. Fuel oils and solvents are common materials that should not, but may be found in switch rooms. The flash point of a solvent or fuel oil is the temperature at which vapor given off will ignite when a spark occurs. Typically, flashpoint of fuel oil is around 60°C, but the ignition temperature, 250°C for light oil, is also an important factor. Carbon disulfide is particularly dangerous in this respect with an ignition temperature of 90°C, while for diethyl ether the ignition temperature is 160°C. Both these solvents could potentially be ignited by hot busbars without the need for a spark. Power factor correction capacitors present a special hazard, because they routinely contain a discharge resistor that can run at 300°C.

Electric motors are rated by their maximum allowable operating temperature, determined by the type of electrical insulation used in the motor [10]. Insulation degrades rapidly when it exceeds its thermal rating. For every 10 K rise over their rating, motor life is reduced by 50%! Ratings for motors are based by the hottest spot allowed on the insulation, which is inside the motor, when the motor is operating in a 40°C ambient environment. Temperatures seen by thermographers on the outside of the motor are usually 20 K less than those on the inside. Table 2 below gives some guidance to acceptable surface temperatures.

Table 2 Guidelines for acceptable surface temperatures.

Maximum temperatures		
External	Class	Internal
85°C	A	105°C
110°C	B	130°C
135°C	F	155°C
160°C	H	180°C

2.2.5. Personal safety and business risk considerations

Electrical thermography is perhaps the most hazardous range of applications. As well as the well-known risk of touching live electrical parts, there is the risk of "arc flash" which is the electrical discharge that occurs when electrical components fail catastrophically. Arc flash risk evaluation will not be covered here. A good description can be found in Ref. [11]. Elsewhere, such as the UK, risk assessment is the regulatory responsibility of the person responsible for the workplace, but regulatory legislation is less prescriptive on safe distances in electrical maintenance. Working near exposed live conductors should be avoided wherever possible, and remote thermal imaging of electrical installations is aimed at increasing safety and reducing risk by removing the necessity for worker proximity to the devices being monitored. The remote sensing can involve design measures such as IR windows, meshes, small observation holes, and arrangement of components so that critical parts can be checked safely.

The risk of electrical installations failing can be reduced by thermal imaging to check for early signs of component failure. The thermographer and the person responsible for the workplace therefore have a duty to develop installations and methods of work that are safe and effective and not introduce any risks to the process being monitored or to the staff involved.

2.3. Building thermography

Building fabric thermography, one of the oldest and largest applications of thermal imaging, involves the detection of temperature differences in the building fabric or structure, which includes the insulation, air layers and pockets, wall plaster and blocks. There is a huge bank of experience that is shared among thermographers primarily through training. Some of the first research in building thermography was conducted in Sweden in 1972 and a report [12] was published giving thermal images of artificially created faults. Following this report, an international standard, ISO 6781 [13], was published. The European and British standards follow this standard closely.

At the time of writing this book, the international standard was being revised and updated. The scope covers testing to identify irregularities but not the determination of transmittance. In fact, it specifically identifies that only differences greater than 50% or at best 25% in transmittance can be detected by the method.

The key to successful thermographic surveys is interpretation of thermal images. This is probably more difficult in building thermography than in other applications because of the range of different fault situations encountered. There is no substitute for good training and experience in diagnosing thermal anomalies in buildings. There are some books that can speed the learning process such as the BINDT thermography handbook [8]. Volume 2 of Ref. [8] contains hundreds of sample thermal images with descriptions of the problems found.

2.3.1. Principles: Why thermal radiation is important in building surveys

About 10% of world energy consumption is used for heating and cooling buildings (37% is used by industry, 20% by transport, and 27% is lost in transmission) [14]. Buildings exchange heat with their environment through walls, roof, floor, and ventilation. Regulations in most countries limit the rate of heat exchange for new buildings, and thermal imaging is used to check compliance with these regulations. Surveys of older buildings are also useful in finding sources of heat gain or loss.

Heat transfer through a wall, for example, is affected by the thermal resistance of the structure, the thermal resistance of the boundary layers, and the environmental temperature on each side, as shown by

$$Q = \frac{A\Delta T}{R_i + R_s + R_o} \tag{8}$$

where Q is the heat transferred, A the area of the surface through which the heat is transferred, R_i the thermal resistance at the interior boundary, R_s the thermal resistance of the combined layers of structural members, and R_o the thermal resistance of the outside boundary. The resistance of the boundary layer depends on radiative and convective heat exchange, so the local air speed has a great influence on its value. On the outside of walls, air speeds are typically much higher than on the inside, $1-10$ m/s outside compared with 0.25 m/s inside.

The difference in environmental temperature occurs across the components of the total thermal resistance in proportion to their resistance values. Typically, $R_i = 0.12$ m^{2}K/W, $R_s = 3.83$ m^{2}K/W, and $R_o = 0.05$ m^{2}K/W, giving a total of 4.0 m^{2}K/W and a U value of $1/Rt = 0.25$ W/m^{2}K. If there is a 10-K temperature difference across the wall, this will result in 0.3 K across the internal surface resistance, 0.125 K

Table 3   Values of R_o at various wind speeds.

Wind speed (m/s)	R_o (m^{2}K/W)
1	0.08
2	0.06
3	0.05
4	0.04
5	0.04
7	0.03
10	0.02

across the outside surface resistance, and the remainder across the structure of the wall. If there is a defect in a part of the wall so that the resistance of the structure is reduced to $0.31\,m^2$K/W, the outer surface temperature will rise 0.92 K to 1.04°C and the inside surface drop 2.2 K to 7.5°C. The difference in surface temperature on the inside is much greater than on the outside, so the thermal image of the inside will show the defect more clearly than that of the outside.

If the wind speed on the outside of the building increases, the outside surface resistance drops so that at wind speeds more than 10 m/s, it is only one quarter of its value at 1 m/s and the same defect would only produce a 0.2 K rise in temperature compared with the good section of the wall. The thermographer would struggle to detect such a small difference.

More detailed treatment of external surface resistance can be found in ISO 6946:2007 [15], which includes the surface resistances listed in Table 3 at different wind speeds. These factors also apply to surface resistance in other thermographic applications. Both electrical and mechanical thermography can be affected by high air speeds, for example, in external electrical and ventilation plant room surveys.

2.3.2. Factors influencing thermal radiation

ISO 6946:2007 [15] states that at external surfaces it is conventional to use the external air temperature, based on an assumption of overcast sky conditions, so that external air and radiance temperatures are effectively equal. This ignores any effect of short-wave solar radiation on external surfaces, dew formation, radiation to the night sky, and the effect of nearby surfaces. Other indices of external temperature, such as radiation-air temperature or sol-air temperature, may be used when such effects are to be allowed.

When thermal imaging is performed on a clear night, the sky radiance temperature can be as low as −50°C with an air temperature well above zero. This has resulted in many buildings being supplied with inadequate heating systems, particularly where the roof forms a large proportion of the total exposed fabric. The roof is typically exposed mainly to the sky, which can have this low radiance temperature. Walls are typically exposed to radiative heat exchange with other walls, and ground, which may have a temperature closer to air temperature. Underestimates of heat loss of up to 10% can be shown in buildings with a large roof area and low air-leakage rate compared with using a more accurate measure of external ambient temperature. Although thermographers rarely try to make quantitative assessments of building heat loss, it is important to understand the heat exchange processes to which their subjects are exposed.

Conversely, solar radiation, both direct and diffuse, must be considered when undertaking building surveys. Most thermographers are aware of this problem and avoid surveys in sunny conditions. It is also important to consider sky temperature in cloudy, diffuse sun conditions.

2.3.3. Insulation faults

Typically, thermal insulation faults show as areas of different surface temperature with a regular shape and relatively sharp edges. With thin skin construction such as insulated double-skin metal cladding, these show as very sharp edged areas as seen in Figure 12.

Figure 12 Thermal image of a typical insulation defect in a double-skin metal roof.

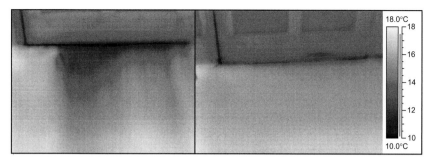

Figure 13 Thermal image of cold air leakage under a door and the same door after draughtproofing.

2.3.4. Air leakage

Air leakage typically shows on the inside of a building as a cold area with irregular shape and graded edges as seen in Figure 13. Cold air entering a building cools the surfaces over which it flows, creating areas that are detectable by thermal imaging. The image on the left is the bottom edge of a door as found with cold air leakage along the floor, and the image on the right is the same door half an hour later after installation of a brush-type draughtproofing strip.

Airtightness of buildings is often tested by pressurizing the building and measuring the air flow required to maintain a specified pressure such as 50 Pa. This test is often a requirement of building codes for new buildings. It provides a measure of the air-leakage quantity, but not the location of air-leakage paths. When buildings fail to reach the required airtightness standard, thermography is often used to detect the location of air leakage.

Air leakage also occurs within structures, and in these circumstances the leakage does not penetrate through the wall or roof and so it does not affect the airtightness of the structure. This type of anomaly sometimes occurs where the outside skin of a structure is not airtight and air penetrates into the thermal insulation layer. If cold, the outside air gets through the insulation, and it may then be in contact with the inner skin of the building. If the inner skin is airtight, the air does not get through into the building. However, it shows as a cold patch on the inside surface, which can be difficult to distinguish from cold patches caused by insulation gaps.

2.3.4.1. Convective air flow.

In heated buildings with air leakage, air enters the building through gaps at low level and leaves through gaps at high level. This means that the air leakage is detectable with thermal imaging at low level inside the building by cold areas on the wall or floor. At high levels, warm air can often be seen leaking out of a building through the roof and under the eaves as seen in Figure 14. Detection is often

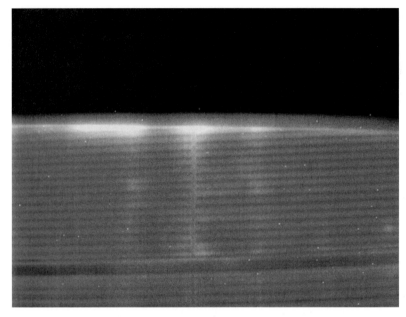

Figure 14   Thermal image of warm air leaking out under roof overhang.

difficult because access to the top of the roof surface may not be possible, but the top of walls where they meet the roof should always be inspected because this is a very common area for air leakage.

2.3.4.2. Forced air flow.   To enhance the detection of air leakage, a building can be pressurized using a method as used for airtightness measurement as seen in Figure 15. If the pressure inside the building is reduced by blowing air out through the typical "blower door" used for airtightness tests, inward air leakage can be forced so that leakage paths are detectable by cold patches on the inside surfaces of buildings.

2.3.5. Precautions for accurate detection

Building thermographers may be aiming to detect thermal insulation faults, thermal bridges, or air leakage. They need to distinguish these, but they also need to identify other extraneous features that are not faults in the building. These include:

- Reflections — identifiable by their apparent movement when the thermographer moves (These can be reflections of hot or cold areas, so the thermographer needs to be aware of temperatures of surrounding surfaces.),

- Heat sources — often identified by a temperature higher than room temperature,
- Intentional variations such as intermediate gutters across the roof of large buildings — designed to have poor thermal insulation so that snow melts in these before the rest of the roof to ensure proper drainage, and
- Areas of enhanced boundary layer resistance such as internal corners — common areas where surface temperature is seen to vary from the flat surface temperature (This is particularly difficult to differentiate from defects because insulation defects also often occur at corners of buildings as seen in Figure 16.).

Figure 15   A typical blower door for airtightness testing of small buildings.

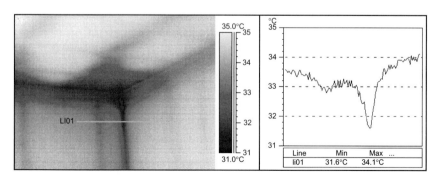

Figure 16   Corner of room showing enhanced boundary layer and genuine faults.

In this image with temperature of internal air at 36°C and external air at 8°C, there is an overall ΔT of 28°C across the wall. The surface temperature drops by 2.5°C at the corner as seen in the temperature profile, resulting in an increase in the temperature drop across the boundary layer from 1°C on the flat wall to 3.5°C in the corner. The boundary layer resistance increases from 3.6% to 12.5% of the total wall resistance. This is normal and not an indication of a fault. However, there is a patch of missing insulation in the corner of the ceiling and a strip across the top of the wall that probably results from thermal bridging by the structural frame.

2.3.6. Quantitative vs. qualitative thermography

Comparative qualitative thermography can locate anomalies quickly. This is all that is required for detection of air-leakage paths. However, if a building is to be surveyed for thermal insulation or thermal bridging, then a measure of severity is required to differentiate serious faults from the trivial anomalies.

There has been little guidance in standards or regulations on acceptable temperature variations in building structures. Although many thermographers have developed their own criteria for assessing building insulation, there has been no incentive to harmonize their methods. The Canadian Building Digest CBD-229 [16] was published in 1983, but this only gives guidance on procedures for identifying thermal deficiencies. The American Society for Testing and Methods (ASTM) Standard C1060-90 [17] offers no criteria either, only a method.

The 2001 Part L2 revision of the Building Regulations for England and Wales [18] introduced a requirement that thermal insulation should be "reasonably continuous" over the whole roof, wall, and floor area of new commercial buildings. Thermography of the completed building is recommended as a means of ensuring this. However, the wording is vague, leaving thermographers, builders, and building owners in doubt about whether a building complies with the requirement. Guidance on meeting these requirements can be found in the Building Services Research and Information Association (BSRIA) guide to building fabric thermography [19].

The UK Government has intimated in The Energy White Paper that the next edition of the Building Regulations will have stricter quality assurance of thermal insulation in buildings. It has become increasingly important to set a rational, uniform standard for acceptance of variations in thermal properties of the building fabric.

2.3.7. Technical difficulties

Heat exchange between walls or roofs and the environment is a complex subject beyond the scope of this chapter. It involves conduction, convection, and radiation to all objects in the hemisphere that they face. It is possible and desirable for the thermographer to approximately measure the appropriate "ambient" or "reflected" temperature in the process of measuring surface temperature. It is this temperature with which the wall or roof exchanges heat by radiation. Thermography routinely measures the surface temperature of the components in the FOV. The air temperature may be less relevant and is often difficult to determine at a distance.

In Figure 12, the black area on the left is a roof light that contains no insulation between the two polycarbonate sheets, but a large patch of insulation is missing to the right of this. The lack of insulation in this defect is confirmed by its temperature being the same as the rooflight.

Clearly a thermal anomaly is more serious if (a) it has a larger area or (b) it has a greater temperature differential. However, for practical purposes it must be possible to assess a building element in different climatic conditions. The criterion must therefore be a dimensionless number, such as a proportion of the difference between the inside and outside temperatures.

2.3.8. Suggested building fabric criteria

Similar problems have been faced by thermographers in cold climates. A paper "Building Thermography in Finland" [20] described the practice in Finland and a new criteria based on a "Thermal Index" TI given by

$$TI = \frac{T_{sp} - T_o}{T_i - T_o} \times 100 \qquad (9)$$

where T_{sp} is the inside surface temperature, T_o the outdoor temperature, and T_i the indoor temperature. In addition, they are recommending that a defect should be defined as a condition in which $TI < 75\%$, and conditions in which $TI < 60\%$ should be considered a health hazard. This definition requires determination of air temperature on each side of the structural element, which may be difficult. The Finnish recommendations are in a government publication on health and housing [21]. In England, the surface temperature factor "f" is defined in the same way, but expressed as a decimal fraction [18].

A slight variation on this is sometimes used, based on surface temperatures, which are easier to measure with IR imaging equipment. This avoids the need to precisely determine ambient temperature, allowing the thermographer more time to collect good quality images.

Acceptance criteria can be defined using this approach. Two examples are:

(a) An anomaly is defined as an area in which surface temperature drops exceed 25% of the difference between inside and outside surface temperatures by

$$T_{sp1} - T_{sp2} > 0.25(T_{sp1} - T_{so}) \qquad (10)$$

where T_{sp1} is the maximum temperature on an area of internal surface, T_{sp2} the minimum temperature in the same area, and T_{so} the outside surface temperature of the same section of wall or roof.

(b) The area of anomalies should not exceed 0.1% of the total exposed fabric area.

This approach has been used successfully by many thermographers in the UK since 2002. Some TIC are now supplied with onboard software to show isotherms for Thermal Index, and these can be used to differentiate "defective" areas at the time of the survey, thus avoiding the need to save large numbers of images of doubtful areas for later analysis.

2.3.9. Personal safety

Building surveys must often be done on new or incomplete buildings. Work on building sites can be dangerous. The site manager should be responsible for ensuring the safety of people working on the site, including the thermographer. Risks include, but are not limited to: falling from unprotected platforms or flat roofs, walking into objects, colliding with site vehicles, being hit by falling objects or masonry chippings, and being bitten by vermin. There should be site rules and procedures, which may include wearing of personal protective equipment (PPE) such as high-visibility jackets, hard hats, steel toecap boots, goggles, and gloves. The thermo-grapher should also conduct a risk assessment and work out a method of conducting the survey while eliminating the risk of injury. In many countries these considerations are legal requirements. The thermographer must be aware of legal and site-specific requirements for the area in which they work.

2.3.10. Aerial surveys

Surveying buildings from the air is appealing because large areas can be covered in a very short time. When surveying a single building or a group of buildings, a helicopter can be a useful means of gaining sufficient height above the roof to cover large areas with each image. Typically flying at 100 m with a conventional camera, an area of 33 m × 50 m can be covered by each image. This is a large enough area to allow comparison of several houses or a small retail unit in each image and larger buildings with a small number of images. To make a "thermal map" of a large area, data collection is carried out by flying transects in a grid pattern. An altitude of over 650 m is normal to comply with regulations and to give better than 1-m resolution. This altitude and resolution using wide angle optics is a compromise of image quality and economy. Higher resolution would either require a lower altitude or a narrower FOV which clearly requires more flight lines and airtime. An overlap of each line prevents "holes" occurring in the map due to possible drifting of the aircraft. The thermal imaging equipment used in all our thermal surveys is of military specification and produces high-resolution imagery. The attitude of the scan head is vertical to the ground and fitted in a stabilized mount. All images are time/date stamped during the survey and can be hidden after processing.

An alternative to aircraft-mounted cameras is the mast-mounted camera. These can reach heights of 50 m above ground level and produce images of 25 m × 16 m areas, big enough for domestic and small commercial surveys. Masts are a relatively low-cost solution that is increasingly popular, because they are easily transported in an ordinary family car or small van and they can be raised in a few minutes with hydraulic actuation powered from the car battery. Cameras can be fitted with pan and tilt heads, and the images can be sent by wire or wireless connection to a monitor in the car. Images can be stitched together after the survey to provide a panoramic view of large buildings as seen in Figure 17.

Figure 17 Composite building image from mast.

Figure 18   Thermal image of two houses with thermal anomalies.

2.3.11. Energy surveys

Thermal imaging can be very useful in identifying areas that are causing unnecessary energy use as a result of their anomalous temperature. A whole house image can show areas of heat loss. But prediction of energy consumption from such images is unreliable because of variations in surface resistance, variations in internal temperature, and local heat sources.

The houses shown in Figure 18 have several thermal anomalies. The roofs appear to be different temperatures, but that may be because they are tiled with different materials or one is better insulated or because the roof space of one is heated. Further information is required before a judgment can be made.

2.3.12. Complementary methods

Air pressurization, temperature measurement, and borescope investigations are all used to assist or confirm thermographic results. For more information, refer to BINDT thermography handbooks [8].

2.4. Mechanical plant thermography

Thermography is used on rotating machinery, pipework systems, process plant, tanks, and vessels. Inspection of mechanical plant most often involves rotating equipment such as compressors, motors, gearboxes, couplings, fans, pumps, and conveyors. But, other plant equipment, such as furnaces, ovens, kilns, boilers, hot water storage tanks, coolers, pipes, valves, and steam traps also generates temperature differences that are symptomatic of faults and detectable with thermography.

2.4.1. Principles: Why thermal radiation is important in mechanical systems

Moving machinery generates heat by friction in all types of bearings. Lubrication is widely used to minimize friction, and thermography is used to show failure of lubrication. Heat is also generated in failed or misaligned bearings. Heat is dissipated from these bearings by convection, conduction, and radiation. Any radiative heat transfer is detected by thermography.

Emissivity and the environmental temperature are particularly important in mechanical thermography, because the equipment is often in plant rooms with other equipment at elevated or reduced temperature. The emissivity of the surface on machines varies greatly with new metal surfaces having values of 0.1 or below and dirty, corroded, or painted surfaces 0.9 or above.

2.4.2. Detectability with IR

If appropriate precautions are taken against variations in air and radiance temperature, emissivity, and normal operating temperature, a wide range of faults can be detected. To find normal operating temperatures, it is recommended that a "baseline" survey is conducted on new equipment as soon as it is known to be operating correctly. Prior to this, thermography can also be used in commissioning new equipment to detect anomalies in operation.

IRT does have some limitations in mechanical inspection in that it cannot always establish the cause of excessive heat; however, its value comes in pinpointing the exact location or the source of overheating quickly, safely, and without interrupting plant and processes. Once a fault is identified, other inspection techniques, such as vibration analysis, may be used if the cause is not readily apparent.

2.4.3. Practical examples

A simple example of use in commissioning is shown in Figure 19 where some of the sections of a new radiant panel heater are blocked so that hot water cannot circulate in them. A temperature profile has been superimposed to confirm the temperatures across the panel.

The inspection of refractory or insulating materials is based on the premise that under normal operating conditions a uniform or an even-temperature profile exists within a vessel and that the external surface temperature will be

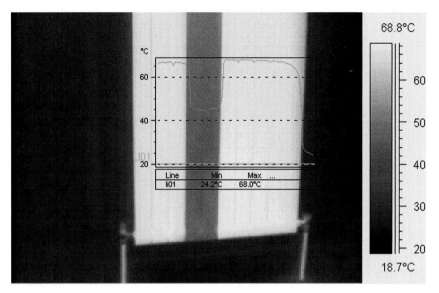

Figure 19 Thermal image of radiant panel heater with some faulty sections and temperature profile superimposed.

a function of the internal temperature. Defects in the refractory lining or moisture-laden areas of insulation will therefore affect the exterior surface temperature uniformity. The usual approach to this kind of inspection would be to thermally map the exterior surfaces and use this as a baseline for subsequent surveys whereby deterioration may be monitored over time through comparison.

Thermography is a very effective method for locating problems in steam systems. Applications include steam trap defects, boiler tube leaks, underground pipe leaks, and insulation defects. Steam leaks are particularly costly because of the high operating pressures involved, such that a release of steam from a relatively small hole or the passing of steam through a defective trap can prove very costly in terms of wasted energy. Thermography provides the means of identifying steam leaks (Figure 20) that would otherwise be undetected due to limitations in conventional inspection methods.

An example of problems with bearings can be seen in Figure 21. The two bearings at either end of a coupling shaft should be at the same temperature, but in this example one is hot, while the other is near to ambient temperature. Further investigation showed a fault in the bearing.

Qualitative thermography is useful in detecting the locations of anomalies, but quantitative thermography helps to rate the severity of anomalies as described in ISO 18434-1 [7].

Figure 20   Steam leak that could not be seen, found with thermography.

Figure 21   Two bearings on a coupling shaft at very different temperatures.

2.4.4. Mechanical plant acceptance criteria

Setting limits for mechanical plant can be much more complicated than for electrical equipment. BSRIA Guidance [22] quotes a range of British and European standards and application guides that give temperature limits or guidance on how to establish temperature limits. ISO 18434-1 [7] gives a table of actions related to temperature rises above established reference temperatures, but states that this is only an example that should be adapted as necessary. Recommendations can be compared in Table 4 to show how different thermographers might interpret the same temperatures in different ways.

Table 4 Examples of mechanical plant temperature criteria.

Degrees above reference	5	10	15	20	25	30	35	40	45	50	55	60
Holst	Normal	Slight	Serious overheating				Excessive overheating				Acute overheating	
ISO18434	Advisory		Intermediate		Serious				Critical			

Holst [1] produced a table for mechanical plant comparable with that published by NETA for electrical equipment. Table 4 is further adapted from this to more closely follow the NETA table. Arguably the temperature rise should refer to the normal reference temperature rather than the ambient, because many mechanical components, such as thrust bearings, operate normally with a temperature rise above ambient due to friction.

2.4.5. Personal safety and business risk considerations

Prior to the commencement of work, minimum safety rules and guidelines should be established in accordance with applicable site rules, local or national standards and regulations and particularly where hazardous environments may exist. An example set of rules is given in ISO18434-1 [7].

These include appropriate qualifications, a site safety induction, use of appropriate PPE, job hazard assessment (JHA), a standby person present with site and plant understanding, and accident, incident, and injury procedures. Unless the necessary qualifications/licenses are held, the thermographer will not perform any plant intervention tasks that are normally done by qualified personnel.

There is a risk of plant failure if thermography is not carried out to identify incipient failures, but there is also a business risk if in doing the thermographic survey some action is taken that triggers a failure. The thermographer and the plant operator/manager should consider the risks on both sides before undertaking the thermographic survey. Often thermographic surveys can be used in place of another condition monitoring or maintenance procedure that involves more plant intervention and thus a greater risk of provoking failures.

2.4.6. Associated assessment methods and calibration

Thermography can usefully be applied as the first stage of a condition monitoring process because so many faults result in thermal changes to the plant or equipment. Thermography may not be sufficient to diagnose the cause of an anomaly such as that shown in Figure 21, but it directs the attention of the maintenance manager to areas where further investigation is required. In this case the cause of overheating could be diagnosed by vibration analysis. In other cases acoustic emission, oil analysis, or wear debris analysis could be used. A trained and experienced thermographer will know what complementary method is appropriate. Training in accordance with ISO18436 Part 7 [23] provides suitable knowledge to thermographers.

Thermographers shall have their TIC calibrated to original equipment manufacturers' guidelines, or established industry practice. Documented calibration checks should be carried out using a traceable blackbody reference in accordance with manufacturer's recommendations, client specifications, or any applicable industry standards. Quick calibration checks should be performed prior to each inspection or survey.

3. THERMAL IMAGING IN FIREFIGHTING

Many fire departments currently employ TIC for firefighting applications. TIC are used for locating victims and other firefighters, determining safe paths through a burning building, and finding possible reignition points after a fire has been extinguished.

The performance of TIC in firefighting applications is often integral to the protection of both lives and property. To aid in first-responder operations, TIC are designed to operate in very harsh conditions, provide users with a wide FOV, and are capable of adjusting signal output to more clearly portray the thermal scene with minimal input from the user. The performance needs of the user are designated by the intended use. For example, a TIC used to determine a fire source in a building may not require the same resolution or sensitivity as a TIC used to locate unconscious victims in a smoky environment. In general, the harsh and varied environments in which TIC are used for emergency response imparts challenges not present in other more traditional applications of thermal imaging.

3.1. Thermal environment

The environment encountered by firefighters varies, depending on the nature of the fire scenario. Decades of fire testing have shown that the gas temperature in a naturally ventilated burning room stratifies, due to buoyancy, into a hot upper layer that contains combustion byproducts and a cooler lower layer composed mainly of ambient air. The time-varying severity of conditions in the room of fire origin and adjacent or nearby rooms will change depending on the type and amount of materials burning, thermal properties of the room surfaces, the ventilation conditions, the size of the room, and a number of other factors [24]. As any firefighter will gladly explain, no two fires are exactly the same.

The nature of the conditions under which TIC are used has in large part influenced the development of imaging technology for first-responder applications. For example, a firefighter may encounter high temperatures, open flames, pools and sprays of water, and thick smoke; therefore, it is important that TIC are capable of seeing in these obstructive conditions with a minimum amount of interference from the surrounding

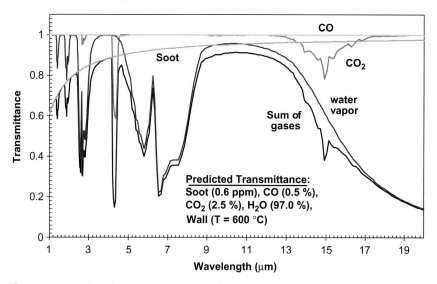

Figure 22   Predicted transmittance through a 12.7-cm (5 in.) measurement pathlength for a simulated combustion atmosphere. The soot, gases, and wall temperatures are at 600°C.

environment. A high-transmittance band approximately in the $8-14\,\mu m$ spectral range is evident in Figure 22, reinforcing the reason why firefighting TIC are designed to operate in this area of the IR spectrum.

Radiance as a function of wavelength is shown in Figure 23. Reflected thermal radiation coming from hot surfaces and flames may interfere with TIC imaging performance. The peak in radiance is much shorter than the spectral range of the TIC. However, it may still interfere with the TIC's ability to see a target in some conditions. The software program RADCAL [25] was used to perform these transmittance and radiance simulations.

The TIC detects thermal radiance from solid surfaces and from gases that radiate in the $8-14\,\mu m$ spectral range. Emissivity affects the radiation in a way that can make the surface or gas appear to have a temperature that is different than it actually is. In general, surfaces that are flat black in color and somewhat rough in texture tend to have high emissivities, and surfaces that are shiny and smooth tend to have low emissivities. Most firefighting TIC are designed to use a constant emissivity value of 0.95 in their internal algorithms used to convert the radiant energy signal to a temperature value. The further away an object's emissivity is from 0.95, the less accurate that object's surface temperature will be. The term "apparent (or radiance) temperature" is used to account for temperature deviations caused by differences in emissivity. TIC for firefighting applications operate between 8 to $14\,\mu m$ because radiant emission intensities for radiating objects is large

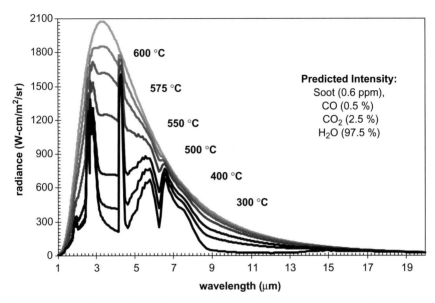

Figure 23 Predicted radiance through a 12.7-cm (5 in.) measurement pathlength for a simulated combustion atmosphere consisting of 0.6-ppm soot, 0.5% CO, 2.5% CO_2, and 97.0% H_2O. The media temperature is held at 600°C and the wall temperature varies from 0°C to 600°C.

enough to detect in this region *and* there is minimal overlap with the spectra of the most common combustion species with the exception of smoke.

The impact of an intervening smoke aerosol on a TIC's view depends on the smoke temperature and concentration as compared to the temperature and emissivity of a target. Smoke may absorb and scatter, but will not emit a significant amount of radiation if its temperature is relatively cool. In a typical compartment fire, smoke may obscure human visibility, but in the IR, attenuation by smoke decreases with wavelength, allowing TIC to "see" through smoke [26].

Chemical compounds that give rise to a dipole moment absorb and emit IR radiation at unique and characteristic frequencies. Band absorption by water vapor is seen at wavenumbers between about 1,300 and 2,000 cm^{-1} (7.7 and 5.0 μm). Figure 24 shows recognizable emission signatures for individual species but not in the spectral region in which TIC are sensitive, reinforcing the reason firefighting TIC are designed to operate in this portion of the IR spectrum. Little band radiation was observed in the portion of the spectrum in which the TIC are sensitive, thus indicating that fuel type is not a dominant parameter when testing TIC imaging performance for typical smoky fuels.

Information regarding the establishment of conditions to test the performance of TIC has been assembled with the objective of developing

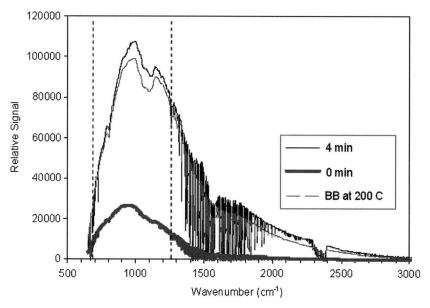

Figure 24 An FTIR was used to measure the relative radiance as a function of wavenumber during the burning of a wood crib over carpet and padding. Results are shown in 0 and 4 min after ignition. Measurement of the radiance of a blackbody source at 200°C (with a standard uncertainty of ±5°C) is also shown.

test conditions that are relevant and representative of fire service operations. Input from users, as well as existing standards, research reports, and test measurements were considered in the development of representative TIC performance testing conditions. The concept of thermal class as a means of categorizing the activities in which TIC are used by firefighters is helpful in defining meaningful bounds on TIC test conditions. The thermal classes link the various requirements of existing standards on related fire service equipment with experimental results from full-scale testing and the scientific literature.

The type of fuel used in full-scale fire experiments was found not to be an important parameter with respect to TIC imaging performance. Radiation emitted from very smoky, poorly characterized fuels, such as a burning wood crib with carpet and padding, does not significantly impact the spectral region employed by fire service TIC.

3.2. Thermal imaging operations in the fire service

Before 2004 in the US, thermal imaging needs for firefighters were not well understood or collected in one document. To assess the thermal imaging needs and activities of firefighters and to facilitate gathering this information, a Workshop on Thermal Imaging Research Needs for First Responders was held at NIST in December 2004. Ties were established between industry, the first-responder community, and NIST that have, over time, helped shape the Draft National Fire Protection Agency (NFPA) 1801 standard [27] and accelerated the somewhat protracted standard development process.

The Workshop on Thermal Imaging Research Needs for First Responders [28] was held to better understand the needs of users, manufacturers, government agencies, and other proponents of TIC technology and to identify barriers that impede advances in the application of thermal imaging technology to emergency response. The issues that were found to be most important to the attendees of the workshop are shown in Figure 25.

Before operating conditions are discussed, it is instructive to list the ways in which TIC are currently being used in the field. This information is provided in Table 5. From this, one can estimate the thermal and spectral environments that impact the functionality and imaging performance of TIC. Table 5 is not a comprehensive list. It was compiled from input from a variety of users and is changing and expanding as the fire service community increasingly accepts TIC technology and discovers the value

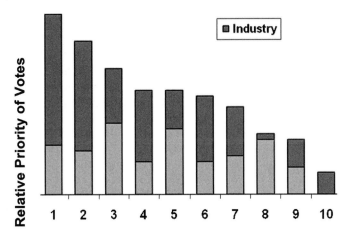

Figure 25 Thermal imaging research needs, as indicated by participants of the Workshop on Thermal Imaging Needs for First Responders in December 2004. The key to the abscissa is: 1. Image quality (research, metrics, standards, contrast, sensitivity); 2. Durability (metric, mean time between failures, ruggedness standard test methods); 3. Training and certification for users; 4. Establishment of minimum/typical test environments; 5. Human factor/ergonomic and human dynamics research; 6. Image display (technology, viewability, metrics); 7. Battery and charger issues (life/maintenance, icons, and self-test improvements); 8. Reduction in TIC cost ("water bottle" sized package priced at $2,000); 9. Standard target (for field test/calibration and emissivity target for turnout gear); and 10. TIC self-test procedure and warning system.

Table 5 Fire service TIC operations.

Activity	Description
Size up	Assess hazard, find fire/heat sources, escape points, and vents
Communication	Lead or direct searches, interface with incident command, account for team members
Search	Locate victims, other firefighters, fugitives, missing persons in dark and/or smoky environments
Tactics	Direct hose stream, check upper layer temperature, check for changing conditions, detect obstacles, passageways, damaged structural members, judge distances, use in rapid intervention teams
Overhaul	Ensure fire is out, look for hidden smoldering and hot spots
Forensics	Identify source of fire, determine fire spread, record video during fire for later use as evidence
Wildland fires	Ground- and air-based search for hot spots and personnel
Hazmat	Determine material levels inside containers, track material movement and spill spread limits
Other	Preventative building maintenance, emergency medical applications, motor vehicle accident investigations

of these tools through successful operations in which TIC played a critical role. Clearly, not all firefighters use TIC for every activity listed in Table 5. Each fire service organization has its own set of operational procedures and needs.

3.3. Design guidelines

Firefighters may rely in part on a TIC to negotiate their way through a burning structure; therefore, most TIC employ a wide FOV in the range of $40-60°$. There are few cases in which a firefighter would need to focus on an object less than 1 m away, which encourages the use of relatively robust and lower cost fixed focus optics that focus from 1 m to infinity. While temperature measurements provide useful information in a fire situation, the accuracy of the measurement is not as critical as it might be in other thermal imaging applications.

The urgency of a fire event also dictates the need to keep the operation of the TIC as simple as possible. Most TIC feature fully automated gain, focus, and iris settings, rarely offering more controls than a large on/off button that can easily be accessed by a firefighter wearing heavy gloves. In some cases the TIC may have one or two added functions, such as a zoom button or a toggle between IR and visible viewing.

There are currently three well-established detector/sensor technologies available for purchase: BST (barium strontium titanate), VOx, and ASi. Manufacturers may utilize a single detector array technology or may choose to offer a selection of TIC models possessing different technologies. While most detectors used for first-responder applications are FPAs, utilizing an array of sensors located at the focal plane of the optics, each specific detector technology is capable of generating different levels of information in the displayed image. This is evident in Figure 26, where TIC representing the three different detector types are viewing an identical thermal scene; all three TIC operate within the same wavelength range. The differences seen in this figure are the result of not only different detector technologies, but also different optical and electronic systems, which can contribute significantly to overall image quality.

The optics package provides an interface between the signal processor and the recorded image and makes an important contribution to the overall performance of the instrument. This fact should be considered when viewing Figure 26. There are as many ways of designing optics and electronics packages as there are TIC form factors and target market prices.

The two typical FPA sensor configurations for fire service applications are 320 rows × 240 columns and 160 rows × 120 columns. The BST and VOx detectors are based on older technology, developed during the 1980s. The BST detectors are solid-state ceramic devices that convert changes in electrical polarization to voltage differences. A thermoelectric cooler provides thermal stability. These are AC-coupled detectors that measure relative levels of IR radiation. Thus, the detector output requires a correction based on reference points provided by a chopper. The VOx and ASi detectors are both microbolometer devices that measure changes in the electrical resistance of the sensor material due to heat from IR radiation. Microbolometers do not require cooling but are relatively sensitive to detector temperature. The output from these DC-coupled detectors is a function of the absolute level of incident IR radiation. A shutter is used to periodically reset the detector's absolute radiation input level and prevent drift among the detector pixels [6]. Initially VOx was used as the sensor material; however, ASi is now also emerging as a viable sensor material.

There are two basic ways in which operating conditions affect TIC: first, the TIC itself must be rugged enough to function in elevated temperatures and humidity, and other adverse conditions; and second, the TIC must be capable of producing images that provide useful information to the user. Functionality tests address the ability of the TIC to operate in a harsh environment. In 1999 the U.S. Navy issued a report that describes a series of tests conducted to determine which of eight commercially available TIC were sturdy enough to function in the harsh environment of a fire onboard a ship [29]. The navy chose 27 criteria ranging from battery life to corrosion resistance by which to judge the performance of the TIC.

Figure 26 These three TIC are viewing the identical thermal scene: a long corridor with a heated mannequin on the floor, and reflective and heated targets mounted on the wall at the end. A fire room is located adjacent to the corridor on the right.

Functionality tests that address many of the same operating conditions have also been incorporated into NFPA 1801. A sampling of these tests is discussed in Section 3.3.1.

Image quality tests address the ability of the TIC to capture an IR scene with sufficient sensitivity and detail to enable the user to perform a particular activity, such as searching for a fire victim. A general overview of these tests is provided in Section 3.3.2.

3.3.1. Design robustness

The nature of the fire service demands that firefighter equipment be rugged. Firefighters may need to use any tool close at hand to help them quickly perform an operation such as venting a room or opening an enclosure. The TIC must perform reliably in high temperatures, flames, low temperatures, in the presence of water, corrosive substances, and electrical interference. The TIC must not develop leaks when exposed to vibrations, drops, and fire hose spray. The TIC must not ignite explosive gas or shock the user. The TIC housing must be fire-hardened, but must not cause the TIC to overheat when operated for extended lengths of time.

3.3.2. Image quality

Image quality is important in its own right but also has implications for TIC design robustness. It is necessary to consider design robustness because image quality may be used as the pass/fail criteria by standards organizations for many design robustness tests. For example, an environmental temperature stress test may require the TIC to be exposed to an elevated temperature of 60°C for 4h, be exposed to a low temperature of −20°C for 4h, and then experience a thermal shock (changing from −20°C to 60°C within 5 min). Immediately after each of these three procedures, the thermal image produced by the TIC may be evaluated to determine whether the procedure had an adverse effect on image quality.

There are four image quality performance metrics identified by the NFPA as being of interest to fire service TIC. They are spatial resolution, effective temperature range (ETR), image nonuniformity, and thermal sensitivity. The tests used to evaluate fire service TIC are described in detail in the following sections.

3.3.2.1. Spatial resolution.

Spatial resolution is a measure of the amount of detail that can be reproduced by a TIC. A firefighter may observe a hotspot, noticeable by a difference in contrast, but not realize that the source is a warm but harmless household item (e.g., a computer, coffee pot, power strip, lamp, etc.). Without enough detail in the image to identify the item of interest, the firefighter may lose valuable time inspecting everything that looks relatively warm. A detailed discussion of the development of spatial resolution as it applies to fire service TIC can be found in Ref. [30]. There are many ways to measure spatial resolution, most of which involve constructing a modulation transfer function (MTF) curve from some sort of IR target having bars, a point, a line, or a slanted edge. A Fourier transform is usually used to convert the blur between relatively light and dark parts of an image to frequencies, from which the MTF curve is derived. For this and all image quality evaluations for fire service TIC, it is recommended that all image processing be performed on images that are captured from the TIC display with a well-characterized high-resolution visible spectrum digital camera.

3.3.2.2. Effective temperature range. The ETR test is intended to measure the range of temperatures at which the TIC provides a useful image to the firefighter. Useful means here that the image quality is adequate for the user to take appropriate action. It is not unusual for a firefighter to use a TIC to search for objects having intermediate surface temperatures, for example, a human victim, another firefighter, or an egress route, when flames are in the FOV. In this situation, the presence of flames or other extreme conditions may cause the TIC to "wash out" or lose contrast in the object of interest. The ETR test measures the temperature range within which objects of intermediate temperature retain enough contrast to be discernable to the user, that is, the temperature range of a useful image. A detailed discussion of the development of ETR as it applies to fire service TIC can be found in Ref. [31].

3.3.2.3. Nonuniformity. Nonuniformity is a measure of the variation of the TIC's response to a uniform test target. There are many ways that the TIC response may be nonuniform: there can be gradients in pixel intensity across the image, various patterns such as narcissus may appear, some degree of broadband noise may be present, blemishes and blotches may be seen, and combinations of these nonuniformities can occur. In Figure 27, images are presented from the same TIC, viewing a uniform target at four temperatures. As can be seen, as the test target temperature increased, nonuniformities in the images became generally more apparent. This TIC shifted to a less thermally sensitive mode between the 100°C and 200°C images, which can reduce the amount of nonuniformity present in images. A detailed discussion of the development of nonuniformity as it applies to fire service TIC can be found in Ref. [32].

3.3.2.4. Thermal sensitivity. Thermal sensitivity is a measure of the smallest temperature difference that the TIC is capable of resolving.

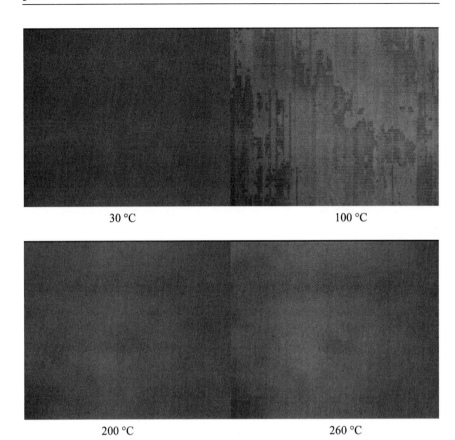

30 °C   100 °C

200 °C   260 °C

Figure 27 Variation in nonuniformity of a TIC with increasing temperature. As the test target temperature increased, nonuniformities became more apparent. The TIC shifted to a less thermally sensitive mode between the 100°C and 200°C images.

The sensitivity of most TIC to temperature differences is a function of the nominal target temperature and generally requires a differential blackbody that can attain differential target temperatures of the order of 5 mK. One way to measure thermal sensitivity is to use a trained observer to indicate when he/she perceives a target shape as the temperature difference increases in an IR target. This method is dependent on subjective human responses. Another approach is to measure pixel intensity changes on the TIC display as the temperature difference increases in an IR target.

For many thermal imaging applications the NETD test is used to determine the thermal sensitivity of the TIC; however, the NETD does not include contributions to image degradation due to the quality of the display. For this reason, the NETD test is not recommended as a method of evaluating image quality for fire service TIC.

3.3.3. Quality assurance/quality control

It is good practice for TIC manufacturers to follow a rigorous quality assurance and quality control process. One such process, of many available processes, is the failure mode and effects analysis (FMEA). With this procedure, the manufacturer identifies and prioritizes critical failures that could have a serious effect on the safety and reliability of a TIC in the anticipated operating environment. The potential failure modes and their effects on the performance of the TIC are described, considering the magnitude of the risk and the probability that it will occur. Steps are then taken to correct the TIC design to reduce the risk and/or the probability to an acceptable level.

4. Standards

There are several standards development organizations that write standards on electronic equipment and other PPE used by the fire service. Standard test methods do exist for TIC used in nonemergency service applications, but there is currently no standard specifically written for TIC used by the fire service.

The following is a brief discussion of some details of existing standards that may have bearing on TIC operating conditions for functional and image quality tests. American National Standards Institute (ANSI) standard ANSI/ISA-12.1201 covers intrinsic safety testing for nonincendive devices [33]. Several ASTM standards are functional tests for exposure to salt spray [34], transmittance of lens [35], flame resistance [36], and liquid penetration [37]. There are also several ASTM test methods for TIC image quality [38–40]; however, these test methods do not include extreme operating environments in their tests. Pertinent publications from the International Standards Organization (ISO) include enclosure integrity [41], electromagnetic compatibility [42], functional safety [43], and resolution measurements for still cameras [44]. The military standard MIL-STD 810F tests resistance to mechanical acceleration [45]. Underwriters Laboratory (UL) produces a large number of standard test methods for many aspects of design for electrical and electronic devices, such as proper grounding and wiring for switches and terminal blocks. The Video Electronics Standards Association (VESA) writes standard test methods for measuring display performance [46]. In addition to the above standards, which are not specifically written for fire service equipment, NFPA produces standards on communications systems [47], protective clothing for structural firefighting [48], protective clothing for wildland firefighting [49], self-contained breathing apparatus [50], and personal alert safety systems [51], which are intended for equipment that might be exposed to thermal environments similar to those experienced by TIC. Note that there are only two NFPA standards, on self-contained breathing apparatus and on personal alert safety systems, which are written specifically for electronic devices used in burning structures. Firefighters sometimes use electronic gas detectors, but these devices have no formal standards for operation or functionality, although some adhere to the ANSI/ISO standard on nonincendive devices. Currently, some TIC manufacturers voluntarily adhere to some of the standards discussed here, but are not required to do so.

5. SUMMARY

This chapter is intended to give the reader a sense of the ways in which IR technology is used in thermography and fire service applications. The relative importance of the various aspects of TIC design and performance is highly dependent on the operating environment and the needs of the user.

ACKNOWLEDGMENTS

The authors gratefully acknowledge the Advanced Technology Program at NIST and the Department of Homeland Security/United States Fire Administration. They would also like to thank the participants of the Workshop on Thermal Imaging Research Needs for First Responders for their generous advice, Dr. Marc Nyden for his assistance with the FTIR measurements, and Dr. Sung Chan Kim, Dr. Yong Shik Han, Roy McLane, and Jay McElroy for their contributions during the full-scale tests. The assistance of the capable crew at the NIST Large Fire Laboratory is very much appreciated as well. The authors would like to thank Ray Faulkner, Andrzej Nowicki, and Norman Walker in the United Kingdom for permission to use some of their thermal images.

REFERENCES

[1] G. C. Holst, *Common Sense Approach to Thermal Imaging*, SPIE Press and JCD Publishing, Winter Park, FL (2000).
[2] P. Muralt, "Micromachined infrared detectors based on pyroelectric thin films," Rep. Prog. Phys. **64**, 1339–1388 (2001).
[3] C. M. Hanson, "Hybrid Pyroelectric–Ferroelectric Bolometer Arrays," in *Uncooled Infrared Imaging Arrays and Systems*, edited by P. W. Kruse and D. D. Skatrud, Academic Press, San Diego, CA, pp. 123–174, Chapter 4 (1997).
[4] D. L. Polla and J. R. Choi, "Monolithic Pyroelectric Bolometer Arrays," in *Uncooled Infrared Imaging Arrays and Systems*, edited by P. W. Kruse and D. D. Skatrud, Academic Press, San Diego, CA, pp. 175–202, Chapter 5 (1997).
[5] C. F. Tsai and M. S. Young, "Pyroelectric Infrared Sensor-Based Thermometer for Monitoring Indoor Objects," Rev. Sci. Instrum. **74**, 5267–5273 (2003).

[6]   P. W. Kruse, "Principles of Uncooled Infrared Focal Plane Arrays," in *Uncooled Infrared Imaging Arrays and Systems*, edited by P. W. Kruse and D. D. Skatrud, Academic Press, San Diego, CA, pp. 17−44, Chapter 2 (1997).

[7]   International Organization for Standardization, "Condition Monitoring and Diagnostics of Machines − Thermography − Part 1: General procedures," ISO 18434-1 (2008).

[8]   N. Walker (ed.), *Infrared Thermography Handbook − Volume 1. Principles and Practice*, The British Institute of Non-Destructive Testing (BINDT), Northampton (2005); A. N. Nowicki, *Infrared Thermography Handbook − Volume 2. Applications*, The British Institute of Non-Destructive Testing (BINDT), Northampton, (2005).

[9]   InterNational Electrical Testing Association, "Standard for Maintenance Testing Specifications for Electrical Power Distribution Equipment and Systems," NETA MTS-2003 (2003), "Acceptance Testing Specification for Electrical Power Distribution Equipment and Systems," NETA ATS-1999, Portage, MI (1999).

[10] J. Nicholas, Thermal Solutions '97, Cleveland, OH (1997).

[11] National Fire Protection Association, "Standard for Electrical Safety in the Workplace," NFPA 70E, Quincy, MA (2009).

[12] I. Paljak and B. Pettersson, "Thermography of Buildings (Termografering av byggnader)," Swedish Council for Building Research (1972).

[13] International Organization for Standardization, "Thermal Insulation − Qualitative Detection of Thermal Irregularities in Building Envelopes − Infrared Method," ISO 6781 (1983).

[14] Energy Information Administration, "International Energy Outlook 2007," DOE/EIA-0484 (2007).

[15] International Organization for Standardization, "Building Components and Building Elements − Thermal Resistance and Thermal Transmittance − Calculation Method," ISO 6946:2007 (2007).

[16] G. A. Chown and K. N. Burn, "Thermographic Identification of Building Enclosure Effects and Deficiencies," National Research Council of Canada, CBD-229 (1983).

[17] American Society for Testing and Materials, "Standard Practice for Thermographic Inspection of Insulation Installations in Envelope Cavities of Frame Buildings," Philadelphia, PA, ASTM C1060-90 (1997).

[18] Building Regulations for England and Wales, Approved Document Parts L1 and L2: Conservation of Fuel & Power. DTLR (2001)

[19] C. Pearson, "Thermal Imaging of Building Fabric − A Best Practice Guide for Continuous Insulation," BSRIA, TN 9/2002 (2002).

[20] T. Kauppinen, "Building Thermography in Finland," in *Thermosense XXV*, edited by K. E. Cramer and X. P. Maldague, SPIE **5073** (2003).

[21] Finland Ministry of Social Affairs and Health, "Asumisterveysohje: Asuntojen ja muiden oleskelutilojen fysikaaliset, kemialliset ja mikrobiologiset tekijät," "Instructions regarding physical, chemical and biological factors in housing," (2003).

[22] N. Barnard and C. C. Pearson, "Guidance to the Standard Specification for Thermal Imaging of Electrical Building Services Installations," FMS 5/1999, BSRIA (1999).

[23] International Organization for Standardization, "Condition Monitoring and Diagnostics of Machines — Requirements for Qualification of Personnel — Part 7: Thermography," ISO 18436-7 (2008).

[24] M. Bundy, A. Hamins, E. L. Johnsson, S. C. Kim, G. H. Ko, and D. B. Lenhert, "Measurements of Heat and Combustion Products in Reduced Scale Ventilation Limited Compartment Fires," NIST Technical Note TN 1483 (2007).

[25] W. L. Grosshandler, "RADCAL: A Narrow-Band Model for Radiation Calculations in a Combustion Environment," NIST TN1402 (1993).

[26]   R. Siegel and J. Howell, *Thermal Radiation Heat Transfer*, Taylor and Francis, New York, 4th ed. (2002).

[27]   National Fire Protection Association, "Standard on Thermal Imagers for the Fire Service," NFPA 1801 (2009).

[28]   F. K. Amon, N. P. Bryner, and A. Hamins, "Thermal Imaging Research Needs for First Responders: Workshop Proceedings," http://fire.nist.gov/bfrlpubs/fire05/PDF/f05036.pdf, NIST SP1040 (2005).

[29]   F. Crowson, B. Gagnon, and S. Cockerham, "Evaluation of Commercially Available Thermal Imaging Cameras for Navy Shipboard Firefighting," Department of the Navy Report No. SS99-001 (1999).

[30]   A. Lock and F. Amon, "Application of Spatial Frequency Response as a Criterion for Evaluating Thermal Imaging Camera Performance," in *Proceedings of SPIE — Infrared Imaging Systems: Design, Analysis, Modeling, and Testing XIX*, edited by G. C. Holst, SPIE **6941**, Orlando, FL (2008).

[31]   F. Amon, A. Lock, and N. P. Bryner, "Measurement of Effective Temperature Range of Fire Service Thermal Imaging Cameras," in *Proceedings of SPIE — Infrared Imaging Systems: Design, Analysis, Modeling, and Testing XIX*, edited by G. C. Holst, SPIE **6941**, Orlando, FL (2008).

[32]   A. Lock and F. Amon, "Measurement of the Nonuniformity of First Responder Thermal Imaging Cameras," in *Proceedings of SPIE — Infrared Imaging Systems: Design, Analysis, Modeling, and Testing XIX*, edited by G. C. Holst, SPIE **6941**, Orlando, FL (2008).

[33]   Instrument Society of America, "Nonincendive Electrical Equipment for Use in Class I and II, Division 2 and Class III, Divisions 1 and 2 Hazardous (Classified) Locations," ANSI/ISA ANSI/ISA-12.12.01-2007 (2007).

[34]   ASTM International, "Standard Practice for Operating Salt Spray (Fog) Apparatus," West Conshohocken, PA, ASTM B117-07a (2007).

[35]   ASTM International, "Standard Test Method for Haze and Luminous Transmittance of Transparent Plastics," West Conshohoken, PA, ASTM D1003-07e1 (2000).

[36]   ASTM International, "Standard Test Method for Flame Resistance of Textiles (Vertical Test)," West Conshohocken, PA, ASTM D6413-08 (1999).

[37]   ASTM International, "Standard Test Method for Liquid Penetration Resistance of Protective Clothing or Protective Ensembles Under a Shower Spray While on a Mannequin," West Conshohocken, PA, ASTM F1359-07 (2004).

[38]   ASTM International, "Standard Test Methods for Measuring and Compensating for Reflected Temperature Using Infrared Imaging Radiometers," West Conshohocken, PA, ASTM E1862-97e1 (2002).

[39]   ASTM International, "Standard Test Methods for Measuring and Compensating for Transmittance of an Attenuating Medium Using Infrared Imaging Radiometers," West Conshohocken, PA, ASTM E1897-97e1 (2002).

[40]   ASTM International, "Standard Test Methods for Radiation Thermometers (Single Waveband Type)," West Conshohocken, PA, ASTM E1256-95 (2007).

[41]   International Organization for Standardization, "Degrees of Protection Provided by Enclosures (IP Code)," Geneva, Switzerland, ANSI/IEC 60529-2004 (2004).

[42]   International Organization for Standardization, "Electromagnetic compatibility (EMC)-Part 3-2: Limits-Limits for harmonic current emissions (equipment input current ≤ 16 A per phase)," Geneva, Switzerland, IEC 61000-3-2 Ed. 3.2 b:2009 (2005).

[43]   International Organization for Standardization, "Functional Safety of Electrical/Programmable Electronic Safety-Related Systems," Geneva, Switzerland, IEC 61508 (2005).

[44] International Organization for Standardization, "Photography-Electronic Still-Picture Cameras-Resolution Measurements," ISO 12233 (2000).

[45] Wright-Patterson Air Force Base, "Environmental Test Methods and Engineering Guidelines," Wright-Patterson Air Force Base, OH, MIL MIL-STD-810E. Method.513.4 (1989).

[46] Video Electronics Standards Association, "Flat Panel Display Measurements (FPDM2)," Milpitas, CA, VESA VESA-2005-5 (2005).

[47] NFPA 1221, "Standard for the Installation, Maintenance, and Use of Emergency Services Communications Systems" National Fire Protection Association, Quincy, MA (2007).

[48] National Fire Protection Association, "Standard on Protective Ensembles for Structural Fire Fighting and Proximity Fire Fighting," Quincy, MA, NFPA 1971 (2007).

[49] National Fire Protection Association, "Standard on Protective Clothing and Equipment for Wildland Fire Fighting," Quincy, MA, NFPA 1977 (1998).

[50] National Fire Protection Association, "Standard on Open-Circuit Self-Contained Breathing Apparatus (SCBA) for Emergency Services," Quincy, MA, NFPA 1981 (2007).

[51] National Fire Protection Association, "Standard on Personal Alert Safety Systems (PASS)," Quincy, MA, NFPA 1982 (2007).

CHAPTER 6

REMOTE SENSING OF THE EARTH'S SURFACE TEMPERATURE

Peter J. Minnett[1] *and* Ian J. Barton[2]

Contents

1. Introduction   334
2. Statement of the Problem   337
   2.1. Land surface temperature   338
   2.2. Sea surface temperature   339
   2.3. Surface renewal theories   341
   2.4. Effects of breaking waves   342
   2.5. Diurnal thermocline effects   342
3. Remote Sensing of Surface Temperature   345
   3.1. Infrared atmospheric correction algorithms and SST   347
   3.2. Remote sensing of land surface temperature   351
   3.3. Ice and snow surface temperature   353
   3.4. Microwave measurements   353
4. Spacecraft Radiometers   354
   4.1. The advanced very high resolution radiometer   355
   4.2. The moderate-resolution imaging spectroradiometer   356
   4.3. The advanced along-track scanning radiometer   357
   4.4. Geostationary meteorological satellites   359
   4.5. Advanced microwave scanning radiometer for the earth
         observing system   359
5. Validation of Surface Temperature Retrievals   361
   5.1. SST validation using radiometers   362
   5.2. SST validation using buoys   365
   5.3. Traceability to temperature standards   367
6. Residual Uncertainties   367
   6.1. AVHRR   368
   6.2. MODIS   369
   6.3. AATSR   371

[1] Meteorology and Physical Oceanography, Rosenstiel School of Marine and Atmospheric Science, University of Miami, 4600 Rickenbacker Causeway, Miami, FL 33149-1098, USA
[2] Marine and Atmospheric Research, Commonwealth Scientific and Industrial Research Organisation, Hobart, Tasmania 7001, Australia

Experimental Methods in the Physical Sciences, Volume 43
ISSN 1079-4042, DOI 10.1016/S1079-4042(09)04306-9

    6.4. Geostationary meteorological satellites   372
    6.5. AMSR-E   372
    6.6. Land surface temperature   373
  7. Applications of Remotely Sensed Surface Temperatures   374
    7.1. Ocean surface fronts, eddies and currents   374
    7.2. Numerical weather prediction   377
    7.3. Severe storm development   377
    7.4. El Niño—Southern Oscillation   378
    7.5. Detection of climate-change signals   378
    7.6. Coral bleaching   379
    7.7. Land surface temperature   379
    7.8. Fire detection   380
  8. Atmospheric Profiles   380
  9. Future Missions   381
    9.1. NPOESS VIIRS   381
    9.2. GMES Sentinel-3 SLSTR   382
    9.3. GCOM-C SGLI   382
  10. Conclusions and Outlook   383
  References   384

1. INTRODUCTION

The surface temperature of the earth is a fundamental and integral parameter within the larger system of the global climate. It is a controlling variable in the exchange of heat, moisture and gases between the surface and the atmosphere. Patterns of sea surface temperature (SST) reveal subsurface oceanic variability, and long-term evolution of the global, regional and seasonal averages of SST are potential indicators of climate change.

For nearly three decades, scanning radiometers on earth observation satellites have provided a time series of consistent measurements of global surface temperatures. Infrared radiometers have been the mainstay of this, although they are now complemented by imaging microwave radiometers. In this chapter, we describe the physical basis of the measurements and the approaches taken to correct for the effects of the intervening atmosphere as this is the major factor determining the accuracy of the retrieved temperatures. How the solutions to the measurement problem influence the technical characteristics of the instruments is discussed, and the techniques used to validate the surface temperatures and thus to determine the residual errors are presented.

The bulk of the sun's energy that is absorbed by the earth's surface arrives at visible and near-infrared wavelengths and the absorbed radiative energy then heats the irradiated surfaces. However, these surfaces also re-radiate infrared energy back to the atmosphere and deep space causing cooling of the surface. It is the balance between the incoming solar (shortwave) radiation and the outgoing infrared (longwave) radiation that determines the equilibrium temperature of the earth—atmosphere system.

Most of the solar energy reaching the water surfaces of the earth is absorbed in the upper layers. Some of this energy is released locally, within the course of the following night, but some heat is retained for longer periods and is moved around the planet by the oceanic surface currents. Subsequent heat release to the atmosphere helps determine the weather and climate patterns around the globe. The seasonal cycle of upper ocean heat content is the integral of the residuals of diurnal heating and cooling while the interannual variability of SST is the manifestation of the imbalance of the heating and cooling cycles from year to year. The perceived warming of the upper ocean in response to the complexities of a changing climate is the long-term integral of the discrepancies of heating and cooling of which the diurnal cycle is the fundamental component.

In contrast to the oceans, solar energy reaching land surfaces heats the surface only and subsurface layers are then heated through heat conduction. Land surfaces can thus show extreme variability with black (highly absorbing) surfaces reaching temperatures well in excess of 60°C during the day, but near freezing during the night especially under clear-sky conditions when the outgoing infrared radiation to space is significant. Land surfaces are usually non-homogeneous with different surfaces having different albedos (reflectance) and thus absorbing a range of solar energy. Surface heating is also dependent on the local topography, shadowing and vegetation. Remote sensing of land surface temperature (LST) using infrared radiation necessarily gives an average surface temperature of the scene covered by the radiometer's field of view. This spatial variability in LST results in poor measurement accuracy and makes it extremely difficult to use *in situ* measurements for validation of any LST measured from space.

The difficulty in making adequate measurements of surface temperature can only be resolved by using satellite radiometers which provide the capability of self-consistent, large-area or global measurements on repeat cycles of hours to days. Satellite measurements of surface temperature, especially SST, are of continued importance and have been at the forefront since the launch of the first Advanced Very High Resolution Radiometer (AVHRR) in 1978. LST is also of vital importance to environmental and energy budget studies, even though the residual uncertainties available are an order of magnitude greater than those for SST. In the past decade, a suite of spacecraft sensors has been launched by the USA and the European Space Agency (ESA) for the accurate measurement of SST. These are the MODerate resolution Imaging Spectroradiometer (MODIS) on *Terra* (launched on 18 December 1999), the MODIS on *Aqua* (launched on

4 May 2002) and the Advanced Along-Track Scanning Radiometer (AATSR) on *Envisat* (launched on 1 March 2002). Additional instruments on *Aqua* measuring SST are the Atmospheric Infrared Sounder (AIRS) and the Japanese-built Advanced Microwave Scanning Radiometer for the Earth Observing System (AMSR-E). This suite of sensors complements the latest models of the AVHRR on the National Oceanic and Atmospheric Administration (NOAA) polar orbiters, and the European Meteorological Operational (MetOp) satellites. The NOAA AVHRRs will be replaced by the Visible Infrared Imager/Radiometer Suite (VIIRS) on the next generation of operational orbiting meteorological satellites to continue the record of satellite-derived surface temperatures into the next decades.

The potential of this long series of satellite measurements in climate research is great [1,2] but, as expounded in the National Research Council Reports [3,4], to make appropriate use of the measurements from these instruments requires a clear understanding of the residual uncertainties in the derived surface temperature fields. One of the primary applications of an extended, well-characterized SST time series is in climate change detection, through improved near-term ($t \leq 50$ years) global climate model simulations. The response of the climate system in general, and SST in particular, to the changing radiative energy flux at the earth's surface is not necessarily a uniform, monotonic increase in temperature. For the ocean, a response to global warming is likely to include changes in patterns of the SST distribution and temporal characteristics, with some regions warming and others cooling [5,6]. The regional and seasonal aspects of the potential climate-change signal in SST underscore the profound need to ensure that satellite-derived SST fields have well-understood uncertainties and that these are not unknowingly contaminated by residual instrumental errors (e.g. related to differential solar heating round the orbit [7]) or geographically or temporally coherent uncertainties in the atmospheric correction algorithms (i.e. related to the distribution of atmospheric water vapor [8] or aerosols), both of which could lead to systematic regional and seasonal errors.

There are many applications of infrared radiometry from space, such as studying the planetary radiation budget from space [9], the properties of clouds and their role in the climate system [10] and the distribution and effects of the cloud-free atmospheric variables [11]. Similarly, surface-based infrared radiometry, broad band or hyperspectral, is used to study the temperature and moisture structure of the atmosphere [12], cloud properties [13] and surface radiative forcing by clouds [14] and aerosols [15]. But while these topics are of interest and importance, we will limit the contents of this chapter to the remote sensing of surface temperatures from polar-orbiting satellites and the discussion of surface-based infrared radiometry is limited to a technique of validating the satellite surface temperature measurements.

The next part of this chapter is a discussion about what is meant by "surface temperature", a concept that is not as obvious as one might think, and this is followed by an introduction to the various remote sensing methods for measuring surface temperature. The emphasis is on using infrared radiometry from polar-orbiting satellites, while the complementary approach of measuring SST using microwave emission is briefly introduced. Short descriptions of the instruments that have contributed to the time series of surface temperature measurements are presented, followed by a discussion of the demonstrated accuracies of the retrieved surface temperatures. The chapter continues with a survey of some applications of the satellite-derived surface temperatures and concludes with a very brief discussion of planned satellite missions that will extend the time series of remotely sensed surface temperature measurements into the coming decade.

2. STATEMENT OF THE PROBLEM

In this section, we review the physical aspects that determine the surface temperature. For land and sea ice, the meaning of the surface temperature is quite straightforward in spite of its inherent variability and difficulty of (*in situ*) measurement. For the ocean, the situation is more complex.

The surface temperature is controlled by the incoming solar energy and the heat exchanges with the overlying atmosphere. The processes of heat exchange comprise of two types: radiative and turbulent. The heat gained by the surface is primarily radiative, being in the form of solar irradiation. Clearly this has a marked diurnal aspect and is strongly dependent on the presence and properties of clouds. The other radiative component, contributing to surface heating, is the infrared exitance from atmospheric gases, aerosols and clouds. It is therefore clear that infrared heating of the surface occurs all the time, not just during the day. The amount of energy absorbed by the surface depends on its reflectivity. Snow- and ice-covered surfaces have a high reflectivity in the visible, so relatively little of the incident insolation (i.e. incident solar radiation) is absorbed by the surface. Conversely, frozen surfaces are strong absorbers of infrared radiation. For the oceans and other open water, nearly all the incident solar radiation and the incident infrared radiation are absorbed. Over land surfaces, the reflectivity for both the visible and infrared spectra varies from location to location. The counterpart of the incident infrared radiative heating at the surface is the cooling by surface emission. This is related to the surface temperature through the Stefan−Boltzmann equation ($\pi L = \varepsilon \sigma T^4$, where ε is the total hemispherical surface emissivity, which depends on the spectral emissivity and surface temperature, σ the Stefan−Boltzmann constant,

T the surface temperature and πL the total emitted energy into a hemisphere per unit area per unit time — see Chapter 3, *Fundamentals*, Vol. 42 of this series). For solid surfaces that are heating or cooling, the subsurface vertical heat transport is accomplished primarily by conduction with heat flowing down a temperature gradient, but for liquid surfaces, the vertical heat flux is more effectively achieved by turbulence in the near-surface layers. Turbulence in the atmospheric boundary layer, arising from natural convention driven by buoyancy effects and from instabilities driven by wind shear, is the mechanism for the non-radiative components of the surface heat budget. These are the sensible heat flux, resulting from a temperature difference between the surface and the atmospheric boundary layer above, and the latent heat flux that is caused by the evaporation of water from the surface. Both turbulent fluxes are strongly dependent on the turbulent energy density in the air above, which is governed by the wind speed and the atmospheric stability. In most situations, the surface is warmer than the air above, and the atmospheric boundary layer is unstable or neutrally stable, and the vertical turbulent fluxes are effective, but in some cases the air is warmer than the underlying surface, and the turbulent fluxes are suppressed. Clearly the latent heat fluxes depend on the availability of water to evaporate, and over the oceans the latent heat fluxes generally exceed the sensible heat fluxes, sometimes by a large margin, but over an arid land surface this is not the case. It is relatively unusual for the latent heat fluxes to heat the surface, but this does occur when there is condensation, resulting in the latent heat of condensation becoming available at the surface, as in dewfall.

The surface temperature is related to the heat fluxes through the thermal capacity and the thermal conductivity of the material constituting the surface, or that immediately below. When a phase change takes place, such as in the melting of snow or ice, the available heat is not used to change the surface temperature. Thus, the surface temperature is a manifestation of the integrated net heat fluxes, and can be very variable in space and time.

2.1. Land surface temperature

The LST is a balance between the incoming solar radiative energy at visible and near-infrared wavelengths, the upward re-emission of energy at longer wavelengths to the atmosphere, clouds and outer space, the slow conduction of heat to subsurface layers, and the sensible and latent heat fluxes into the atmosphere. The sensible heat is simply the transport of heat into the atmosphere through local turbulence, while the latent heat refers to the evaporation of moisture and the transfer of water vapor into the atmosphere. The surface temperature is the main controlling parameter for the exchange of energy between the surface and the atmosphere, and accurate measurements are thus required for detailed local, regional and global numerical models of weather and climate, as well as many other applications discussed later.

Over small spatial scales, it is difficult to take account of the effects of land surface heterogeneity — the surface substance (trees, grassland, soil, agricultural areas, etc.), surface roughness, color, shadowing and topography all affect the local distribution of surface temperature. However, over larger scales it is possible to use infrared radiometers to average the surface temperature variability and derive temperature estimates that can have useful applications in many areas. Of course, the local variability of surface temperature also makes accurate validation of a remote sensing temperature estimate rather difficult. To this end, there is a continual search for large homogeneous land surfaces that can be used to confirm the accuracy of any LST estimate. Price [16] has discussed early attempts at LST measurements using the AVHRR on NOAA-7 while Prata [17] has reported on the use of two Australian sites to validate LST estimates from both AVHRR and the Along-Track Scanning Radiometers (ATSR).

As with the measurement of SST, it is also necessary to take account of the atmospheric absorption of the surface-emitted infrared radiation before it reaches a satellite-borne radiometer. Again, the usual method is to use the differential absorption of the atmosphere at two or more infrared wavelengths to account for the atmospheric absorption. The derivation of LST algorithms is discussed in a later section.

2.2. Sea surface temperature

The exchange of heat between the sea surface and the atmosphere modifies the temperatures in the immediate vicinity of the surface. Given the relative transparency of water to solar radiation in the visible part of the electromagnetic spectrum (the source of heat at the ocean surface), the opacity of water to infrared radiation and the complexities of dynamic coupling between the surface and atmosphere, the SST is not simple to define.

The surface temperature of the ocean is often called the "skin SST" and this is the source of the radiance measured in space by infrared radiometers. The temperature of the body of water below the surface, the temperature measured using *in situ* contact thermometers, is referred as the "bulk or seawater temperature". The near-surface temperature gradients, as illustrated in Figure 1, result from three distinct processes: the absorption of insolation, the heat exchange with the atmosphere and levels of subsurface turbulent mixing. In conditions of low wind speed, the heat generated in the upper few meters of the water column by the absorption of solar radiation is not mixed through the surface layer, but causes thermal stratification and temperature differences between the uppermost layer of the ocean and

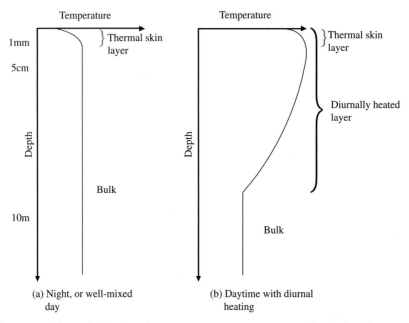

Figure 1   Schematic showing the upper ocean temperature profiles during the (a) night time, or daytime with good vertical mixing in the upper layer, and (b) daytime during conditions conducive to the formation of a diurnal warm layer. The depth scale is non-linear and some illustrative depths are marked.

the water below. There is a strong diurnal component to the magnitude of these temperature gradients, as well as a dependence on cloud cover, which modulates the insolation, and wind speed, which influences the turbulent mixing [18,19]. The surface, skin layer of the ocean, less than 1 mm thick [20,21], is nearly always cooler than the underlying water because the heat flux is nearly always from the ocean to the atmosphere. The heat flow, supplying energy for both the turbulent and radiant heat loss to the atmosphere, is accomplished by molecular conduction through the aqueous side of the interface and this is associated with a temperature gradient in the surface skin layer. The relationship between skin and bulk SSTs just below the surface (measurable at ~ 5 cm) is reasonably well behaved in the mean, that is, on time scales of tens of seconds and longer and spatial scales of tenths of a meter and greater. Figure 2 illustrates the wind-speed dependence of the temperature drop across the thermal skin layer, the colors indicating the time of day these open-ocean measurements were taken. The dark line is the wind-speed parameterization published by Donlon et al. [22] based on measurements of multiple investigators taken at night, to avoid the complicating effects of possible diurnal heating, and the light line is a similar parameterization based on data including those taken

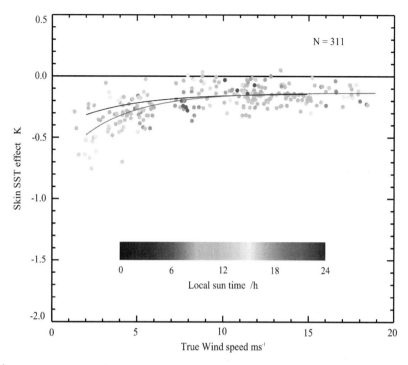

Figure 2   Wind-speed dependence of the temperature difference across the skin layer, measured radiometrically by the M-AERI (see text below) and a subskin, bulk temperature measured by a surface following float at a depth of about 5 cm. The shades of gray represent the local time of day at which the measurements were taken during a research cruise off New Zealand. The dark solid line depicts the parameterization of Donlon et al. [22] based on night-time measurements alone, and the light line the parameterization derived from measurements that include many taken during the day. After Ref. [23].

during the daytime, but in which the bulk temperature was measured at a depth of ~ 5 cm from a surface-following float [23].

2.3. Surface renewal theories

The usual direction of heat transfer from the surface of the ocean to the overlying atmosphere results in the ocean being a fluid cooled at the interface, and since the density of water is a function of temperature, the heat loss tends to destabilize the surface giving rise to free convection in the water immediately beneath the skin layer. By continuity, there is a compensating upward motion in the water that brings warmer fluid to the underside of the skin layer and this supports the heat conduction through the skin layer to the air above. This process is referred as "surface renewal". The horizontal scales of the vertical turbulent motion in the water are centimeters, and thermal imaging radiometers reveal the resulting temperature patterns at the water surface. Figure 3 shows images taken during an experiment in a wind-wave tank at the University of Miami. The patterns represent warmer water brought to the underside of the skin layer and the horizontal convergence of the fluid, driven by the subsurface convection, concentrates the water that has been cooled by the heat loss to the atmosphere. When the wind blows above the surface, the natural convection is enhanced by shear-flow instabilities and the length scales of the temperature patterns at the surface become smaller (Figure 3(b)). The analytical expressions for the behavior of the thermal skin later are complex [24], but have been addressed through numerical modeling with some success [25].

The scales of the surface renewal processes are many orders of magnitudes smaller than the satellite resolution at the sea surface and so a spatially averaged skin layer is considered when dealing with remote sensing of SST from space.

2.4. Effects of breaking waves

The thermal skin layer of the ocean is a very robust natural feature, but it can be removed by breaking waves. This mechanical rupture can be visualized using infrared imagers and causes a warm patch of ocean surface. In the cases where the breaking wave results in a whitecap or foam patch, the position of the warm surface expression is initially related to the foam, but as the foam moves downwind, the warmer signature remains in its wake [26]. The ruptured skin layer can be an efficient conduit for the flow of heat and gases from the ocean to the atmosphere [27] but the thermal skin must begin re-establishing itself immediately as the heat flow to the atmosphere is still conductive and requires a temperature gradient across the skin layer. The time for the thermal scar of the ruptured skin takes several seconds to disappear, and appears to be linearly related to the crest speed of the breaking wave which is indicative of the subsurface turbulent energy density. Faster moving waves generate more turbulence, leading to longer skin-layer recovery times [26].

In terms of satellite remote sensing of SST in the infrared, the rupture of the skin layer by breaking waves is not a significant issue. The areas of the ruptures comprise a small fraction of the surface, and are not resolved by the kilometer scales of the satellite radiometers, and additionally the thermal variability resulting from the wave breaking is relatively small.

2.5. Diurnal thermocline effects

Better understanding of the physics of the diurnal heating of the upper ocean is basic to improving the representation of the interactions between

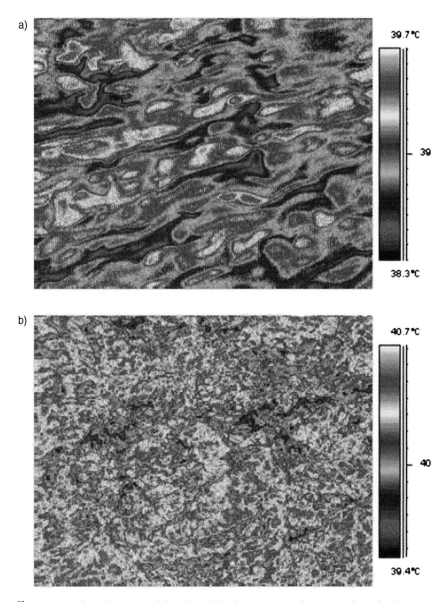

Figure 3 Infrared images of the skin SST taken in a wind-wave tank at the University of Miami. The airflow was from left to right (with a small component in the direction to the top of the figures). The air–sea temperature difference was $-15\,\mathrm{K}$, which is large for most natural situations, but was chosen to provide a clear temperature signal in the skin layer. The size of each image is approximately $30\,\mathrm{cm} \times 20$ cm. (a) Wind speed of $2\,\mathrm{m/s}$, without any wave breaking, and (b) wind speed of $9\,\mathrm{m/s}$, with wave breaking.

ocean and atmosphere in numerical simulations and forecasting of weather and climate. The global sampling offered by satellites provides the best approach to study this component of the earth's climate system.

Radiometric skin temperature measurements provide sampling of the diurnal warming at the ocean surface. Thermal stratification of the top few meters of the ocean occurs during the day, in clear-sky and calm conditions ([19,28] and the references therein). The existence of this diurnal warm layer was first described in 1942 [29] and has been studied extensively since [18,30,31]. Surface temperature deviations greater than 3 K, referenced to subsurface temperatures below the extent of surface heating, are not uncommon [32−34] and may persist for many hours. Rarely, diurnal heating in excess of 6 K is observed [35]. Failure to account for a diurnal cycle in SSTs can lead to errors in determining surface fluxes for numerical weather prediction (NWP) and climate models [36,37]. Tropical atmospheric circulation is sensitive to relatively small changes in SST [38−40].

As implied in Figure 1, the vertical distribution of the diurnal warming and its surface expression are determined by the absorption of radiation at differing depths (influenced by the absorptivity of the water), the vertical diffusion of heat, the vertical stratification in the upper ocean "mixed" layer and the rate of turbulent mixing within the mixed layer. During the night, the ocean loses heat to the atmosphere and deep space, thus cooling the surface layer. The cooler, denser water at the surface becomes gravitationally unstable and results in free convection, overturning within the mixed layer and possibly causing the layer to deepen. The stratification may also be decreased by surface wind stress-induced mechanical mixing, by shear-flow instabilities within the stratified layers or by breaking surface and internal waves.

The relationship with deeper bulk temperature, at depths of half to several meters where many bulk SST measurements are taken, is the same on average during the night, and during the day for wind speed greater than about 6 m/s [22]. But under low winds, the relationship is very variable − vertically, horizontally and temporally [33,34]. The difference between the skin temperature and that measured by a bulk, *in situ* thermometer is quite variable and highly dependent on the depth of the bulk measurement, as illustrated in Figure 1.

Use of the bulk temperature in applications where a temperature of the surface is really required introduces an uncertainty. When bulk temperatures are used to validate the satellite skin temperature retrievals, these near-surface gradients appear in the uncertainty budget of the satellite retrieval and lead to an over-estimate of the uncertainties [41]. While physical models of the growth and decay of the diurnal thermocline [18,30,42,43] require high temporal resolution forcing fields to produce reliable predictions, which is a limitation on their use in several applications, empirically based models based on satellite measurements themselves may hold promise [44].

3. Remote Sensing of Surface Temperature

Satellite radiometers measure the electromagnetic radiation emerging from the "top of the atmosphere" at the orbital height. In its passage through the atmosphere, the radiance field undergoes modification by molecular absorption, re-emission and scattering. The spectrum of the top-of-atmosphere emission is related to an equivalent temperature through Planck's equation, referred as the brightness temperature. This is a spectrally varying quantity as it depends not only on the thermal radiation emitted and reflected from the surface, but also on the transmission and emission properties of the atmosphere. The upper panel of Figure 4 illustrates atmospheric transmission spectra through three representative atmospheres, corresponding to polar (cold and dry), mid-latitude and tropical (warm and moist) conditions. The lower panel of Figure 4 shows Planck's function calculated for four temperatures in the range covered by the earth's surface, along with the relative spectral response functions of five of the spectral bands used to derive surface temperatures (see below).

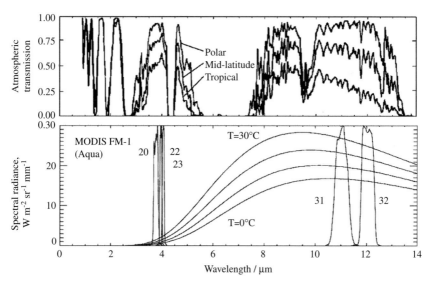

Figure 4 Spectral dependence of the atmospheric transmission for wavelengths of electromagnetic radiation from about 1 to 14 μm, for three characteristic atmospheres (above), and (below) the blackbody emission for temperatures of 0, 10, 20 and 30°C, and the relative spectral response functions of the bands MODIS (Flight Model 1) on *Aqua* used to derive SST.

The most common satellite radiometers used for the measurement of surface temperature measure the spectral radiance, $L_{b,\lambda}$, within distinct wavelength intervals of the electromagnetic spectrum. From these measurements, the brightness temperatures can be calculated using the Planck equation:

$$L_{b,\lambda}(T) = \frac{2hc_0^2\lambda^{-5}}{e^{hc_0/(\lambda kT)} - 1} \tag{1}$$

where h is Planck's constant, c_0 the speed of light in vacuum, k the Boltzmann's constant, λ the wavelength in vacuum and T the temperature of the surface. This expression for the emission is for a blackbody, a perfect emitter, and in reality this is modified by the value of the emissivity, $\varepsilon(\lambda)$, which always has a value less than unity. For a detailed discussion of Planck's law, see Chapter 3, *Fundamentals*, Vol. 42 of this series.

For surface temperature measurements, the spectral intervals (wavelengths) are chosen where three conditions are met: (a) the surface emits a measurable amount of radiant energy, (b) the atmosphere is sufficiently transparent to allow some of the energy to propagate to the spacecraft and (c) current technology exists to build radiometers that can measure the energy to the required level of accuracy within the bounds of size, weight and power consumption imposed by the spacecraft. In reality, these constrain the instruments to two relatively narrow regions of the infrared part of the spectrum and to low-frequency microwaves. The infrared regions, the so-called atmospheric windows, are situated between wavelengths of 3.5−4.1 and 8−13 μm (Figure 4); the microwave measurements are made at frequencies of 6−12 GHz.[1] The individual spectral intervals for each radiometer are referred as "channels" or "bands" and these have finite widths and distinct shapes, insofar as having a transmission that is wavelength dependent. This is called the relative spectral response function, $\Phi_i(\lambda)$, and ideally has unit transmission within the desired band-pass, and zero transmission elsewhere. In reality, this is not the case and determining the spectral shape of $\Phi_i(\lambda)$ for each band and instrument prior to launch is very important but often very difficult, especially in quantifying the "out-of-band" response. The requirement for accurate knowledge of $\Phi_i(\lambda)$ is driven by the conversion of the measured signal, a spectral radiance, to a temperature, using Planck's function. For some purposes, using the central wavelength, or the centroid of the $\Phi_i(\lambda)$ is adequate, but for accurate radiometry the wavelength integral of Planck's function modulated by the relative spectral response function

---

[1] In microwave radiometry, it is usual to refer to spectral intervals in terms of frequency, whereas in the visible and infrared, wavelength is most commonly used.

for the channel has to be used:

$$L_i(T) = \frac{\int L_{b,\lambda}(\lambda, T)\Phi_i(\lambda)\mathrm{d}\lambda}{\int \Phi_i(\lambda)\mathrm{d}\lambda} \tag{2}$$

where $L_i(T)$ is spectral radiance measured at band i for a target temperature T. This relationship is usually evaluated prior to launch for temperatures in the dynamic range of the sensors to generate "look-up tables" for use in the procedures to process the radiometer data.

As the electromagnetic radiation propagates through the atmosphere, some of it is absorbed and scattered out of the field of view of the radiometer, thereby attenuating the original signal. If the attenuation is sufficiently strong, none of the radiation from the sea reaches the height of the satellite, and such is the case when clouds are present in the field of view of infrared radiometers. Even in clear-sky conditions, a significant fraction of the sea surface emission is absorbed in the infrared windows. This energy is re-emitted, but at a temperature characteristic of that height in the atmosphere. Consequently, the brightness temperatures measured through the clear atmosphere by a spacecraft radiometer are cooler than would be measured by a similar device just above the surface. This atmospheric effect, frequently referred as the temperature deficit, must be corrected accurately if the derived surface temperatures are to be used quantitatively.

3.1. Infrared atmospheric correction algorithms and SST

The peak of the Planck function for temperatures typical of the surface is around $10\,\mu\text{m}$, close to the long-wavelength infrared window, which is therefore well suited to surface temperature measurement (Figure 4). However, the main atmospheric constituent in this spectral interval that contributes to the temperature deficit is water vapor, which is very variable in both space and time. Other molecular species that contribute to the temperature deficit are quite well mixed throughout the atmosphere, and therefore cause a relatively constant temperature deficit that is easier to correct.

For surface temperature retrievals from a well-calibrated and well-characterized spacecraft infrared radiometer, the limit of the accuracy is imposed by the effectiveness of the correction of the atmospheric effects. The essence of the atmospheric correction algorithm is the differential absorption (and re-emission) of infrared radiation by atmospheric con-stituents at different spectral intervals (the "channels" or "bands"). This leads to different brightness temperatures at satellite altitudes for collocated measurements in different bands, and the correct combination of these results in a pixel-by-pixel correction for the effect of the intervening atmosphere. This approach can only be applied to parts of the images that have been screened for the presence of cloud, as, when thick or extending over the entire pixel, these prevent the infrared radiation propagating from the surface to the satellite radiometer. When thin or obscuring only part of the pixel, the cloud radiances contribute to the measurements and contaminate the SST retrieval. The "atmospheric windows", in the 3.5−4.1 and 10−13 μm spectral intervals (Figure 4), are referred as the mid-infrared window and the thermal infrared window, respectively. The shorter wavelengths are contaminated by sunlight reflected from both land and sea surfaces during the day ("sun glitter" or "sun glint"), and so such measurements are usually confined to LST and SST retrievals during the night-time part of each orbit. Because of the rather small number of channels available, and the wide range of clear-sky atmospheric variability, especially of water vapor, the atmospheric correction is subject to uncertainty, even within the limits imposed by the inherent noise in the radiometer measurements, which have a very strong temperature dependence and consequently the detectors of satellite infrared radiometers are cooled to temperatures in the range of 77−105 K. The residual errors in the atmospheric correction make significant contributions to the satellite surface temperature retrieval uncertainties.

The variability of water vapor requires an atmospheric correction algorithm based on the information contained in the measurements themselves. The current atmospheric correction algorithms are the product of over three decades of research, and while some approximations invoked to render the problem tractable may appear unjustified, especially given the complexity of the processes they represent and the wide ranges of environmental variability with which they have to contend, the ultimate justification is in the magnitudes of the uncertainties in the derived SST fields. The uncertainties are determined by comparison with independent measurements and methodologies and results are discussed in subsequent parts of this chapter.

By examining the behavior of the atmospheric transmission spectra for different atmospheric conditions (see Figure 4), the hypothesis can be formulated that the difference in the brightness temperatures measured in two channels, i and j, is related to the temperature deficit in one of them. The atmospheric correction algorithm for SST retrieval can be formulated thus:

$$\text{SST} - T_i = f(T_i - T_j) \tag{3}$$

where SST is the derived SST, and T_i and T_j the brightness temperatures in channels i and j, respectively.

Further, by assuming that the atmospheric attenuation is small in these channels, so that the radiative transfer can be linearized, and that the channels are spectrally close so that Planck's function can also be linearized, the algorithm can be expressed in the very simple form:

$$\text{SST} = a_0 + a_i T_i + a_j T_j \tag{4}$$

where a_0, a_i and a_j are coefficients [45]. These are determined by regression analysis either of coincident satellite and *in situ* measurements, such as from drifting buoys, or of simulated satellite measurements derived by radiative transfer modeling of the propagation of the infrared radiation from the sea surface through a representative set of atmospheric profiles [46,47].

This simple empirical algorithm has been applied for many years in the operational derivation of the sea surface from measurements of the AVHRR (see below), the product of which is called the Multi-Channel SST (MCSST), where i refers to channel 4 ($\lambda \approx 10.8\,\mu m$) and j to channel 5 ($\lambda \approx 12.0\,\mu m$).

More complex forms of the algorithms have been developed to compensate for some of the limitations of the linearization and the shortcomings of the underlying assumptions. One such widely applied algorithm takes the form:

$$\text{SST} = b_0 + b_1 T_i + b_2 (T_i - T_j)\text{SST}_r + b_3 (T_i - T_j)(\sec\theta - 1) \tag{5}$$

where SST_r is a reference SST (or first-guess temperature) and θ the zenith angle to the satellite radiometer, measured at the sea surface. When applied to AVHRR data, with i and j referring to channels 4 and 5, the derived SST is called the Non-Linear SST (NLSST) [48]. A refinement is called the Pathfinder SST (PFSST) [49] in the research program designed to post-process AVHRR data over nearly two decades to provide a consistent data set for climate research. In the PFSST, the coefficients are derived by comparison with measurements from buoys on a monthly basis for two different atmospheric regimes, distinguished by the value of the $T_4 - T_5$ differences being above or below 0.7 K.

For the MODIS [50,51], the same formulation of the algorithm is used and i and j denote bands 31 and 32 ($\lambda \approx 11$ and $12\,\mu m$, respectively).

MODIS has three bands in the mid-infrared atmospheric transmission window allowing an alternative scheme that exploits the greater temperature sensitivity of the infrared emission on the short wavelength side of the peak of the Planck function. However, as noted above, because of the effects of sun glitter, this approach can only be used at night. The algorithm, using two bands in the $4\,\mu m$ atmospheric window, is:

$$\text{SST}_4 = c_0 + c_1 T_{3.9} + c_2 (T_{3.9} - T_{4.0}) + c_3 (\sec\theta - 1) \tag{6}$$

Other variants of the algorithms are possible, such as squared terms in the brightness temperature differences, or an explicit dependency on the columnar water vapor terms, but to date such algorithms have not produced a systematic improvement to the SST retrieval accuracy.

   This approach of using collocated and contemporaneous match-ups with *in situ* measurements can only be adopted after launch, while the numerical simulations using a radiative transfer model can be done prior to launch and thus be used to study the likely error characteristics of the retrievals while the instrument is being specified, designed, built and characterized.

   The success of the numerical simulations depends on several factors and how faithfully these represent the environmental conditions. These include knowledge of the spectroscopy of the atmospheric constituents that interact with the infrared radiation, the spatial and temporal distributions of these (especially water vapor); the atmospheric thermal state, including the air—sea temperature difference; and the surface conditions, specifically the angular and wind-speed dependence of the surface emissivity in the spectral intervals of interest. A very important factor is the completeness of the instrument model, such as the noise characteristics of the detectors, the reflectivities (angular and spectral) of the optical surfaces, optical cross-talk, relative spectral response of the channels (including out-of-band response), the characteristics of the electronic components (e.g. digitization intervals, preferences of the digitizers, electronic cross-talk, etc.), scattered light sources and paths (which includes the temperature distributions and changes within the instrument and of external sources) and the characteristics of the on-board calibration sources (blackbody calibration target(s), cold space view).

   An entirely different approach sometimes referred as the "forward solution" uses radiative transfer simulations given the state of the atmosphere at the time of the measurement. This can be derived from NWP fields or retrievals of the atmospheric temperature and humidity structure derived from satellite instruments on the same satellite. In principle, this is a very desirable option as it removes the sources of uncertainties inherent in the statistical approach. However, it is prone to uncertainties of its own, resulting from inaccuracies in the specification of the atmospheric state, or in the atmospheric spectroscopy. In the past, these drawbacks have outweighed the actual benefits, but as the accuracies in the NWP fields and in the satellite atmospheric state retrievals improve, this approach may become feasible [52].

   All atmospheric correction algorithms work effectively only in the clear atmosphere. The presence of clouds in the field of view of the infrared radiometer contaminates the measurements so that these pixels must be identified and removed from the SST retrieval process. It is not necessary for the entire pixel to be obscured; even so small a portion as 3—5%, dependent on cloud type and height, can produce unacceptable errors in the SST measurement. Thin, semi-transparent cloud, such as cirrus, can have similar effects to subpixel geometric obscuration by optically thick clouds. Consequently, great attention must be paid in the SST derivation to the identification of measurements contaminated by radiance from even a small amount of clouds. This is the principal disadvantage to SST measurement by space-borne infrared radiometry. Since there are large areas of cloud cover over the open ocean, it may be necessary to composite the cloud-free parts of many images to obtain a complete picture of the SST over an ocean basin.

Similarly, aerosols in the atmosphere can introduce significant errors in SST measurement. Volcanic aerosols injected into the cold stratosphere by violent eruptions produce unmistakable signals that can bias the retrieved SST too cold by several degrees. A more insidious problem is caused by less readily identified aerosols at lower, warmer levels of the atmosphere that can introduce systematic errors of much smaller amplitude.

An alternative approach to correcting the effects of the intervening cloud-free atmosphere is to make measurements of the same area of sea surface through two different atmospheric path lengths. The pairs of such measurements must be made in quick succession, so that the surface temperature and atmospheric conditions do not change in the time interval. This approach is used by the two ATSRs on the European satellites ERS-1 and ERS-2, and the AATSR on *Envisat*. The differences in the brightness temperatures between the slanted atmospheric path and nadir path are a direct measurement of the effect of the atmosphere and permit a more accurate determination of the SST. For ATSR series, the atmospheric correction algorithms take the form:

$$SST = d_0 + \sum d_{n,i} T_{n,i} + \sum d_{s,i} T_{s,i} \tag{7}$$

where the subscripts n and s refer to measurements from the nadir and slant path views, i indicates two or three atmospheric window channels and the set of d is coefficients. The coefficients, derived by radiative transfer simulations, have an explicit latitudinal dependence.

3.2. Remote sensing of land surface temperature

All the satellite-borne radiometers designed to give estimates of SST are also capable of providing estimates of LST, although with much reduced accuracy. As with SST estimation, the effects of atmospheric absorption and surface emissivity are of paramount importance and must be taken into account with any analysis of infrared data obtained over land surfaces, but a major difficulty with this is the inhomogeneity and roughness of the surfaces. Space-borne radiometers designed to give global measurements of surface temperatures have a footprint in the order of 1 km − some have improved resolution of better than 100 m (e.g. the Thematic Mapper on Landsat-5) while others give data with a 5−20 km footprint (the network of geostationary meteorological satellites) and most land surfaces do not have uniform vegetation, or other cover, on these scales. During daytime, and under the cloud-free conditions required for LST estimation using infrared radiometers, rough surfaces provide a wide range of temperatures depending on the shadowing effects on solar heating. Objects in full view of the sun can have a surface temperature that is more than 20 K warmer than those in the shade. A measurement from space will then be an areal average of the infrared radiation emitted from these surfaces. Even an effective surface emissivity over these large areas will be an average value, and is thus extremely difficult to estimate and to validate.

Early LST measurements using the AVHRR on the NOAA-7 satellite were reported by Price [16]. He used a simple split-window algorithm with coefficients that were dependent on surface emissivity to estimate the temperature over agricultural areas of the central USA.

LST is now an approved product from both USA and European environmental satellites, including the infrared measurements from the MODIS instruments on *Terra* and *Aqua*. The Institute for Computational Earth Systems Science at the University of California, Santa Barbara hosts the MODIS Land Surface Temperature Group [53]. As reported on their URL, two LST algorithms were developed, one being the generalized split-window algorithm [54] and the other is the physics-based day/night LST method [55].

The first LST algorithm uses MODIS data in bands 31 and 32 in the atmospheric transmission window (at wavelengths of 11 and 12 μm), and is suitable for land-cover types such as dense evergreen canopies, lake surfaces, snow and most soils, that have a stable emissivity that is known to within 0.01. The infrared emissivities in MODIS bands 31 and 32 are inferred from the land-cover types based on thermal infrared bi-directional reflectance distribution function (BRDF) models [56,57] that simulate the scene emissivity from the proportions, surface structures and spectral emissivities of the components in the scene. Examples of spectral emissivities of terrestrial materials can be found in the MODIS UCSB Emissivity Library [58].

The day/night method retrieves LST and band emissivities simultaneously from pairs of daytime and night-time MODIS data in seven thermal infrared bands. Further information on these LST algorithms can be found in the MODIS LST Algorithm Theoretical Basis Document [59].

Early work on the derivation of LST in Europe has been reported by Sobrino et al. [60] and Prata [17,61] who both compared AVHRR LST retrievals with ground-based measurements over uniform land sites. More recently, ESA has been deriving an operational LST product from the AATSR infrared data [62]. For the ESA product, the emissivity is not specified but is built into a vegetation type that must be included in the LST algorithm. Vegetation types are currently restricted to a set of 13 types, but more are planned for future applications. The ESA Algorithm Theoretical Basis Document [63] gives further details.

3.3. Ice and snow surface temperature

The SST and LST infrared atmospheric correction algorithms are generally optimized for the global range of atmospheric conditions, and when such algorithms are used in a set of conditions that exhibit limited atmospheric variability, especially when those are at climatological extremes, significant errors can result in the derived surface temperatures [64]. This is the case in the measurement of ice and snow surface temperatures which, in addition to sometimes being very cold, often occur in conditions of low atmospheric water vapor content such as in polar regions or in high mountain ranges. Similarly, SST and LST temperatures in polar regions are likely to be biased. One approach to improve the surface temperature retrievals is to optimize the coefficients in a standard multi-channel surface temperature algorithm for the cold surface temperatures and low water vapor atmospheres [65]. Another approach, demonstrated for the SST measurement in polar regions, is to use a very simple, single channel algorithm [66,67]. This is based on the recognition that the brightness temperatures measured by the channels that share the thermal infrared window are particularly well correlated in dry polar atmospheres and introducing the second channel in the algorithm does not add significantly more information while it does add noise to the retrieval.

3.4. Microwave measurements

Microwave radiometers use a similar measurement principle as infrared radiometers, having several spectral channels to provide the information to correct for extraneous effects, and blackbody calibration targets to ensure the accuracy of the measurements. The suite of channels is selected to include sensitivity to the parameters interfering with the SST measurements, such as rain and surface wind speed; this requires measuring microwave emission at higher frequencies. The sensitivity of the microwave emission from the sea surface is dependent not only on the temperature, but also on the emissivity, and this is strongly spectrally dependent. The microwave emissivity of the sea surface is much lower than in the infrared, and the frequencies that are used for sea–surface temperature determination are limited to the $6-10\,\mathrm{GHz}$ band [68]. A simple combination of the brightness temperatures, such as in Equation (3), can be used to retrieve the SST [69].

The infrared emissivity of open water is very high, but that of snow is yet higher — while snow is one of the brightest natural surfaces in the visible part of the spectrum (reflectivity $>90\%$), it is one of the darkest in the infrared (reflectivity $<1\%$; emissivity >0.99) [65], so the brightness temperatures (i.e. those determined radiometrically) of snow, ice and water are very close to their respective thermodynamic temperatures. That is not the case in the microwave part of the spectrum, where the emissivity of sea water is low, and that for ice is high [70]. The emissivity is dependent on polarization, frequency and ice type and, while this variability makes the derivation of ice surface temperature somewhat uncertain, it has been exploited to derive the ice cover of polar oceans [71,72]. Given that the water emissivity is low, the microwave brightness temperatures of open water are much lower than those of ice and snow, even though the thermodynamic temperature is higher. Analyses of microwave brightness temperatures over the polar regions give maps of sea ice extent rather than surface temperature, which can better be derived using infrared data [73]. Indeed most of our knowledge of the advance and retreat of the seasonal ice cover, and its long-term decline, is based on satellite microwave radiometry. Because of the large footprints of the microwave radiometers at the surface, and the high spatial variability of the surface leading to mixed surface types in the individual footprints, the algorithms are based on the identification of the fractions of each surface type in each retrieval cell. This requires a good knowledge of the microwave characteristics of the "end members" of the mixing lines, that is, the signals that would be measured if a uniform surface type were to fill a footprint [74,75]. One problem, particular to microwave radiometers and absent in their infrared counterparts, is that the spatial resolution at the surface is a strong function of the frequency of the emission being measured, with the higher frequencies having the better spatial resolution. Techniques have been developed that make use of iterative calculations of the fractions of each surface type in the footprints of each frequency to retrieve surface ice type and cover and water cover at the resolution of the highest frequency [76].

4. SPACECRAFT RADIOMETERS

In this section, we describe some of the characteristics of the radiometers used to measure surface temperature from polar-orbiting satellites. These spacecraft complete an orbit of the earth at an altitude of about 800 km above the surface in about 100 min. The orbital planes of the satellites are inclined to the earth's equatorial plane at an angle of about 98°, and the satellites pass over close to the poles on each revolution of the orbit. This orbit type is sometimes referred as "sun synchronous" as the satellites pass overhead at about the same time each day. Earth-observing satellites generally cross the equator within a few hours of local noon, and thus half of each orbit is in daylight and the other half covers the night-time part of the planet. In the infrared, the radiometers have several aspects in common: not only in the selection of the spectral channels, but also in their spatial resolution at the surface. Other aspects have evolved with successive generations of the same sensor series, or in new radiometer design.

4.1. The advanced very high resolution radiometer

The satellite instrument series that has produced the longest time series of surface temperature measurements is the AVHRR. This first flew on Television Infrared Observation Satellite-N (TIROS-N) launched in late 1978. AVHRRs have flown on successive operational satellites of the NOAA series from NOAA-6 to NOAA-18, launched on 20 May 2005. The operational requirement means that there are generally two AVHRRs in use at any given time. The NOAA satellites are in a near-polar, sun-synchronous orbit at a height of about 850 km above the earth's surface and with an orbital period of close to 102 min. The equator-crossing times of the two NOAA satellites are about 2.30 a.m. and p.m. and about 7.30 a.m. and p.m. local time. The AVHRR has five channels: 1 and 2 at $\sim$ 0.65 and $\sim 0.85\,\mu$m are responsive to reflected sunlight and are used to detect clouds and identify coastlines in the images, and to monitor the health of vegetation using measurements from the daytime part of each orbit. Channels 4 and 5 are in the atmospheric window close to the peak of the thermal emission from the surface and are used primarily for the measurement of surface temperature. Channel 3, positioned at the shorter wavelength atmospheric window, is responsive to both surface emission in the infrared and reflected sunlight.[2] For the daytime part of each orbit, the NLSST algorithm is used operationally to derive SST from brightness temperature measurements from channels 4 and 5; see Equation (5). But during the night-time part of each orbit, channel 3 brightness temperatures can also be used in variants of the atmospheric correction algorithm to determine SST. One such night-time SST algorithm used operationally by NOAA in the USA takes the form:

$$\mathrm{SST} = e_0 + e_1 T_4 + e_2(T_3 - T_5) + e_3(\sec\theta - 1) \tag{8}$$

where T_3, T_4 and T_5 are brightness temperatures measured in channels 3, 4 and 5 of the AVHRR, and e_i are coefficients.

The presence of reflected sunlight during the daytime part of the orbit prevents much of these data from being used for SST measurement. Because of the roughening of the sea surface by the wind, some individual facets of the surface are tilted in such a way as to fulfill the conditions of specular reflection of the sun into the radiometer. The area contaminated by reflected sunlight (sun glitter) can be quite extensive, and is dependent on the local surface wind speed.

---

[2] The first AVHRR had only four channels, and the telemetry system replicated channel 4 data in channel 5. The current versions, starting with the instrument on NOAA-15, are six-channel radiometers, but only five are telemetered to the ground. The channel 3 telemetry is split between the conventional 3.7 μm measurements being transmitted at night, but measurements in a new channel at $\lambda = 1.6\,\mu$m being optionally transmitted during the daytime part of each orbit.

The images in each channel, digitized to 10 bits, are constructed by scanning the field of view of the AVHRR across the earth's surface by a mirror inclined at 45° to the direction of flight. The spatial resolution of the radiometer at nadir, the subsatellite point, is a nominal 1 km², increasing by beam spreading along a longer path length to about 4 km × 8 km at the edges of the swath. The rate of rotation of the scan mirror, 6.67 Hz, is such that successive scan lines are contiguous at the surface directly below the satellite. The width of the swath ($\sim$ 2,700 km), which is symmetrical about the subsatellite point, means that the swaths from successive orbits overlap so that the whole earth's surface is covered without gaps twice each day.

The calibration of the infrared channels is done by a two-point measurement: one of deep space (zero radiance) and the other of a "blackbody" calibration target that is on the base-plate of the instrument. Non-linearities in the response of the detectors are measured before launch and are assumed to be stable over the lifetime of the radiometer. The calibration target temperature is measured by five embedded thermometers. However, the calibration target is not shielded and exhibits significant spatial temperature gradients which change around the orbit, along with the average temperature [7]. This introduces an uncertainty into the calibration of the AVHRR brightness temperatures.

The first of a new series of AVHRRs was launched on 19 October 2006, on the first of three planned MetOp satellites which form a new European remote sensing program to improve weather forecasts and to support climate monitoring. The MetOp program is run by the ESA and the European Organisation for the Exploitation of Meteorological Satellites (EUMETSAT). This commitment means that AVHRRs will continue to provide surface temperature measurements over a time span of more than three decades.

4.2. The moderate-resolution imaging spectroradiometer

The MODIS is a 36-band imaging radiometer on the NASA Earth Observing System (EOS) satellites *Terra* [77], launched on 18 December 1999, and *Aqua* [78], launched on 4 May 2001. MODIS [79,80] is much more complex than other radiometers designed for surface temperature measurement, but uses the same atmospheric windows. In addition to the usual two bands in the 10−13 μm spectral interval, MODIS has three bands in the 3.7−4.1 μm window, which, although limited by sun-glitter effects during the day, provide more accurate measurement of surface temperature during the night. Several of the other 31 bands of MODIS contribute to the SST measurement by better identification of residual cloud and aerosol contamination. The swath width of MODIS, at 2,330 km, is somewhat narrower than that of AVHRR with the result that a single day's coverage is not entire, but the gaps from one day are filled in on the next. The spatial resolution of the infrared window bands is 1 km at nadir.

MODIS introduced several innovations to wide-swath infrared radiometry including a well-shielded blackbody calibration target that is inside the instrument in a well-controlled thermal environment, 10 detectors per channel, and a "paddle-wheel" scan mirror geometry. The axis of rotation of the scan mirror is aligned with the flight vector of the satellite, with the entrance aperture of the instrument in the normal plane. This means that the angle of incidence on the scan mirror changes across the swath, but it also means that both sides of the mirror can be used to generate successive sets of scans. The mirror rotation rate, 0.338 Hz, is much slower than for AVHRR while maintaining the same spatial resolution. This results in the integration time for each pixel being longer and therefore leads to an improved signal-to-noise ratio. However, the reflectivity of the scan mirror is dependent on the angle of incidence, and this is accentuated in the thermal infrared by a multi-layer coating on the mirror surface designed to improve the polarization response in the visible part of the spectrum. Corrections for these effects have been developed based on pre-launch characterization measurements and empirical post-launch analyses. The MODIS measurements are digitized to 12-bit resolution.

The MODIS Characterization Support Team (MSCT) monitors the performance of the MODISs on *Terra* and *Aqua* to ensure the on-board calibration procedures post-launch continue to produce high-quality measurements. For the infrared data used for surface temperature retrievals, this includes analysis of the performance of the blackbody calibration target and ensuring that the corrections for instrumental artifacts are current and appropriate. For example, for the Terra MODIS, the blackbody temperature variations are less than ± 0.30 mK on a scan by scan basis and the changes in the responses of the thermal emissive bands are $< 0.7\%$ on an annual basis [81]. The URL http://www.mcst.ssai.biz/mcstweb/index.html gives many details of the MSCT activities and their results.

4.3. The advanced along-track scanning radiometer

In 1991, ESA launched the first of their ATSRs that were specifically designed to measure SST with residual uncertainties of less than 0.3 K. These radiometers use a dual-view technique as well as having three infrared channels. An offset conical scan gives near nadir measurements beneath the satellite as well as a forward view with a zenith angle (of the satellite from the earth's surface) of close to 55° [82]. The field of view of the (A)ATSRs sweeps out a curved path on the surface, beginning at the point directly below the satellite, moving out sideways and forwards. Half a mirror revolution later, the field of view is about 900 km ahead of the subsatellite point in the center of the forward view. The path of the field of view returns to the subsatellite point, which, during the period of the mirror rotation, has moved 1 km ahead of the starting point. Thus, the pixels forming the successive swaths through the nadir point are contiguous. The orbital motion of the satellite means that the nadir point overlays the center of the forward view after about 2 min. These two views provide more information of the effect of the atmosphere and facilitate an improved atmospheric correction to the measured brightness temperatures. One benefit of having dual-view measurements at the same wavelength is to reduce any inherent errors due to the presence of significant aerosols in the atmosphere [83].

Careful pre-launch characterization of the spectral and optical pro-perties of the radiometers has been crucial to the development of multi-channel algorithms for deriving surface temperature. A high-quality atmospheric radiative transfer model has been used to derive the operational algorithms for the ATSR instruments. The model is used with a large number of measurements of atmospheric profiles over a wide range of marine atmospheric conditions to produce a synthetic data base of satellite brightness temperatures each with an associated SST. Using a multiple linear regression technique that includes noise estimates in the synthetic brightness temperatures, linear algorithm coefficients to give SST can then be derived. For the first two instruments, the model used was developed at the Rutherford Appleton Laboratory in the UK [84] but for AATSR, an improved model, based on the original Závody model, has been developed by Merchant and LeBorgne [85] and now provides the most recent algorithms for SST derivation. This latest version of the model is also being used to provide new algorithms for the earlier instruments in a program to re-process the long-term data archive provided by this series of satellites.

The ATSR instruments, with their two views and three different wavelengths (six "channels"), can provide SST using several different sets of algorithms. In all cases, a simple linear algorithm is used with the selected brightness temperatures; see Equation (7). During the daytime, the two main algorithms use either the nadir 11 and 12 μm brightness temperatures (in the same manner as the AVHRR and MODIS) or both views at these two wavelengths in a four-channel derivation. During the night, the algorithms commonly used include a three-channel nadir view only, or a dual-view version with all six channels.

Accurate calibration of the brightness temperatures is achieved by using two on-board blackbody cavities, situated between the apertures for the nadir and forward views such that they are scanned on each rotation of the mirror. One calibration target is at the spacecraft ambient temperature while the other is heated, so that the measured brightness temperatures of the earth's surface are straddled by the calibration temperatures. The values of the wavelength-dependent emissivity of the blackbodies of ATSR are 0.9986 at 3.75 μm, 0.9993 at 10.85 μm and 0.9992 at 12.0 μm, and these have been determined with sufficient accuracy that the contribution to the uncertainty in the ATSR brightness temperature measurements from the blackbodies is < 35 mK (3σ) [86]. The ATSR-2 and AATSR each have similar calibration performance [87]. The limitation of the simple scanning geometry of the ATSR is a relatively narrow swath width of 512 km. This disadvantage is offset by the improvement in absolute accuracy of the atmospheric correction, and of its better insensitivity to aerosol effects.

4.4. Geostationary meteorological satellites

A system of geostationary satellites located above the equator is maintained by the meteorological agencies of the USA, Europe, India, Japan and China. These satellites all carry visible and infrared radiometers that can provide regular (better than hourly) measurements of the SST and LST in their region. The spatial resolution of these radiometers in the infrared is less than that for the polar-orbiting satellites (typically 5 km) due to their greater distance from the earth's surface of about 36,000 km. However, the hourly data provide better opportunities for cloud-free monitoring of the surface temperatures. Nearly all of the radiometers on geostationary satellites have a split-window capability and simple multi-channel algorithms are used to derive surface temperatures in the same manner as for the polar-orbiting satellites. There are two distinct types of geostationary satellite: the "spin stabilized" satellites, such as the European Meteosat series, rotate at 100 rpm about an axis parallel to that of the earth, and "three-axis stabilized" satellites such as the US Geostationary Operational Environmental Satellite (GOES) series rotate once per sidereal day to keep a single face of the satellite pointing toward the earth throughout its orbit. The three-axis stabilized approach has some inherent advantages in terms of the amount of time that the radiometers are pointed toward the earth and therefore collecting useful information, but the daily variation in solar heating and entry of sunlight into the sensor around midnight, when the sun passes behind the earth (from the satellite's viewpoint), lead to problems with the internal thermal balance of the radiometers.

4.5. Advanced microwave scanning radiometer for the earth observing system

The design of the optical path through infrared radiometers relies on the use of reflective surfaces, with the only exception usually being the transmission band-pass filters which define the spectral response of each channel. The calculation of the instantaneous field of view of the radiometer is relatively simple as the principles of geometric optics can be used, and the spatial resolution is independent of the wavelength of the radiation being measured. This is feasible because the dimensions of the apertures (typically centimeters to tens of centimeters) are very much greater than the wavelength of the electromagnetic radiation (typically 10 μm or less). For microwave radiometers, a similar requirement for the use of reflective surfaces holds, but the wavelength of the radiation is only an order of magnitude or so smaller than the primary scan mirror, usually an off-axis parabolic antenna. This means the spatial resolution of the instrument is diffraction limited, and depends on the ratio of the antenna diameter to the wavelength of the electromagnetic radiation. Given the constraints on the size of an antenna that can be accommodated on a spacecraft, generally a meter or less, the spatial resolution of a microwave radiometer at the earth's surface is dependent on the frequency of the emission being measured in each channel.

Microwave radiometers use a similar measurement principle to infrared radiometers, having several spectral channels to provide the information to correct for extraneous effects, and blackbody calibration targets to ensure the accuracy of the measurements. A critical component to the SST retrieval is the value of the surface emissivity which is strongly dependent on the emission angle as well as frequency. The effects of the angular dependence can be reduced by using a conical scan approach, so the angle of intersection of the radiometer field of view with the earth's surface is a constant. There remains a dependency on the surface roughness as the tilts of facets of the sea surface are dependent on the local wind (and this provides a mechanism for the measurement of wind speed by microwave radiometry), and this can be reduced in the SST retrieval by using measurements from multiple channels. The sensitivity of the microwave emission is more pronounced at low frequencies, around 7 GHz. A simple combination of the brightness temperatures in different channels can again be used to retrieve the SST.

The AMSR-E [88] was developed in Japan, and is part of the instrument payload on the NASA *Aqua* satellite [78]. AMSR-E is a six-frequency dual-polarized passive microwave radiometer with the largest main reflector of its kind (1.6 m diameter). The SST measurement capability is through channels close to 6.9 GHz which have a spatial resolution of ~56 km. Because of the diffraction effects, a significant fraction of the energy received by microwave radiometers comes not from the main lobe of the antenna pattern, that is, where the antenna points the field of view, but from side lobes. In coastal regions, this leads to a significant contamination of the oceanic signal, which is small because of the low emissivity of sea water, from the land, where the emissivity is relatively large. This means that accurate SST measurements are not feasible within a hundred kilometers or so of land. An advantage of microwave measurements results from the fact that the microwave emission from the sea surface propagates through the atmosphere, including clouds, with relatively little attenuation. Thus, SSTs can be derived in cloudy conditions. This is a significant advantage over infrared SST measurements, and in many applications this outweighs the inherent disadvantages of poor spatial resolution and side-lobe contamination.

An important recent development is the merging of SST fields derived from MODIS and AMSR-E, both on *Aqua*. In cloud-free conditions, and close to coasts, the infrared retrievals dominate, whereas in cloudy conditions the microwave measurements are used. Finding the optimum approach to melding data from different sensors is currently an active field of research.

5. VALIDATION OF SURFACE TEMPERATURE RETRIEVALS

The utility of a satellite-derived surface temperature field depends on the knowledge of its accuracy and the characteristics of the residual uncertainties. For example, the assimilation of a surface temperature measurement in an NWP operation requires an estimate of the uncertainty of the measurement. A temperature measurement that is believed to be relatively inaccurate is given less weight than one that is believed to have smaller uncertainties. Similarly, the analysis of a time series of surface temperature measurements made with the intention of revealing signatures of climate change will not lead to a convincing result if the uncertainties associated with the measurements are larger than the anticipated signal.

The approach to determine the uncertainties in satellite-derived surface temperatures is to compare them with coincident measurements from an independent source. This is called "validation" as it leads to a confirmation of the calibration procedures that take place in the instrument on the satellite, and the processing steps required to derive a surface temperature from the radiometric measurements. Of particular importance are the corrections of any instrumental artifacts in the raw measurements and of the effects of the intervening atmosphere. Ideally, the reference measurements should be more accurate than the satellite measurements, and they should share as many of the same characteristics as possible, such as spatial coverage, type of measurement (e.g. a radiometric validation measurement is preferred over a contact thermometer) and time of measurement, so that "like is compared with like" and additional sources of uncertainty caused by the near-surface temperature gradients (Figure 1) can be rendered negligible. In reality, such a set of reference measurements is not always available or even feasible, and an error budget has to be constructed that takes into account the contributions of imperfections in the process.

For some applications of satellite-derived surface temperatures, such as monitoring the positions and evolution of the surface expressions of thermal fronts in the ocean, absolute accuracy is not of great significance and a good precision, or relative accuracy, is more important. However, for many purposes the accuracy of the temperatures derived from satellite data is of prime importance. Perhaps the most demanding application of satellite-derived surface temperatures is in climate research where a multi-decadal time series of global surface temperatures is required to detect small changes that are expected to reveal the response of the climate to changing forcing. Such time series are often referred as "Climate Data Records" (CDRs) and a publication of the US National Academy of Sciences [3] defines a CDR as "a data set designed to enable study and assessment of long-term climate change, with 'long-term' meaning year-to-year and decade-to-decade change. Climate research often involves the detection of small changes against a background of intense, short-term variations". To derive CDRs from satellite data, "calibration and validation should be considered as a process that encompasses the entire system, from the sensor performance to the derivation of the data products". Given the likely sources of uncertainties in the surface temperature retrievals, an important aspect of any validation exercise is ensuring that the full range of orbital and atmospheric conditions is sampled. The orbital aspect is important as the thermal conditions on the spacecraft change markedly around the orbit, with thermal shock being experienced as the satellite enters and leaves the shadow of the earth [7], and the performance of a radiometer is dependent, at least to some degree, on its thermal environment. Sampling the full range of atmospheric conditions is important as the uncertainties in the surface temperatures are likely to be dominated by the residual, uncorrected effects of the atmosphere, especially in the infrared. As stated in the National Academy of Sciences publication [3], it is important to continue validation efforts over the lifetimes of the spacecraft sensors to ensure that the effects of degradation of the instruments in orbit are not misinterpreted as being caused by environmental signals. In generating time series of surface temperatures that span several satellite missions, the role of validation includes providing the necessary continuity in the derived fields.

There are several approaches to validating satellite-derived surface temperatures that use different instruments. Some are mounted on aircraft, others on ships or buoys and others at fixed validation sites on land.

5.1. SST validation using radiometers

The approach using aircraft is particularly attractive as it utilizes radiometers that can be constructed to replicate quite closely the measurements of the satellite radiometer. The aircraft radiometer, in common with the satellite instrument, is required to have internal calibration to provide accurate measurements, but unlike the satellite sensor, the calibration procedure of the aircraft instrument can be carefully checked before and after each flight or deployment. The speed of the aircraft permits measurements over distances that provide a spatial average comparable to the footprint of the satellite instruments, and the aircraft can fly to areas that are cloud-free at the time of the satellite overpass, if feasible. There are several flight options for validation, including high-altitude transects underneath the satellite, whereby the aircraft sensor is used to replicate the satellite measurement, but a source of residual uncertainty is in the effects of the atmosphere between the aircraft and satellite. An alternative approach is to fly at a low level so that the aircraft instrument is being used to measure the surface temperature, but here the source of uncertainty is in the effects of the atmosphere beneath the aircraft. The major drawback, though, of using aircraft for surface temperature validation is cost.

For SST validation, an alternative approach is to mount infrared radiometers on ships. While lacking the speed of aircraft to maneuver into cloud-free areas, long-term deployment of instruments is quite feasible, and even if up to about 90% of possible overpasses are contaminated by clouds [49], those that remain can make a significant contribution to the validation of the satellite SST retrievals. The ship-based radiometers have to be mounted so as to have a clear view of the sea surface ahead of the bow wave of the ship, so they are measuring the skin SST undisturbed by the presence of the ship, as far as is feasible. Since the emissivity of the sea surface is not unity, a small component of the signal measured by the radiometer viewing the sea surface is reflected sky radiance, and a measurement of the incident atmospheric emission is needed to correct for this. Consequently, the validating instrument must be able to view the sky at the same angle to zenith as the sea view is inclined to nadir, as illustrated in Figure 5. The radiometers can be mounted at heights of a few meters to tens of meters above the sea surface, so the effects of the atmosphere below are negligible or easily corrected to good accuracy [89,90]. As with radiometers on spacecraft and aircraft, a ship-based instrument must be calibrated throughout the field deployment using internal calibration targets, and the calibration procedure should be checked using laboratory facilities before and after each deployment.

To date, the ship-based instruments are of two types: filter radiometers where the band-pass (the relative spectral response) of the instrument is determined by an optical filter and hyperspectral interferometers which measure a broad spectral range, a section of which is selected for the skin SST measurement.

Examples of well-calibrated filter radiometers developed for the validation of satellite-derived SSTs include the Infrared Scanning Autonomous Radiometer (ISAR) [91], the Calibrated InfraRed *in situ* Measurement System (CIRIMS) [92,93] and the Scanning Infrared Sea Surface Temperature Radiometer (SISTeR; described in Ref. [94]). These are autonomous instruments that can be installed on ships to operate for weeks or months at a time with little or no operator intervention.

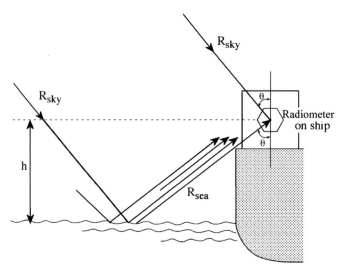

Figure 5 The viewing geometry of a ship-board radiometer measuring the skin
SST. The sky view is necessary to provide a correction for the small component in
the sea-view measurements that is reflected infrared emission from the sky. The
intersection of the radiometer field of view and the sea surface should be ahead of
the bow wave to minimize the local influence of the ship.

For example, an ISAR is mounted on the M/V *Val de Loire* and M/V *Pride
of Bilbao* [95] and another on the M/V *Jingu Maru*. In the past, a CIRIMS
has been installed on the research vessels *Thomas G. Thompson* and the
Ronald H. Brown, and the US Coast Guard icebreaker *Polar Sea* [92]. Data
are recorded internally or on a local computer and transmitted to shore
via satellite telemetry link. The ISAR and CIRIMS use commercial
radiometers with band-pass filters covering the 9.6−11.5 μm wavelength
range. The SISTeR has a filter wheel with three filters corresponding to the
mid- and thermal infrared bands of the AVHRRs and (A)ATSRs. All three
have two small internal blackbody calibration targets, for real-time
calibration in the field. These radiometers, and others, are described in
Barton et al. [94].

The hyperspectral validation data are derived from the Marine-
Atmospheric Emitted Radiance Interferometer (M-AERI) [89]. This is a
Fourier Transform Infrared (FTIR) interferometric spectroradiometer
that operates in the range of infrared wavelengths from ∼3 to ∼18 μm
and measures spectra with a resolution of ∼0.5 cm^{-1}. It was developed
specifically for the validation of MODIS skin SST retrievals and includes
very accurate real-time calibration using two internal blackbody cavities.
Two infrared detectors are used to achieve this wide spectral range, and
these are cooled to ∼78 K (close to the boiling point of liquid nitrogen)
by a Stirling cycle mechanical cooler to reduce the noise equivalent

temperature difference to levels well below 0.1 K. A scan mirror, which is programmed to step through a pre-selected range of angles, directs the field of view from the interferometer to either of the internal blackbody calibration targets or to the environment from nadir to zenith. The interferometer integrates measurements over a pre-selected time interval, typically $60-100$ s, to obtain a satisfactory signal-to-noise ratio, and a typical cycle of measurements including two view angles to the atmosphere, one to the ocean, and calibration measurements, takes about $5-10$ min. The M-AERI is equipped with pitch and roll sensors so that the influence of the ship's motion on the measurements can be determined.

The absolute accuracy of the infrared spectra produced by the M-AERI is determined by the effectiveness of the blackbody cavities as calibration targets. The absolute accuracy of the spectral measurements of the M-AERI is better than 0.03 K [89]. The absolute uncertainties of the retrieved skin SST, determined by operating two M-AERIs side-by-side and by comparing M-AERI measurements with those from other well-calibrated radiometers, are less than 0.05 K [89,94] which are sufficiently small to give confidence in the use of such data in the validation of satellite SST retrievals. Since the launch of *Terra*, over 40 M-AERI cruises have been undertaken on research vessels. In addition, an M-AERI and suite of atmospheric sensors have been deployed on the cruise ship *Explorer of the Seas*, operated by Royal Caribbean International, which was equipped at construction with scientific laboratories for oceanographic and atmospheric research. This vessel has become one of the most valuable single sources of measurements for validation of satellite retrievals of SST. Other radiometer cruises have spanned a wide range of climatological regimes, from polar regions to the tropics. In addition to sampling the full range of SST, from freezing to the high temperatures in the Red Sea and Tropical Western Pacific, validation data have been taken in a wide range of marine atmospheres.

One advantage of taking validation measurements from moving ships is that the integration along the ships' tracks provides a spatially averaged measurement that is closely akin to the spatially averaged measurement within the spacecraft radiometer footprint, at least on the kilometer scales of the retrievals in the infrared. This approach assumes that a one-dimensional average along the ships' tracks is a good approximation of the two-dimensional average of the spacecraft radiometer.

5.2. SST validation using buoys

The radiometric measurements of skin temperatures provide the most physically appropriate data sets for the validation of satellite-derived SSTs, but they are relatively small in number compared to those provided by the network of drifting buoys. The buoys have an advantage of numbers which helps to offset the disadvantage of uncertain calibration [96] and the additional source of uncertainty resulting from near-surface temperature gradients (see above). The large numbers of the buoy data set render it too valuable a resource for SST validation for it to be neglected, but the quality control of the buoy data sets has to be thorough. These devices are rarely retrieved at the end of their lifetime to assess calibration drift, sensor damage or contamination.

Moored buoys in the deep tropical Pacific and Atlantic Oceans are also used as a source of validating data. They generally have a thermometer at a depth of about 3 m, and while these are recovered for post-deployment re-calibration [97] the disturbance to the upper ocean by the presence of the mooring introduces a source of uncertainty into the comparison that is difficult to quantify.

The distribution of collocated (within 0.1° of latitude and longitude) and coincident (within 30 min) measurements from the *Aqua* MODIS and drifting and moored buoys is shown in Figure 6 for the year 2003. There are 12,536 match-ups in the plot and only those satellite retrievals with the best quality flag (i.e. confidently cloud-free and satellite zenith angle <45°) are shown, but even with this large number there are ocean areas that are poorly sampled.

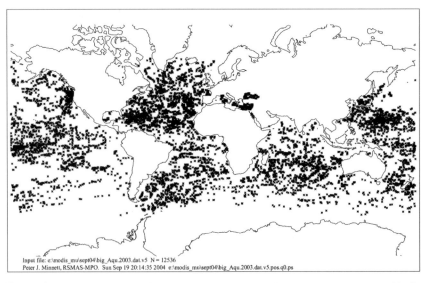

Input file: e:\modis_mu\sept04\big_Aqu.2003.dat.v5 N = 12536
Peter J. Minnett, RSMAS-MPO. Sun Sep 19 20:14:35 2004 e:\modis_mu\sept04\big_Aqu.2003.dat.v5.pos.q0.ps

Figure 6   The distribution of the match-ups between the MODIS on *Aqua* and bulk SST measurements from drifting and moored buoys, for 2003. There are 12,536 match-ups with the highest quality flag (QF = 0), indicating the highest level of confidence in the effectiveness of the cloud screening. Generally, less than 10% of all match-ups between buoys and satellite data are QF = 0.

5.3. Traceability to temperature standards

An important aspect to the generation of CDRs of SST, especially since they span several satellite missions, is ensuring that the surface temperatures derived from the satellite measurements are traceable to national temperature standards. As part of the pre-launch characterization of the satellite instruments, the radiometers are carefully calibrated in thermal-vacuum chambers to replicate the conditions on orbit. But the instruments are not recovered at the end of the mission for re-calibration and re-characterization. The path to national temperature standards, therefore, is through the sensors used to validate the satellite retrievals. In the USA, the national reference standards are maintained by the National Institute of Standards and Technology (NIST) and to ensure traceability to NIST standards, an infrared calibration facility has been established at the Rosenstiel School of Marine and Atmospheric Science at the University of Miami. Three international workshops have been held at which many of the ship-board radiometers used to validate satellite-derived SSTs were calibrated using a water-bath blackbody calibration target, built to a NIST design [98]. This calibration target consists of a black-painted, thin-walled, hollow, tapered, copper cone surrounded by a water bath, the temperature of which can be very accurately controlled. The water-bath temperature is monitored by two thermometers, with calibrations traceable to NIST standards. The radiometers are calibrated by pointing them into the cone. The radiation emerging from the cone depends not only on its temperature, as given by the thermometers in the water bath, but also on its emissivity. The emissivity was determined, and hence the calibration system characterized, by the NIST Transfer Radiometer (TXR) [99], which is the infrared radiometric standard for the NASA EOS program [100]. The TXR was also used to characterize the laboratory blackbody calibrators used elsewhere to check the internal calibration of the ship-deployed radiometers [101]. The comparative performance of the radiometers in conditions such as they experience in the field was determined by mounting them together on a short cruise of the R/V *F.G. Walton Smith* [94]. The outcome of this exercise is that the radiometers with internal calibration and an effective correction for the reflected sky radiance are capable of measuring the skin SST with uncertainties <0.1 K and they can be used to help generate CDRs of SST. Other at-sea comparisons between radiometers of different design have taken place on a more *ad hoc* basis, with similar results [93].

6. Residual Uncertainties

Improvements to the accuracy of satellite-derived SSTs continue to be made as our understanding of the behavior of the instruments, and hence of the instrumental artifacts in the data stream, becomes more complete and refinements to the atmospheric correction algorithms are found, often as a result of analyses of the validation data sets. The residual uncertainties in the retrievals are often expressed as a mean error, or bias, and a scatter, or standard deviation about the mean; this terminology assumes a Gaussian error distribution. Sometimes the bias and scatter are combined in a root mean square (rms) error estimate. However, the reduction of the uncertainty fields to a few numbers does not indicate the complexity of the information that is required for some applications, such as the assimilation of the data in NWP and Ocean Forecasting models [102]. Some of the uncertainties result from systematic dependences on some of the parameters that determine the conditions at the time of the satellite measurement, such as the satellite zenith angle (through the slant path length through the atmosphere and the surface emissivity), the atmospheric water vapor and temperature distribution, the characteristics of aerosols and the surface wind speed. And, of course, the residual uncertainties depend on the details of the processing algorithms, in particular the cloud screening approaches, and the choice of coefficients in the atmospheric correction algorithm. The coefficients are generally optimized in the sense of producing minimum errors on a global basis, and application of the resulting SST fields in regionally or seasonally constrained analyses can result in different error characteristics [8,103–105].

The following summary of accuracies of the SSTs derived from various sensors should be interpreted with these caveats in mind. They are determined by comparison with independent measurements and generally the discrepancies are ascribed to the satellite measurement. Since the independent measurements have their own uncertainties, and the method of comparison introduces additional uncertainties, the SST retrievals from the satellite-based measurements are likely to be more accurate than indicated. A recent study involving collocated measurements of three different sources types of SSTs provides sufficient information to enable the standard deviation of error on each observation type to be derived, thus allowing a more appropriate apportioning of the uncertainties [106].

6.1. AVHRR

The statistics shown in Table 1a are for the AVHRR PFSST fields [49] and show a near-zero mean error and standard deviation of about half a degree when compared with independent measurements as described above. The coefficients in the atmospheric correction algorithms used in the Pathfinder project are derived by robust regression between the AVHRR brightness temperatures and *in situ* measurements from drifting buoys, so the retrieved SSTs are a "bulk SST" and they are related to the skin SST, which is more directly related to the radiance measured in space, through a mean skin

Table 1a Uncertainties in the Pathfinder AVHRR SST retrievals.

Data	All
Mean (K)	0.02
Standard deviation (K)	0.53

From cloud-free comparisons between all AVHRRs in the period 1985–1998 and drifting buoys. After Ref. [49].

Table 1b Uncertainties in the AVHRR SST retrievals vs. MAERI.

Data	Mean (K)	Standard deviation (K)	N	Excluding Arctic cruise		
				Mean (K)	Standard deviation (K)	N
All	0.14	0.36	299	0.07	0.31	219
Day	0.18	0.40	142	0.00	0.24	62
Night	0.10	0.33	157	0.10	0.33	157

After Ref. [41]. © American Meteorological Society.

temperature difference. Using a radiometric SST from the M-AERI [89] reduces the error (Table 1b), with the exception of polar regions [41] where the multi-channel atmospheric correction algorithm is known to be prone to additional uncertainties [66]. Given that the AVHRR was originally designed to be a cloud imager with little emphasis on absolute radiometry, these statistics are remarkable.

6.2. MODIS

The error statistics of the MODIS SST retrievals from the instruments on *Terra* and *Aqua* are shown in Tables 2a–d, derived from comparisons with buoys and radiometers. The processing scheme is V5 for both instruments. The periods covered are from the start of each of the missions to mid-2007. The first thing to note is that the numbers of valid comparisons with buoys is very much greater than with radiometers (mainly from M-AERIs). The statistics are shown for daytime and night-time retrievals using the $11 - 12\,\mu m$ measurements and the night-time retrievals using the $4\,\mu m$ data. The MODIS SST retrievals are of a skin temperature and the mean errors (biases) that result from the comparison with the buoys are a manifestation of the thermal skin effect, which is typically in the range of $0.15 - 0.20\,K$ [22]. The bias errors approximate to zero in the comparisons with the radiometer data. The standard deviations of the uncertainties are higher during the day than at night, reflecting the larger contribution of the variability in diurnal heating between the surface and the depth of the buoy

Table 2a Uncertainties in the *Terra* MODIS SST retrievals vs. buoys.

Data	11 µm SST			4 µm SST		
	Mean (K)	Standard deviation (K)	N	Mean (K)	Standard deviation (K)	N
All	−0.15	0.61	254834			
Day	−0.11	0.64	149893			
Night	−0.20	0.55	104941	−0.18	0.47	114562

Table 2b Uncertainties in the *Aqua* MODIS SST retrievals vs. buoys.

Data	11 µm SST			4 µm SST		
	Mean (K)	Standard deviation (K)	N	Mean (K)	Standard deviation (K)	N
All	−0.17	0.49	243826			
Day	−0.15	0.55	151814			
Night	−0.22	0.48	92012	−0.22	0.42	99986

Table 2c Uncertainties in the *Terra* MODIS SST retrievals vs. M-AERI.

Data	11 µm SST			4 µm SST		
	Mean (K)	Standard deviation (K)	N	Mean (K)	Standard deviation (K)	N
All	0.04	0.53	4751			
Day	0.09	0.58	1999			
Night	0.01	0.53	2752	−0.02	0.43	3056

Table 2d Uncertainties in the *Aqua* MODIS SST retrievals vs. M-AERI.

Data	11 µm SST			4 µm SST		
	Mean (K)	Standard deviation (K)	N	Mean (K)	Standard deviation (K)	N
All	0.00	0.56	2093			
Day	0.04	0.59	832			
Night	−0.02	0.53	1261	−0.06	0.45	1399

measurements during the day [28]. The day—night differences in the scatter are much smaller in the radiometer comparisons. The night-time 4 μm SST retrievals show markedly smaller scatter than the 11−12 μm SST, which reflects not only the inherent superiority of the measurements in the shorter wavelength atmospheric window, but also the better instrument performance. The dual-sided paddle-wheel scan mirror has a multi-layer interference coating that introduces a dependence of the reflectivity on the scan angle for wavelengths greater than about 8 μm [81]. Imperfect corrections for this effect introduce an additional uncertainty in the 11−12 μm SST retrievals. The uncertainty characteristics in the *Aqua* MODIS retrievals are better than for the *Terra* MODIS and this presumably results from improvements in the instrument construction and pre-launch characterization of the second instrument, as some of the lessons learned in the development of the *Terra* MODIS could be applied to the MODIS on *Aqua*.

6.3. AATSR

The inherent advantage of atmospheric correction based on the dual view of the AATSR is demonstrated in the residual uncertainties when compared to buoys and radiometers (Tables 3a and b). The disadvantage of the narrow swath (∼500 km) is also revealed in the relatively small number of comparisons. The night-time algorithm uses all three infrared channels at both views, while the daytime algorithm uses both views of

Table 3a Uncertainties in the *Envisat* AATSR SST retrievals.

Data	Buoy comparisons			Radiometer comparisons		
	Mean (K)	Standard deviation (K)	N	Mean (K)	Standard deviation (K)	N
Day	0.02	0.39	∼5500	0.11	0.33	18
Night	0.04	0.28	∼5500	−0.06	0.20	12

From information in Ref. [95].

Table 3b Uncertainties in the *Envisat* AATSR SST retrievals vs. M-AERI skin SST in the Caribbean Sea, measured from the *Explorer of the Seas*.

Data	Mean (K)	Standard deviation (K)	N
Day	0.16	0.36	32
Night	0.04	0.26	84

From information in Ref. [107].

the two channels in the thermal infrared atmospheric window, but not the mid-infrared channel at 3.7 μm wavelength because of the contamination by sun glitter. The buoy comparisons are taken from the 12-month period from 19 August 2002, and amount to about 30 per day, with collocation and coincidence criteria of 10 min of latitude and longitude, and 3 h [95]. Given that the AATSR SST retrieval is a skin temperature, the smaller bias errors derived from the comparison to buoys to that with radiometers are curious and probably indicate a warm bias in the SSTs, which is apparent in the daytime radiometer comparisons. The smaller night-time bias in the radiometer comparison, along with the smaller standard deviation compared to the daytime retrievals, indicates the benefit of including the 3.7 μm measurements in the derivation.

An independent validation exercise using M-AERI data from the *Explorer of the Seas* in the Caribbean Sea [107] supports the evidence of a small warm bias in the AATSR skin SST data with a clear day–night difference (Table 3b). The standard deviations revealed in all of the AATSR validation are notably smaller than for any other infrared radiometer.

6.4. Geostationary meteorological satellites

In principle, the accuracies of the surface temperatures derived from the infrared channels on radiometers on geostationary satellites should be comparable to those on polar orbiters, given that the atmospheric corrections are essentially the same. However, the standard deviations of the SST uncertainties, when compared to temperatures measured from buoys, are in general greater than for those derived from radiometers on polar-orbiting satellites [108]. For the GOES SST retrievals, part of the elevated scatter (0.6–0.9 K) is caused by errors that are systematic in the time of day [108], and are therefore likely to be related to instrumental artifacts with a diurnal character.

6.5. AMSR-E

The near-real time validation statistics of the AMSR-E SST retrievals are shown as daily means against a variety of subsurface temperature measurements at http://www.ssmi.com/amsr/amsre_sst_validation_statistics.html. These go back to 1 June 2002, close to the start of the mission, and extend to the present. Table 4a gives the AMSR-E SST uncertainties as a mean and standard deviation when compared to drifting buoy measurements up to 2 July 2008. Table 4b gives the uncertainties when compared to the M-AERI on the *Explorer of the Seas* in the Caribbean Sea. The bias errors in the AMSR-E retrievals when compared with skin and bulk SSTs are physically reasonable given that the emission depth of microwave radiation is greater than for the infrared, and thereby contains

Table 4a Uncertainties in the *Aqua* AMSR-E SST retrievals vs. drifting buoys.

Data	All
Mean (K)	−0.03
Standard deviation (K)	0.58
N	1646488

Derived from data from http://www.ssmi.com/amsr/amsre_sst_validation_statistics.html for the period 1 June 2002−2 July 2008.

Table 4b Uncertainties in the *Aqua* AMSR-E SST retrievals vs. M-AERI on the *Explorer of the Seas* data from July 2002 to July 2005.

Data	Night+day
Mean (K)	0.18
Standard deviation (K)	0.59
N	139

emission from below the thermal skin layer of the ocean, which results in the anticipation of microwave measurements of SSTs being somewhat warmer than the skin SST. However, given the instrumental issues related to the temperature knowledge of the AMSR-E hot-load calibration target, and the empirical approach to its correction [109], the characteristics of the bias errors in the AMSR-E SSTs are likely to be governed by this correction for an instrumental artifact rather than the geophysical processes at the sea surface.

6.6. Land surface temperature

The MODIS LST algorithms have been validated using both ground-based and air-borne data from intensive field campaigns. Details are given at www.icess.ucsb.edu/modis/modis-lst_valid.html. Validation of the LST products from the AATSR on *Envisat* has been reported by Sobrino et al. [62]. The wide range of different vegetation covers and their lack of homogeneity over satellite radiometer fields of view mean that tables of LST accuracy show a wide variability compared to those reported for SST. The accuracies predicted by Prata [63] in his AATSR LST Algorithm Theoretical Basis Document are of the order of 2.5 K during the day and close to 1.0 K during the night. Many of the reports of LST validation taken over a wide range of land surfaces suggest that these values are representative of the derivations from the MODIS, AVHRR and AATSR instruments. In some cases the accuracies are somewhat better than these values, but in others the estimates show larger deviations from the predictions.

7. APPLICATIONS OF REMOTELY SENSED SURFACE TEMPERATURES

Remote sensing of surface temperatures is a relatively mature science with many applications. Some rely on the unique viewpoint of earth-orbiting satellites to provide spatial information that is not obtainable by any other means, while others exploit the accuracy of the surface temperature retrievals, and some require both. Here we provide a brief overview of some applications. It is not an exhaustive list and neither is it detailed, but is intended to give a flavor of the wide range of applications. The interested reader can find more details in the cited references.

7.1. Ocean surface fronts, eddies and currents

One of the most striking features apparent in infrared images of SST is ocean fronts, which delineate the boundaries between dissimilar surface water masses (Figure 7). The most conspicuous cases are western boundary currents, such as the Gulf Stream in the Atlantic Ocean and the Kuroshio in the Pacific Ocean, both of which transport warm surface water poleward and away from the western coastlines. In the Atlantic Ocean, the flow of the warm surface water of the Gulf Stream can be seen in SST images to extend into the Norwegian Sea, and into the Arctic Ocean, where the heat in the surface layer prevents winter ice formation in the Barents Sea and to the west of Svalbard. The surface water in this current system provides

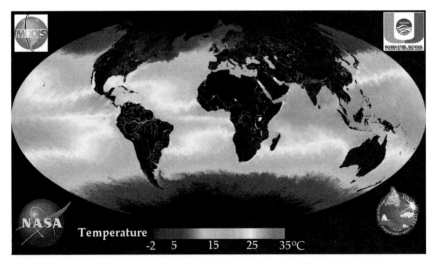

Figure 7 Global cloud-free composite of skin SST derived from the measurements of MODIS on *Terra* taken during May 2001.

heat to the atmosphere, moderating the climate in Western Europe and producing a marked zonal difference in the conditions on the opposite coasts of the Atlantic and Greenland–Norwegian Seas. In the southern hemisphere, the boundary currents off Australia and South America are also apparent in infrared imagery. Off South Africa, the SST images provide a view of the flow of the Aghulas Current, the western boundary current of the southern Indian Ocean, meeting the Atlantic Ocean. Here the Aghulas Current is often forced back to the south and east, giving rise to eddies of water with a warm surface temperature, with diameters of several hundreds of kilometers. These subsequently propagate into the South Atlantic. The near-zonal fronts in the Southern Ocean are apparent in the surface temperature images.

Instabilities like those seen in the Aghulas Current are a common feature of all fronts and are revealed in detail in satellite-derived images of the SST, and the meanders of the Gulf Stream and Kuroshio are particularly pronounced. These frequently give rise to detached eddies on both sides of the main current flow. Another notable feature in SST images is instabilities along the boundaries of the Equatorial current system in the Pacific Ocean. The extent and structure of these features were first described by analysis of satellite SST images [110]. Similar but less pronounced equatorial instability waves are also seen in the Atlantic Ocean.

Kelly and Strub [111] published one of the first papers on the use of successive satellite images to estimate ocean surface currents by tracking ocean surface features. Their technique has been refined by successive researchers including Bowen et al. [112] and Barton [113] to provide operational ocean current estimates from brightness temperature images from the operational AVHRR instruments. For the correlation analysis, the infrared images are preferred as they are available for both night and day and, as surface variability only is required, infrared brightness temperatures are again preferred over SST fields. In the Maximum Cross-Correlation technique, two infrared images are first mapped onto a common geographic grid maintaining the high spatial resolution of close to 1 km. These images need to be separated in time by between 2 and 14 h — shorter times do not allow significant movement of any surface features while longer times allow their dispersion to restrict any pattern matching. For each location for which a surface movement vector is required, a search window or template is selected in the first image and then scanned over the second image until a maximum cross-correlation is found giving an estimate of the surface movement between the two images. Search windows with dimensions of between 20 and 40 km are common and the search area is usually restricted to within 50 km of the template location.

Figure 8 shows an analysis of ocean currents from two AVHRR images off the western Australian coast. The figure also shows that this technique can produce vectors (currents) that are obviously incorrect.

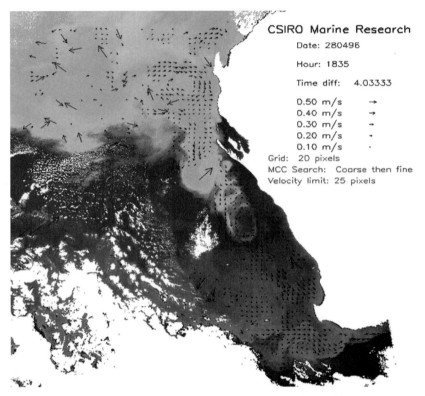

Figure 8 An example of the surface current field (arrows) derived from the Maximum Cross-Correlation method, between the NOAA-14 AVHRR channel 4 brightness temperatures ($\lambda = 11\,\mu m$) shown in the image and those from the NOAA-12 AVHRR 4 h later. The images cover a large area of $1{,}600 \times 1{,}600$ pixels including the Leeuwin Current off the west coast of Australia. The white area to the right (east) is Australia, and other white areas are clouds.

Several techniques have been advanced to eliminate these "rogue" vectors from the analysis. These include nearest neighbor matching [112], restrictions on the minimum cross-correlation coefficient and derived displacement vectors, comparisons with altimeter-derived currents [114] and a reciprocal filtering technique described by Barton [113].

With a minimum of two AVHRR radiometers operating at any one time, it is possible to derive surface currents from the infrared imagery on an operational basis. However, as with SST estimation, currents can only be derived in areas where both images are cloud-free. A combination of altimeter data and MCC currents provides a valuable tool for oceanographic research and monitoring. These ocean current fields can be used in weather and ocean-state forecasting, ship routing, pollution monitoring, search and rescue, fisheries and many other applications.

7.2. Numerical weather prediction

The earth's surface is the lower boundary of the atmosphere and correct specification of its temperature is an important step toward improving the skill of numerical weather forecasts. This is most relevant over the ocean where the surface temperature distribution is particularly dynamic and the coupling between the ocean and atmosphere in terms of exchange of heat and moisture is particularly important. In the past several years, an international program, the Global Ocean Data Assimilation Experiment (GODAE) High Resolution SST Pilot Project (GHRSST-PP), has focused on providing satellite-derived SST fields from a variety of sensors, along with error estimates for each individual pixel, in real-time for use in NWP centers around the world [102]. To be of use in an NWP environment, the processed SSTs have to be available within 6 h of the measurement being taken. The SST fields are also being used in the relatively new applications of ocean forecasting.

7.3. Severe storm development

Accurate forecasting of the path and intensity of land-falling severe storms is very important. These are called hurricanes in the Atlantic and eastern Pacific Oceans, typhoons in the western north Pacific Ocean and cyclones in the western south Pacific and Indian Oceans. Storm intensification is driven by the heat available in the upper ocean of which SST is a good indicator. As a useful rule, regions of SSTs of 26°C and higher appear to be necessary for severe storm intensification. To forecast hurricanes in the Atlantic, satellite-derived SST maps are used in the prediction of the development of storm propagation from the area off Cape Verde where atmospheric easterly waves spawn the nascent storms. Closer to the USA and Caribbean, the SST field is important in determining the sudden intensification that can occur just before landfall. An important case is the sudden intensification of the hurricanes in the Gulf of Mexico that can be strongly influenced by the position of the Loop Current. In reality, the SST field is an approximation for the heat content of the upper ocean, and a better approach makes use of satellite-derived SSTs in conjunction with estimates of the mixed layer depth, that can be derived from satellite altimeter surface elevation measurements and a simple upper ocean model [115].

In addition to the role of SST in severe storm intensification, there is an opposite effect that can cause a sudden weakening of the storm. After a storm has passed, it often leaves a wake of cooler water at the surface that is readily identifiable in the satellite-derived SST fields. A well-documented example of the decay of a hurricane occurred with Hurricane Danielle in August 1998. A few days earlier, Hurricane Bonnie had intensified over the warm waters north of the Bahamas, and, by extracting heat from the ocean, reduced the surface temperature leaving a cold wake along her path. As Hurricane Danielle encountered this cool pool of surface water, she quickly lost her strength, only to recover as she ventured out over warmer water to the north [116]. Without the evidence of the cold wake derived from satellite measurements, in this case from microwave data, the sudden decay of the second hurricane would have been inexplicable, at least until detailed post-event analysis and modeling had shed light on the situation.

7.4. El Niño—Southern Oscillation

Considering forecasting severe events on a longer time scale, the El Niño—Southern Oscillation (ENSO) is an important feature, probably the largest anomaly in the ocean—atmosphere system to occur on time scales of many months with irregular, interannual recurrence [117]. "El Niño" is the name given to an anomaly in the tropical eastern Pacific Ocean off Ecuador and Peru that is associated with an increase in the surface temperature, frequently occurring around Christmas, hence the appellation. It has since been recognized as an anomaly that extends across the entire tropical Pacific Ocean and is related to a perturbation in the surface air pressure distribution, the Southern Oscillation. The normal situation is for the western Pacific to be warmer at the surface than the eastern. This large area of warm surface water, called the Warm Pool, heats the air above, contributing to the forcing of the large-scale patterns of winds over the globe, including the Trade Winds. At irregular and unpredictable intervals of several years, the Warm Pool extends to the east, warming the eastern Pacific Ocean and disturbing the atmospheric circulation. This has consequences around most of the globe by changing the positions and intensities of storms and rainfall patterns, causing drought and floods that disrupt agriculture and lead to loss of life and damage to property. The prediction of ENSO events, and their consequences, is very difficult and not well developed, so the monitoring of SSTs in the tropical Pacific to give early indication of an incipient ENSO event is a valuable component to forecasting the wide-ranging effects.

7.5. Detection of climate-change signals

The average surface temperature of the oceans has been suggested to be a global thermometer that might indicate the initial effects of climate change [118]. However, because of the complex structure of the ocean currents, the way they influence the surface temperatures, the large-scale perturbations of the ENSO on interannual time scales and other features, the anticipated heating trend expected to result from global warming is very difficult to identify clearly. The anticipated global change signal in SSTs is a few tenths of a degree per decade and to detect this requires a long time series of satellite-derived SSTs [1]. A recent study using 20 years of AVHRR PFSSTs has indeed revealed trends in SST of (0.18 ± 0.04) and (0.17 ± 0.05) K per decade from daytime and night-time data [2]. Clearly, more confidence in these trends, and information about the climate response to increasing levels of greenhouse gases, will result from continued measurements of the global SSTs from satellite.

7.6. Coral bleaching

A final example of the application of satellite-derived SSTs is the now routine monitoring of elevated temperatures in the tropics where living coral reefs can be harmed. When the temperatures exceed the local average summertime maximum for several days, the symbiotic relationship between the coral polyps and their algae, which is necessary for the corals to thrive, breaks down and mortality of the reef-building animals can result. This leads to extensive areas where the coral reef is reduced to the skeletal structure without the living and growing tissue, giving the reef a white appearance [119]. Time series of AVHRR-derived SST have been shown to be valuable predictors of reef areas around the globe that are threatened by warmer than usual water temperatures [120]. Although it is not possible to alter the outcome, SST maps have been useful in determining the scale of the problem and identifying threatened or vulnerable reefs. An example of real-time application of SSTs to monitoring potential coral bleaching events is to be found at http://www.osdpd.noaa.gov/PSB/EPS/SST/climohot.html.

7.7. Land surface temperature

In the same manner as for SST, there are several essential applications of LST estimates in modeling for climate and weather prediction, and regular use of these data is made by most meteorological services. Estimates of SST are essential for forecasting the paths and intensities of tropical storms, but once landfall is made surface LST can be used to assess the transition of storms from hurricanes to local rain depressions.

Another application of LST is in agriculture for the prediction and measurement of frost damage to crops. During the life cycle of wheat, there is a two-week period following anthesis (flowering) when the plant is extremely vulnerable to frost damage. Prediction of likely frost occurrences allows farmers to take measures that will limit the damage to their crops. Other uses in agriculture include the forecasting of crop growth, the estimate of crop yields and the evaluation of the best time for harvest. The difference between the LST measured at night and during the day (the thermal inertia) can give a reasonably accurate estimate of local soil moisture that has applications in agriculture and in surface energy balance studies. Monitoring of frost potential also allows road traffic warnings to be issued by local authorities when appropriate.

Infrared mapping at times of floods can also give a better measurement of flooded areas than can be gained by other means, and it has been shown that daytime measurements, when the difference between the temperature of the floodwaters and the surrounding land is greatest, gives the best estimates of which areas are flooded [121].

7.8. Fire detection

Satellite-borne infrared radiometers have the capability of detecting bushfires through the detection of pixels that have a higher brightness temperature than expected from normal land surfaces [122,123]. In most cases if a simple signal threshold is exceeded, then it is assumed to indicate the presence of a wildfire. Several countries and regions have now established operational systems to alert emergency agencies of fire locations and intensities. In the USA, NOAA developed a Hazard Mapping System Fire and Smoke Product [124] using data from the GOES, MODIS and AVHRR imagery, and the US Department of Agriculture Forest Service has a web-based service based on MODIS measurements [125]. In Australia, the Sentinel bushfire monitoring system was developed by CSIRO, but is now an operational product managed by Geoscience Australia [126]. Again, bushfire detection is derived from AVHRR and MODIS infrared signals. A threshold system is used for the brightness temperature difference between the 4 and 11 µm channels with a separate threshold value for day and night observations. The 4 µm channel is most sensitive to hot-spot detection as the temperature/radiance gradient at this wavelength is significantly larger than that at 11 µm. Systems have also been established in Europe and Africa.

8. ATMOSPHERIC PROFILES

Although the main subject of this chapter is the measurement of surface temperature from satellites, in which the presence of the variable atmosphere is primarily a source of error, the infrared measurements from satellites can also be used to determine vertical air temperature profiles as well as the abundances of some atmospheric constituents. In contrast to the surface temperature measurements, which are made in spectral intervals where the atmosphere is relatively transparent, the measurements of atmospheric properties use spectral regions where the infrared signal at the top of the atmosphere is dominated by atmospheric emission.

For deriving the atmospheric temperature profile, the emitted infrared radiation should come from a constituent that is well mixed and has a strong thermal infrared spectral absorption feature (e.g. the 4.3 μm absorption band of CO_2). Infrared radiometers designed for "atmospheric sounding" have multiple, narrow spectral bands situated on the wing of well-defined emission features, where the atmospheric transmissivity varies strongly with wavelength. The measured brightness temperatures are related to the radiation emitted from different height ranges in the atmosphere. Brightness temperatures measured near the center of the spectral features indicate conditions close to the top of the atmosphere, while others, derived from where there is greater transmission, are related to temperatures lower in the atmosphere. Thus, it is possible to retrieve a profile of air temperature for cloud-free conditions. Similarly, profiles of the concentrations of gases that are not well mixed through the atmosphere, such as water vapor and ozone, can be derived. Measurements from the TIROS Operational Vertical Sounder (TOVS), a suite of three radiometers — one infrared and two microwave — that flies along with the AVHRR on the NOAA operational meteorological satellites, produce vertical profiles of temperature using the 4.3 and 15 μm absorption features of CO_2, the spectral intervals at 6.3 μm for water vapor and at 9 μm for ozone.

9. FUTURE MISSIONS

At the time of writing, mid-2008, plans are well developed for three of the instruments that will continue the time series of infrared-derived surface temperatures into the next decades. These are the VIIRS being developed in the USA, the Sea and Land Surface Temperature Radiometer (SLSTR) in Europe and the Second-generation Global Imager (SGLI) in Japan. These are described briefly below.

9.1. NPOESS VIIRS

The long time series of SSTs from the AVHRRs on the NOAA polar-orbiting satellites is coming to an end as the NOAA satellite series will be superseded by the National Polar-orbiting Operational Environmental Satellite System (NPOESS) [127]. But AVHRRs will continue to be flown on the European MetOp satellites until 2020. On the NPOESS satellites, SST will be measured in the infrared by the VIIRS [128]. At 0.65 km, the spatial resolution of VIIRS at nadir will be somewhat better than most of the previous infrared imagers, but it retains the standard infrared atmospheric window channels: two in the mid-infrared window

($\lambda = 3.7$ and $4.05\,\mu m$) and two in the thermal infrared window ($\lambda = 10.8$ and $12.0\,\mu m$). The VIIRS design takes some of the innovative aspects introduced in MODIS, including the internal blackbody calibration target and multiple detectors per spectral band, but to avoid the sources of uncertainty in the MODIS SSTs derived from measurements in the thermal infrared window introduced by the paddle-wheel scan mirror, the VIIRS optical path includes a rotating telescope assembly instead. The first of the VIIRS series is planned for launch in 2011 on the NASA NPOESS Preparatory Project (NPP) satellite.

9.2. GMES Sentinel-3 SLSTR

The European Global Monitoring for Environment and Security (GMES) [129] program comprises five series of satellites, named Sentinel-1 to -5. The Sentinel-1, -2 and -3 series will be launched to take measurements of the earth's surface from polar orbit. The Sentinel-4 is planned for geostationary orbit, and to focus on atmospheric chemistry, as will Sentinel-5 from polar orbit. Accurate infrared radiometry for the measurement of surface temperature in the $2012-2020$ period will be done from the two satellites planned for the Sentinel-3 series with the SLSTR [130], which is a development continuing the (A)ATSR series of dual-view radiometers. It will have a different scan mirror mechanism that will permit a wider nadir swath of 1,675 km, but the measurement geometry restricts the slant path swath to about 750 km. SLSTR is currently planned to have three infrared channels at 3.7, 10.8 and $12.0\,\mu m$ with a nadir spatial resolution of 1 km, and a further six channels for other purposes. It is anticipated that the excellent radiometric calibration of the (A)ATSR series using two internal blackbody calibration targets will be maintained in the SLSTRs. A dual-view, multi-channel atmospheric correction scheme will be used where the nadir and slant-path swaths overlap, and beyond that a single-view, multi-channel algorithm will be applied.

9.3. GCOM-C SGLI

The Japan Aerospace Exploration Agency (JAXA) is preparing two series of polar-orbiting satellites, designated Global Change Observation Mission-Water (GCOM-W) and Global Change Observation Mission-Climate (GCOM-C). Three satellites with a lifetime of five years are planned for each mission. Thus, the program is expected to run for 13 years or longer, with the first launch of GCOM-W in 2012 and of GCOM-C in 2014. The instruments on GCOM-C will comprise a suite of radiometers covering the wavelength range from the ultraviolet ($\lambda = 0.375\,\mu m$) to the thermal infrared ($\lambda = 12.0\,\mu m$), which are collectively known as the Second-SGLI. The infrared scanner (IRS) is currently planned to have two channels in the thermal infrared ($\lambda = 10.8$ and $12.0\,\mu m$) with a spatial resolution at nadir of $0.5\,km$ and a swath width of $1,400\,km$.

The companion series of satellites, GCOM-W will carry the Advanced Microwave Scanning Radiometer 2 (AMSR2) that will have characteristics similar to AMSR-E, including a good SST measurement capability, but with a better spatial resolution of $35\,km \times 62\,km$ at $6.9\,GHz$, that results from a larger primary scanning antenna of $2.0\,m$ diameter.

10. Conclusions and Outlook

The accurate measurement of the surface temperature of the earth on global scales, yet with sufficient spatial and radiometric resolution to reveal structure on the scales of kilometers, is a triumph of satellite remote sensing. The images of eddies and fronts in the ocean have revolutionized our perception of the upper ocean energetics. The residual uncertainties in the surface temperature retrievals have been demonstrated to be sufficiently small for their widespread use in a variety of research and operational applications, including NWP. It is remarkable that a consistent time series of nearly three decades has been compiled and analyzed to reveal the very small temperature trends likely to be associated with the response of the climate to changing greenhouse gas forcing. The generation of "CDRs" from satellite data has been demonstrated to be feasible for surface temperatures measured from space.

The complementarity of the SSTs derived from microwave and infrared radiometry, and their comparable error characteristics, implies that substantial benefits are to be gained by merging these fields to exploit the microwave measurements through clouds with the higher spatial resolution of the infrared retrievals in cloud-free conditions. This is a topic of current research.

A number of operational and experimental satellite radiometers are currently approaching the end of their planned lifetimes. Replacement instruments of new and innovative design are being constructed, or planned, but there is a risk that failure of some of the instruments currently in space will cause a reduction in the amount or quality of the satellite-derived surface temperatures.

One of the greatest strengths of satellite remote sensing is the combined retrieval of several related variables in the climate system, such as surface temperature, atmospheric water vapor, cloud properties and trace gases. Some variables are accessible to remote sensing using infrared radiation while others are measured using reflected sunlight in the ultraviolet and visible parts of the spectrum and others using microwave emission. Another branch of satellite remote sensing employs active sensors: radars and lidars. In this chapter, we have focused on passive infrared radiometry for the measurement of surface temperature, as it is a well-defined, measurable variable and is a prime variable in the climate system. The interested reader can find expanded details of this and other aspects of satellite remote sensing in a number of textbooks, such as those by Robinson [131], and Kidder and Haar [132], or more general surveys of the field, such as those edited by Gurney et al. [133] and by King et al. [134].

REFERENCES

[1] M. R. Allen, C. T. Mutlow, G. M. C. Blumberg, J. R. Christy, R. T. McNider, and D. T. Llewellyn-Jones, "Global Change Detection," Nature **370**, 24−25 (1994).

[2] S. A. Good, G. K. Corlett, J. J. Remedios, E. J. Noyes, and D. T. Llewellyn-Jones, "The Global Trend in Sea Surface Temperature from 20 Years of Advanced Very High Resolution Radiometer Data," J. Climate **20**, 1255−1264 (2007).

[3] NRC, *Issues in the Integration of Research and Operational Satellite Systems for Climate Research: II. Implementation*, National Academy of Sciences, Washington, DC (2000).

[4] NRC, "*Climate Data Records from Environmental Satellites*," Washington, DC (2004).

[5] K. S. Casey and P. Cornillon, "A Comparison of Satellite and in situ-based Sea Surface Temperature Climatologies," J. Climate **12**, 1848−1863 (1999).

[6] A. E. Strong, E. J. Kearns, and K. K. Gjovig, "Sea Surface Temperature Signals from Satellites − An Update," Geophys. Res. Lett. **27**, 1667−1670 (2000).

[7] O. B. Brown, J. W. Brown, and R. H. Evans, "Calibration of Advanced Very High Resolution Radiometer Infrared Observations," J. Geophys. Res. **90**, 11667−11677 (1985).

[8] P. J. Minnett, "The Regional Optimization of Infrared Measurements of Sea-Surface Temperature from Space," J. Geophys. Res. **95**, 13497−13510 (1990).

[9] B. A. Wielicki, B. R. Barkstrom, E. F. Harrison, R. B. Lee, G. Louis Smith, and J. E. Cooper, "Clouds and the Earth's Radiant Energy System (CERES): An Earth Observing System Experiment," Bull. Am. Meteorol. Soc. **77**, 853−868 (1996).

[10] S. Platnick, M. D. King, S. A. Ackerman, W. P. Menzel, B. A. Baum, J. C. Riédi, and R. A. Frey, "The MODIS Cloud Products: Algorithms and Examples from Terra," IEEE Trans. Geosci. Remote Sensing **41**, 459−473 (2003).

[11] M. D. King, W. P. Menzel, Y. J. Kaufman, et al., "Cloud and Aerosol Properties, Precipitable Water, and Profiles of Temperature and Humidity from MODIS," IEEE Trans. Geosci. Remote Sensing **41**, 442−458 (2003).

[12] M. Szczodrak, P. J. Minnett, N. R. Nalli, and W. F. Feltz, "Profiling the Lower Troposphere over the Ocean with Infrared Hyperspectral Measurements of the Marine-Atmosphere Emitted Radiance Interferometer," J. Oceanic Atmos. Technol. **24**, 390−402 (2007).

[13] A. Mahesh, V. P. Walden, and S. G. Warren, "Ground-Based Infrared Remote Sensing of Cloud Properties over the Antarctic Plateau. Part I: Cloud-Base Heights," J. Appl. Meteorol. **40**, 1265−1278 (2001).

[14] J. A. Hanafin and P. J. Minnett, "Cloud Forcing of Surface Radiation in the North Water Polynya," Atmos. Ocean **39**, 239−255 (2001).

[15] A. M. Vogelmann, P. J. Flatau, M. Szczodrak, K. Markowicz, and P. J. Minnett, "Observations of Large Aerosol Infrared Forcing at the Surface," Geophys. Res. Lett. **30**, 1655 (2003).

[16] J. C. Price, "Land Surface Temperature Measurements from the Split Window Channels of the NOAA-7 Advanced Very High Resolution Radiometer," J. Geophys. Res. **89**, 7231−7237 (1984).

[17] A. J. Prata, "Land Surface Temperatures derived from the Advanced Very High Resolution Radiometer and the Along-Track Scanning Radiometer 2. Experimental Results and Validation of AVHRR Algorithms," J. Geophys. Res. **99**, 13025−13058 (1994).

[18] C. Fairall, E. Bradley, J. Godfrey, G. Wick, J. Edson, and G. Young, "Cool-Skin and Warm-Layer Effects on Sea Surface Temperature," J. Geophys. Res. **101**, 1295−1308 (1996).

[19] J. F. Price, R. A. Weller, and R. Pinkel, "Diurnal Cycling: Observations and Models of the Upper Ocean Response to Diurnal Heating, Cooling and Wind Mixing," J. Geophys. Res. **91**, 8411−8427 (1986).

[20] J. A. Hanafin and P. J. Minnett, "Profiling Temperature in the Sea Surface Skin Layer using FTIR Measurements," in *Gas Transfer at Water Surfaces*, edited by M. A. Donelan, W. M. Drennan, E. S. Saltzmann, and R. Wanninkhof, American Geophysical Union, Washington, DC, pp. 161−166 (2001).

[21] J. A. Hanafin, "On Sea Surface Properties and Characteristics in the Infrared," PhD Thesis, Meteorology and Physical Oceanography, University of Miami (2002).

[22] C. J. Donlon, P. J. Minnett, C. Gentemann, T. J. Nightingale, I. J. Barton, B. Ward, and J. Murray, "Toward Improved Validation of Satellite Sea Surface Skin Temperature Measurements for Climate Research," J. Climate **15**, 353−369 (2002).

[23] P. J. Minnett, M. Smith, and B. Ward, "Measurements of the Oceanic Thermal Skin Effect," Deep Sea Res. II (2009), submitted for publication.

[24] A. V. Soloviev and P. Schlüssel, "Parameterization of the Cool Skin of the Ocean and of the Air−Ocean Gas Transfer on the basis of Modeling Surface Renewal," J. Phys. Oceanogr. **24**, 1339−1346 (1994).

[25] R. I. Leighton, G. B. Smith, and R. A. Handler, "Direct Numerical Simulations of Free Convection beneath an Air−Water Interface at Low Rayleigh Numbers," Phys. Fluids **15**, 3181−3193 (2003).

[26] A. T. Jessup, C. J. Zappa, M. R. Loewen, and V. Hesany, "Infrared Remote Sensing of Breaking Waves," Nature **385**, 52−55 (1997).

[27] W. Eifler and C. J. Donlon, "Modeling the Thermal Surface Signature of Breaking Waves," J. Geophys. Res. **106**, 27, 163−27, 185 (2001).

[28] C. L. Gentemann and P. J. Minnett, "Radiometric Measurements of Ocean Surface Thermal Variability," J. Geophys. Res. **113**, C08017 (2008).

[29] H. U. Sverdrup, M. W. Johnson, and R. H. Fleming, *The Oceans: Their Physics, Chemistry, and General Biology*, Prentice-Hall, Englewood Cliff, NJ (1942).

[30] J. D. Woods and W. Barkmann, "The Response of the Upper Ocean to Solar Heating. I: The Mixed Layer," Q. J. R. Meteorol. Soc. **112**, 1−27 (1986).

[31] P. Schluessel, W. J. Emery, H. Grassl, and T. Mammen, "On the Bulk-Skin Temperature Difference and its Impact on Satellite Remote Sensing of Sea Surface Temperatures," J. Geophys. Res. **95**, 13, 341−13, 356 (1990).

[32] R. Yokoyama, S. Tanba, and T. Souma, "Sea Surface Effects on the Sea Surface Temperature Estimation by Remote Sensing," Int. J. Remote Sensing **16**, 227−238 (1995).

[33] P. J. Minnett, "Radiometric Measurements of the Sea-Surface Skin Temperature − The Competing Roles of the Diurnal Thermocline and the Cool Skin," Int. J. Remote Sensing **24**, 5033−5047 (2003).

[34] B. Ward, "Near-Surface Ocean Temperature," J. Geophys. Res. **111**, C02005 (2006).

[35]   C. L. Gentemann, P. J. Minnett, P. LeBorgne, and C. J. Merchant, "Multi-Satellite Measurements of Large Diurnal Warming Events," Geophys. Res. Lett. **35**, L22602 (2008).

[36]   P. J. Webster, C. A. Clayson, and J. A. Curry, "Clouds, Radiation, and the Diurnal Cycle of Sea Surface Temperature in the Tropical Western Pacific," J. Climate **9**, 1712−1730 (1996).

[37]   J. D. Woods, W. Barkmann, and A. Horch, "Solar Heating of the Oceans − Diurnal, Seasonal, and Meridional Variations," Q. J. R. Meteorol. Soc. **110**, 633−686 (1984).

[38]   S. S. Chen and R. A. Houze, "Diurnal Variation and Lifecycle of Deep Convective Systems over the Tropical Pacific Warm Pool," Q. J. R. Meteorol. Soc. **123**, 357−388 (1997).

[39]   T. N. Palmer and D. A. Mansfield, "Response of Two Atmospheric General Circulation Models to Sea-Surface Temperature Anomalies in the Tropical East and West Pacific," Nature **310**, 483−485 (1984).

[40]   J. Shukla, "Predictability in the Midst of Chaos: A Scientific Basis for Climate Forecasting," Science **282**, 728−731 (1998).

[41]   E. J. Kearns, J. A. Hanafin, R. H. Evans, P. J. Minnett, and O. B. Brown, "An Independent Assessment of Pathfinder AVHRR Sea Surface Temperature Accuracy using the Marine-Atmosphere Emitted Radiance Interferometer (M-AERI)," Bull. Am. Meteorol. Soc. **81**, 1525−1536 (2000).

[42]   A. Schiller and J. S. Godfrey, "A Diagnostic Model of the Diurnal Cycle of Sea Surface Temperature for Use in Coupled Ocean−Atmosphere Models," J. Geophys. Res. **110**, C11014 (2005).

[43]   C. L. Gentemann, P. J. Minnett, and B. Ward, "Profiles of Ocean Surface Heating (POSH): A New Model of Upper Ocean Diurnal Thermal Variability," J. Geophys. Res. **114**, C07017 (2009).

[44]   C. L. Gentemann, C. J. Donlon, A. Stuart-Menteth, and F. J. Wentz., "Diurnal Signals in Satellite Sea Surface Temperature Measurements," Geophys. Res. Lett. **30**, 1140−1143 (2003).

[45]   L. McMillin, "Estimation of Sea-Surface Temperatures from Two Infrared Window Measurements with different Absorption," J. Geophys. Res. **80**, 5113−5117 (1975).

[46]   D. T. Llewellyn-Jones, P. J. Minnett, R. W. Saunders, and A. M. Závody, "Satellite Multichannel Infrared Measurements of Sea-Surface Temperature of the N.E. Atlantic Ocean using AVHRR/2," Q. J. R. Meteorol. Soc. **110**, 613−631 (1984).

[47]   F. Chevallier, S. D. Michele, and A. P. McNally, "Diverse Profile Datasets from the ECMWF 91-Level Short-Range Forecasts," Tech. Rep. No. NWPSAF-EC-TR-010, Satellite Application Facility for Numerical Weather Prediction (2006).

[48]   C. C. Walton, W. G. Pichel, J. F. Sapper, and D. A. May, "The Development and Operational Application of Nonlinear Algorithms for the Measurement of Sea Surface Temperatures with the NOAA Polar-Orbiting Environmental Satellites," J. Geophys. Res. **103**, 27999−28012 (1998).

[49]   K. A. Kilpatrick, G. P. Podestá, and R. H. Evans, "Overview of the NOAA/NASA Pathfinder Algorithm for Sea Surface Temperature and associated Matchup Database," J. Geophys. Res. **106**, 9179−9198 (2001).

[50]   W. E. Esaias, M. R. Abbott, I. Barton, et al., "An Overview of MODIS Capabilities for Ocean Science Observations," IEEE Trans. Geosci. Remote Sensing **36**, 1250−1265 (1998).

[51]   C. O. Justice, J. R. G. Townshend, E. F. Vermote, et al., "An Overview of MODIS Land Data Processing and Product Status," Remote Sensing Environ. **83**, 3−15 (2002).

[52] C. J. Merchant, P. LeBorgne, A. Marsouin, and H. Roquet, "Optimal Estimation of Sea Surface Temperature from Split-Window Observations," Remote Sensing Environ. **112**, 2469−2484 (2008).

[53] http://www.icess.ucsb.edu/modis/modis-lst.html

[54] Z. Wan and J. Dozier, "A Generalized Split-Window Algorithm for retrieving Land-Surface Temperature from Space," IEEE Trans. Geosci. Remote Sensing **34**, 892−905 (1996).

[55] Z. Wan and Z.-L. Li, "A Physics-Based Algorithm for Retrieving Land-Surface Emissivity and Temperature from EOS/MODIS Data," IEEE Trans. Geosci. Remote Sensing **35**, 980−996 (1997).

[56] W. C. Snyder, Z. Wan, Y. Zhang, and Y.-Z. Feng, "Classification-Based Emissivity for Land Surface Temperature Measurement from Space," Int. J. Remote Sensing **19**, 2753−2774 (1998).

[57] W. Snyder and Z. Wan, "BRDF Models to predict Spectral Reflectance and Emissivity in the Thermal Infrared," IEEE Trans. Geosci. Remote Sensing **36**, 214−225 (1998).

[58] http://www.icess.ucsb.edu/modis/EMIS/html/em.html

[59] Z. Wan, *MODIS Land-Surface Temperature Algorithm Theoretical Basis Document (LST ATBD) Version 3.3*, Institute for Computational Earth System Science, University of California, Santa Barbara (1999).

[60] J. A. Sobrino, Z.-L. Li, M. P. Stoll, and F. Becker, "Improvements in the Split-Window Technique for Land Surface Temperature Determination," IEEE Trans. Geosci. Remote Sensing **32**, 243−253 (1994).

[61] A. J. Prata, "Surface Temperatures derived from the Advanced Very High Resolution Radiometer and the Along Track Scanning Radiometer. 1. Theory," J. Geophys. Res. **98**, 16689−16702 (1993).

[62] J. A. Sobrino, G. Sòria, and A. J. Prata, "Surface Temperature Retrieval from Along Track Scanning Radiometer 2 Data: Algorithms and Validation," J. Geophys. Res. **109**, D11101 (2004).

[63] A. J. Prata, "Land Surface Temperature Measurement from Space: AATSR Algorithm Theoretical Basis Document," CSIRO Division of Atmospheric Research (2002).

[64] P. J. Minnett, "A Numerical Study of the Effects of Anomalous North Atlantic Atmospheric Conditions on the Infrared Measurement of Sea-Surface Temperature from Space," J. Geophys. Res. **91**, 8509−8521 (1986).

[65] J. R. Key, J. B. Collins, C. Fowler, and R. S. Stone, "High-Latitude Surface Temperature Estimates from Thermal Satellite Data," Remote Sensing Environ. **61**, 302−309 (1997).

[66] R. F. Vincent, R. F. Marsden, P. J. Minnett, K. A. M. Creber, and J. R. Buckley, "Arctic Waters and Marginal Ice Zones: A Composite Arctic Sea Surface Temperature Algorithm using Satellite Thermal Data," J. Geophys. Res. **113**, C04021 (2008).

[67] R. F. Vincent, R. F. Marsden, P. J. Minnett, and J. R. Buckley, "Arctic Waters and Marginal Ice Zones: Part 2 − An Investigation of Arctic Atmospheric Infrared Absorption for AVHRR Sea Surface Temperature Estimates," J. Geophys. Res. **113**, C08044 (2008).

[68] T. T. Wilheit, A. T. C. Chang, and A. Milman, "Atmospheric Corrections to Passive Microwave Observations of the Ocean," Boundary-Layer Meteorol. **18**, 65−77 (1980).

[69] F. J. Wentz, "A Well-Calibrated Ocean Algorithm for SSM/I," J. Geophys. Res. **102**, 8703−8718 (1997).

[70]   D. T. Eppler, L. D. Farmer, A. W. Lohanick, et al., "Passive Microwave Signatures of Sea Ice," in *Microwave Remote Sensing of Sea Ice*, edited by F. D. Carsey, American Geophysical Union, Washington, DC, pp. 47−71 (1992).

[71]   J. C. Comiso, D. J. Cavalieri, C. L. Parkinson, and P. Gloersen, "Passive Microwave Algorithms for Sea Ice Concentration − A Comparison of Two Techniques," Remote Sensing Environ. **60**, 357−384 (1997).

[72]   K. Steffen and A. Schweiger, "NASA Team Algorithm for Sea Ice Concentration Retrieval from Defense Meteorological Satellite Program Special Sensor Microwave Imager: Comparison with Landsat Satellite Imagery," J. Geophys. Res. **96**, 21971−21987 (1991).

[73]   J. C. Comiso, "Warming Trends in the Arctic from Clear Sky Satellite Observations," J. Climate **16**, 3498−3510 (2003).

[74]   D. J. Cavalieri, P. Gloersen, and W. J. Campbell, "Determination of Sea Ice Parameters with the Nimbus-7 SMMR," J. Geophys. Res. **89**, 5355−5369 (1984).

[75]   J. C. Comiso, "Arctic Multiyear Ice Classification and Summer Ice Cover using Passive Microwave Satellite Data," J. Geophys. Res. **95**, 13411−13422 (1990).

[76]   T. Markus and B. A. Burns, "A Method to estimate Subpixel-Scale Coastal Polynyas with Satellite Passive Microwave Data," J. Geophys. Res. **100**, 4473−4487 (1995).

[77]   Y. J. Kaufman, D. D. Herring, K. J. Ranson, and G. J. Collatz, "Earth Observing System AM1 Mission to Earth," IEEE Trans. Geosci. Remote Sensing **36**, 1045−1055 (1998).

[78]   C. L. Parkinson, "Aqua: An Earth-Observing Satellite Mission to Examine Water and other Climate Variables," IEEE Trans. Geosci. Remote Sensing **41**, 173−183 (2003).

[79]   W. L. Barnes, T. S. Pagano, and V. V. Salomonson, "Prelaunch Characteristics of the Moderate Resolution Imaging Spectroradiometer (MODIS) on EOS-AM1," IEEE Trans. Geosci. Remote Sensing **36**, 1088−1110 (1998).

[80]   B. Guenther, X. Xiong, V. V. Salomonson, W. L. Barnes, and J. Young, "On-Orbit Performance of the Earth Observing System Moderate Resolution Imaging Spectroradiometer; First Year of Data," Remote Sensing Environ. **83**, 16−30 (2002).

[81]   X. Xiong, C. Kwo-Fu, W. Aisheng, W. L. Barnes, B. Guenther, and V. V. Salomonson, "Multiyear On-Orbit Calibration and Performance of Terra MODIS Thermal Emissive Bands," IEEE Trans. Geosci. Remote Sensing **46**, 1790−1803 (2008).

[82]   A. J. Prata, R. P. Cechet, I. J. Barton, and D. T. Llewellyn-Jones, "The Along-Track Scanning Radiometer for ERS-1-Scan Geometry and Data Simulation," IEEE Trans. Geosci. Remote Sensing **28**, 3−13 (1990).

[83]   S. J. Brown, A. R. Harris, I. M. Mason, and A. M. Závody, "New Aerosol Robust Sea Surface Temperature Algorithms for the Along-Track Scanning Radiometer," J. Geophys. Res. **102**, 27973−27989 (1997).

[84]   A. M. Závody, C. T. Mutlow, and D. T. Llewellyn-Jones, "A Radiative Transfer Model for Sea-Surface Temperature Retrieval for the Along Track Scanning Radiometer," J. Geophys. Res. **100**, 937−952 (1995).

[85]   C. J. Merchant and P. LeBorgne, "Retrieval of Sea Surface Temperature from Space, based on Modeling of Infrared Radiative Transfer: Capabilities and Limitations," J. Atmos. Oceanic Technol. **21**, 1734−1746 (2004).

[86]   I. M. Mason, P. H. Sheather, J. A. Bowles, and G. Davies, "Blackbody Calibration Sources of High Accuracy for a Spaceborne Infrared Instrument: The Along Track Scanning Radiometer," Appl. Opt. **35**, 629−639 (1996).

[87] D. L. Smith, J. Delderfield, D. Drummond, T. Edwards, C. T. Mutlow, P. D. Read, and G. M. Toplis, "Calibration of the AATSR Instrument," Adv. Space Res. **28**, 31−39 (2001).

[88] T. Kawanishi, T. Sezai, Y. Ito, et al., "The Advanced Microwave Scanning Radiometer for the Earth Observing System (AMSR-E), NASDA's Contribution to the EOS for Global Energy and Water Cycle Studies," IEEE Trans. Geosci. Remote Sensing **41**, 184−194 (2003).

[89] P. J. Minnett, R. O. Knuteson, F. A. Best, B. J. Osborne, J. A. Hanafin, and O. B. Brown, "The Marine-Atmospheric Emitted Radiance Interferometer (M-AERI), a High-Accuracy, Sea-Going Infrared Spectroradiometer," J. Atmos. Oceanic Technol. **18**, 994−1013 (2001).

[90] W. L. Smith, R. O. Knuteson, H. E. Revercomb, et al., "Observations of the Infrared Radiative Properties of the Ocean − Implications for the Measurement of Sea Surface Temperature via Satellite Remote Sensing," Bull. Am. Meteorol. Soc. **77**, 41−51 (1996).

[91] C. Donlon, I. S. Robinson, M. Reynolds, W. Wimmer, G. Fisher, R. Edwards, and T. J. Nightingale, "An Infrared Sea Surface Temperature Autonomous Radiometer (ISAR) for Deployment aboard Volunteer Observing Ships (VOS)," J. Atmos. Oceanic Technol. **25**, 93−113 (2008).

[92] A. T. Jessup and R. Branch, "Integrated Ocean Skin and Bulk Temperature Measurements using the Calibrated Infrared in situ Measurement System (CIRIMS) and Through-Hull Ports," J. Atmos. Oceanic Technol. **25**, 579−597 (2008).

[93] R. Branch, A. T. Jessup, P. J. Minnett, and E. L. Key, "Comparisons of Shipboard Infrared Sea Surface Skin Temperature Measurements from the CIRIMS and the M-AERI," J. Atmos. Oceanic Technol. **25**, 598−606 (2008).

[94] I. J. Barton, P. J. Minnett, C. J. Donlon, S. J. Hook, A. T. Jessup, K. A. Maillet, and T. J. Nightingale, "The Miami2001 Infrared Radiometer Calibration and Inter-Comparison: 2. Ship Comparisons," J. Atmos. Oceanic Technol. **21**, 268−283 (2004).

[95] G. K. Corlett, I. J. Barton, C. J. Donlon, et al., "The Accuracy of SST Retrievals from AATSR: An Initial Assessment through Geophysical Validation against in situ Radiometers, Buoys and other SST Data Sets," Adv. Space Res. **37**, 764−769 (2006).

[96] W. J. Emery, D. J. Baldwin, P. Schlüssel, and R. W. Reynolds, "Accuracy of in situ Sea Surface Temperatures used to Calibrate Infrared Satellite Measurements," J. Geophys. Res. **106**, 2387−2405 (2001).

[97] H. P. Freitag, M. E. McCarty, C. Nosse, R. Lukas, M. J. McPhaden, and M. F. Cronin, "COARE Seacat Data: Calibrations and Quality Control Procedures," NOAA Technical Memorandum ERL PMEL-115 (1999).

[98] J. B. Fowler, "A Third Generation Water Bath based Blackbody Source," J. Res. Natl. Inst. Stand. Technol. **100**, 591−599 (1995).

[99] J. P. Rice and B. C. Johnson, "The NIST EOS Thermal-Infrared Transfer Radiometer," Metrologia **35**, 505−509 (1998).

[100] J. P. Rice and B. C. Johnson, "A NIST Thermal Infrared Transfer Standard Radiometer for the EOS Program," The Earth Observer **8**, 31 (1996).

[101] J. P. Rice, J. J. Butler, B. C. Johnson, et al., "The Miami2001 Infrared Radiometer Calibration and Intercomparison: 1. Laboratory Characterization of Blackbody Targets," J. Atmos. Oceanic Technol. **21**, 258−267 (2004).

[102] C. Donlon, I. Robinson, K. S. Casey, et al., "The Global Ocean Data Assimilation Experiment High-Resolution Sea Surface Temperature Pilot Project," Bull. Am. Meteorol. Soc. **88**, 1197−1213 (2007).

[103] A. Kumar, P. J. Minnett, G. Podesta, and R. H. Evans, "Error Characteristics of the Atmospheric Correction Algorithms used in Retrieval of Sea Surface Temperatures from Infrared Satellite Measurements; Global and Regional Aspects," J. Atmos. Sci. **60**, 575−585 (2003).

[104] S. C. Shenoi, "On the Suitability of Global Algorithms for the Retrieval of SST from the North Indian Ocean using NOAA/AVHRR Data," Int. J. Remote Sensing **20**, 11−29 (1999).

[105] F. Eugenio, J. Marcello, A. Hernández-Guerra, and E. Rovaris, "Regional Optimization of an Atmospheric Correction Algorithm for the Retrieval of Sea Surface Temperature from the Canary Islands−Azores−Gibraltar Area using NOAA/AVHRR Data," Int. J. Remote Sensing **26**, 1799−1814 (2005).

[106] A. G. O'Carroll, J. R. Eyre, and R. W. Saunders, "Three-Way Error Analysis between AATSR, AMSR-E, and in situ Sea Surface Temperature Observations," J. Atmos. Oceanic Technol. **25**, 1197−1207 (2008).

[107] E. J. Noyes, P. J. Minnett, J. J. Remedios, G. K. Corlett, S. A. Good, and D. T. Llewellyn-Jones, "The Accuracy of the AATSR Sea Surface Temperatures in the Caribbean," Remote Sensing Environ. **101**, 38−51 (2006).

[108] G. A. Wick, J. J. Bates, and D. J. Scott, "Satellite and Skin-Layer Effects on the Accuracy of Sea Surface Temperature Measurements from the Goes Satellites," J. Atmos. Oceanic Technol. **19**, 1834−1848 (2002).

[109] F. Wentz, C. L. Gentemann, and P. Ashcroft, "On-Orbit Calibration of AMSR-E and the Retrieval of Ocean Products," in *12th Conference on Satellite Meteorology and Oceanography, American Meteorological Society*, Long Beach, CA (2003).

[110] R. Legeckis, "Long Waves in the Eastern Equatorial Pacific Ocean: A View from a Geostationary Satellite," Science **197**, 1179−1181 (1977).

[111] K. A. Kelly and P. T. Strub, "Comparison of Velocity Estimates from Advanced Very High Resolution Radiometer in the Coastal Transition Zone," J. Geophys. Res. **97**, 9653−9668 (1992).

[112] M. M. Bowen, W. J. Emery, J. L. Wilkin, P. C. Tildesley, I. J. Barton, and R. Knewtson, "Extracting Multiyear Surface Currents from Sequential Thermal Imagery using the Maximum Cross-Correlation Technique," J. Atmos. Oceanic Technol. **19**, 1665−1676 (2002).

[113] I. J. Barton, "Ocean Currents from Successive Satellite Images: The Reciprocal Filtering Technique," J. Atmos. Oceanic Technol. **19**, 1677−1689 (2002).

[114] J. L. Wilkin, M. M. Bowen, and W. J. Emery, "Mapping Mesoscale Currents by Optimal Interpolation of Satellite Radiometer and Altimeter Data," Ocean Dynamics **52**, 95−103 (2002).

[115] L. K. Shay, G. J. Goni, and P. G. Black, "Effects of a Warm Oceanic Feature on Hurricane Opal," Monthly Weather Rev. **128**, 1366−1383 (2000).

[116] F. J. Wentz, C. Gentemann, D. Smith, and D. Chelton, "Satellite Measurements of Sea-Surface Temperature through Clouds," Science **288**, 847−850 (2000).

[117] S. G. Philander (ed.), *El Niño, La Niña, and the Southern Oscillation*, Academic Press, San Diego, CA (1989).

[118] J. E. Harries, D. T. Llewellyn-Jones, P. J. Minnett, R. W. Saunders, and A. M. Zavody, "Observations of Sea-Surface Temperature for Climate Research," Philos. Trans. R. Soc. A: Math. Phys. Eng. Sci. A **309**, 381−395 (1984).

[119] R. Berkelmans, "Time-Integrated Thermal Bleaching Thresholds of Reefs and their Variation on the Great Barrier Reef," Mar. Ecol. Prog. Ser. **229**, 73−82 (2002).

[120] M. W. Gleeson and A. E. Strong, "Applying MCSST to Coral Reef Bleaching," Adv. Space Res. **16**, 151−154 (1995).

[121] I. J. Barton and J. M. Bathols, "Monitoring Floods with AVHRR," Remote Sensing Environ. **30**, 89−94 (1990).

[122] J. Dozier, "A Method for Satellite Identification of Surface Temperature Fields of Subpixel Resolution," Remote Sensing Environ. **11**, 221−229 (1981).

[123] L. Giglio, J. Descloitres, C. O. Justice, and Y. J. Kaufman, "An Enhanced Contextual Fire Detection Algorithm for MODIS," Remote Sensing Environ. **87**, 273−282 (2003).

[124] http://www.ssd.noaa.gov/PS/FIRE/hms.html

[125] http://activefiremaps.fs.fed.us/

[126] http://sentinel.ga.gov.au

[127] http://www.ipo.noaa.gov/index.html

[128] http://www.ipo.noaa.gov/Technology/viirs_summary.html

[129] http://www.esa.int/esaLP/SEMZHM0DU8E_LPgmes_0.html

[130] http://esamultimedia.esa.int/docs/GMES/GMES_Sentinel3_MRD_V2.0_update.pdf

[131] I. S. Robinson, *Satellite Oceanography: An Introduction for Oceanographers and Remote-Sensing Scientists*, Ellis Horwood, Chichester, UK (1985).

[132] S. Q. Kidder and T. H. V. Haar, *Satellite Meteorology: An Introduction*, Academic Press, San Diego, CA (1995).

[133] R. J. Gurney, J. L. Foster, and C. L. Parkinson (eds.), *Atlas of Satellite Observations related to Global Change*, Cambridge University Press, Cambridge, UK (1993).

[134] M. D. King, C. L. Parkinson, K. C. Partington, and R. G. Williams (eds.), *Our Changing Planet. The View from Space*, Cambridge University Press, New York, NY (2007).

CHAPTER 7

INFRARED AND MICROWAVE MEDICAL THERMOMETRY

E. Francis J. Ring[1], Jürgen Hartmann[2], Kurt Ammer[3], Rod Thomas[4], David Land[5] and Jeff W. Hand[6]

Contents

1. Introduction		394
2. Infrared Ear Thermometers for Clinical Thermometry		394
	2.1. Principle of the IR thermometry of the human ear	395
	2.2. The practice of ear thermometry	396
	2.3. Reliability of ear thermometers	398
3. Infrared Thermal Imaging in Medicine		400
	3.1. Background	400
	3.2. Thermal imaging systems for medicine	403
	3.3. Standardization of the imaging process and digital image processing	406
	3.4. Summary	410
4. Pulsed Photo-Thermal Radiometry (PPTR)		411
	4.1. Background	411
	4.2. Typical applications	412
	4.3. Laser−tissue interaction	413
	4.4. Optimizing laser therapies using thermal imaging	414
	4.5. Conclusion	427
5. Microwave Radiometry for Medical Applications		427
	5.1. Principles of medical microwave radiometry	427
	5.2. Microwave properties of tissues	428

[1] Medical Imaging Research Unit, Faculty of Advanced Technology, University of Glamorgan, Pontypridd CF37 1DL, UK
[2] Physikalisch-Technische Bundesanstalt, Braunschweig and Berlin, Germany
[3] Institute for Physical Medicine and Rehabilitation, Hanusch Krankenhaus, Heinrich Collinstrasse 30, Vienna A1140, Austria
[4] Faculty of Applied Design and Engineering, Swansea Metropolitan University, Mount Pleasant, Swansea, SA1 6ED, UK
[5] Department of Physics and Astronomy, University of Glasgow, Glasgow G12 8QQ, UK
[6] Radiological Sciences Unit, Hammersmith Hospital, DuCane Road, London W12 0HS, UK

Experimental Methods in the Physical Sciences, Volume 43
ISSN 1079-4042, DOI 10.1016/S1079-4042(09)04307-0

5.3.	Microwave radiometry measurement technique	429
5.4.	Signal temperature measurement requirements	432
5.5.	Radiometer configurations	434
5.6.	Equivalent temperature fluctuation of radiometer measurements	437
5.7.	Performance of practical clinical radiometers	438
5.8.	Multifrequency radiometry for temperature profile estimation	438
5.9.	Clinical applications	440
6. Summary		442
References		442

1. INTRODUCTION

From the earliest days of medicine, the link between health, disease, and temperature has been recognized. However, the measurement of human body temperature is relatively recent, the greatest advances being realized in the last century. Most of the early studies of temperature in man and animals have been based on contact thermometry, but in the last 50 years there has been a dramatic improvement and availability of radiometric systems, bringing the opportunity to achieve reliable noncontact body temperature measurement. Current technology allows us to measure temperature in the ear with a handheld radiometer, and to image skin surface temperature in the infrared (IR) spectrum without contact. High-speed thermal imaging brings new opportunities in dynamic testing of skin temperature response, as shown in the section on pulse wave photo-thermal imaging. Finally, the techniques for microwave detection of subsurface body temperature continue to be developed. While the energy levels of microwaves emitted from the human body are much lower than IR, there are some advantages, since bone and hair that will prevent heat transfer in the IR, may be partially transmitted at these frequencies. There is now a wide range of radiometric techniques for the study of human body temperature. These are, of course, subject to limitations, and inevitably some are very operator dependant. Nevertheless, continual development of sensors and methods to improve stability and certainty in measurement will doubtlessly increase the scope and uptake of radiometric temperature measurement in medicine and health science.

2. INFRARED EAR THERMOMETERS FOR CLINICAL THERMOMETRY

The core temperature of the human body has been used since the times of Hippocrates as an indicator of the health status of an individual [1,2].

A distinct temperature indicating the state of health cannot be given in general, but a temperature of 37°C is thought to be the human core body temperature in normal health [3]. However, human body temperature strongly depends on the location of the measurement. One problem of human body temperature measurement is to assess the internal temperature of the body in a simple way and without the use of invasive methods. In intensive care medicine, human body temperature is measured using the bladder, the pulmonary artery, or the esophagus as locations for a contact temperature sensor. However, all of these locations require a more or less invasive procedure, and cannot be used for routine human body temperature measurements. Therefore, for more routine medical practice, external sites located at the skin surface or in easily accessible visceral cavities, for example, mouth, armpit, or rectum, are used. Measurement in the mouth, by locating the thermometer probe below the tongue, can produce a variation of normal body temperature from 33.2°C to 38.2°C, while measuring in the armpit, that is axilla, ranged from 35.5°C to 37.0°C, and for the rectum, that is rectal measurement, from 34.4°C to 37.8°C [4]. However, as these temperatures are measured at the outside of the human body these are not core temperatures and have to be correlated to the human core body temperature. This is because for all medical purposes the measurement of the human − but also the animal − core body temperature is required [5]. Historically, contact thermometers and in particular the mercury-in-glass thermometer have been used for that purpose. However, as contact thermometers have to thermally equate with the location or measurement site, a considerable time span has to be allowed before a contact thermometer will accurately express the temperature of that location. As described elsewhere in this book, noncontact measurement of temperature using emitted thermal radiation is achieved in a very short time. However, the emissivity of the location to be measured has to be known in order to calculate the real body temperature. Radiometric measurement, therefore, requires knowledge of the body site emissivity. To avoid effects from emissivity and from the environment, the IR ear thermometer makes use of the quasi-blackbody behavior of the human ear for detecting body core temperature measurements.

2.1. Principle of the IR thermometry of the human ear

A long cavity made of a distinct material, due to its geometry has a higher emissivity than the bare material from which it is made; see Chapters 5 and 6 of the companion volume on fundamentals (Vol. 42 of this serial). The emissivity of the cavity is close to unity as long as the diameter d of the cavity is much smaller than its length l. With this fact in mind, noncontact measurement of temperature using emitted thermal radiation is advantageous if a long cavity can be used. In fact, the human and animal ear canal is nearly ideal in representing a blackbody and offers a near emissivity corrected temperature measurement. Additionally, the tympanic membrane of the human ear canal is closely located to the hypothalamus (about 3.5 cm distant) and the major brain arteries [6,7]. The hypothalamus is the organ that regulates human body temperature. Therefore, the tympanic membrane and the inner walls of the ear canal are close to the human body core temperature. The ear canal is, therefore, one human body location that is accessible and effectively noninvasive for core temperature measurement.

To measure the temperature of the tympanic membrane (the ear drum) the radiation thermometer, that is, the IR ear thermometer, should have a direct sight onto the tympanic membrane. For an infant, the auricle (i.e., the outer ear) must be pulled back horizontally, while for an adult the auricle must pulled back and upwards to properly adjust the ear canal for a direct sight on to the tympanic membrane [8].

There are also other radiation thermometric devices, designed to measure human body temperature by sensing the thermal radiation emitted by human skin. However, using such devices a correction has to be applied for the emissivity of the human skin. Additionally, as the temperature of the human skin is easily affected by the temperature of the surrounding ambient and by its own physiological status (i.e., sweat or cosmetics), so such devices suffer from additional uncertainties, besides that of the uncertain skin emissivity. This section only deals with IR ear thermometers for determining human body core temperature by sensing thermal radiation from the ear canal.

2.2. The practice of ear thermometry

The application of an IR ear thermometer can be seen in Figure 1. To measure thermal radiation emitted by the ear, specific detectors and optical designs must be used. In principle, two types of IR detectors can be applied: quantum and thermal detectors. However, the application of quantum detectors for IR radiation detection usually implies cryogenic cooling, making them rather impractical. Thus, thermal detectors are best suited for such handheld devices [9]. Usually thermopile and pyro–electric detectors are used in IR ear thermometers. The optical probe of an IR ear thermometer is shown in Figure 2.

The radiation emitted by the ear canal is transmitted by the front window, and is partially absorbed and guided to the sensor by the reflective waveguide. Due to the partly absorbed radiation, the IR window is heated, generating an undesired time-dependent error. Thermal detectors in general suffer from drift due to the change in the sensor's surface temperature. This drift can be dealt with using different approaches. The first approach is a reference radiation source of known and stable radiance

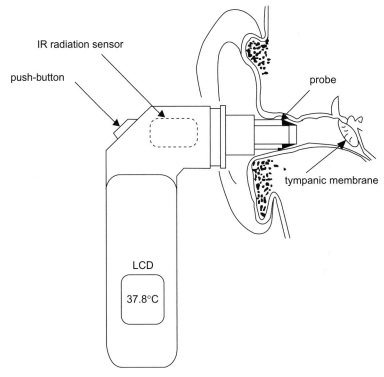

Figure 1 Measurement of the far IR flux from the ear canal [10]. The equivalent network contains thermal resistances between various parts of the system. Temperature changes across resistances are shown as an example.

temperature, which is put in the line of sight of the IR sensor prior to every measurement. As long as the surface temperature of the sensor does not change between the reference and the tympanic temperature measurements, the effect of the sensor's surface temperature can be corrected for. Such a device utilizing a pyro-electric thermal sensor is shown in Figure 3. In the second approach, an internal contact thermometer measures the temperature of the sensor and the reading is corrected for this temperature, as illustrated in Figure 4. These two approaches can also be used for correcting the reading for a temperature drift of the sensor due to changing environmental conditions.

To prevent contamination of the IR window (e.g., by ear wax) and to ensure a sanitary barrier, probe covers are usually used with the IR ear thermometers. Such probe covers may be disposable or reusable. For most IR ear thermometers, the probe covers are made of polymers with a thickness of $20-30\,\mu m$. Also the materials used for these probe covers, such as polypropylene and polyethylene, are reasonably transparent in the far-IR region of the spectrum. It is essential that the variation of transmission from

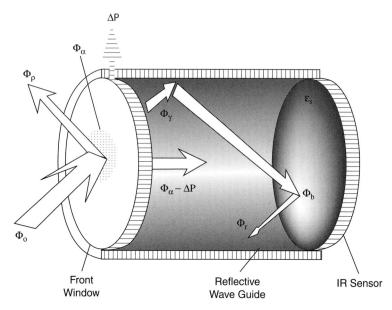

ΔP

Φ_α

Φ_ρ

Φ_γ

ε_s

$\Phi_\alpha - \Delta P$

Φ_b

Φ_r

Φ_o

Front
Window

Reflective
Wave Guide

IR Sensor

Figure 2   Optical probe for an IR ear thermometer [9].

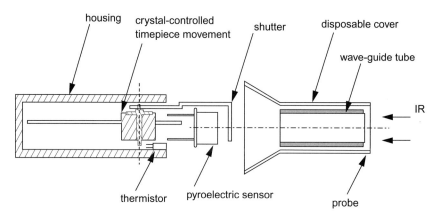

housing   crystal-controlled
timepiece movement   shutter   disposable cover

wave-guide tube

IR

thermistor   pyroelectric sensor   probe

Figure 3   Simplified diagram of a pyro-electric IR ear thermometer using a shutter
[10].

probe to probe be very low to maintain accurate temperature measurement
performance of the thermometer [11].

2.3. Reliability of ear thermometers

Looking at a detailed geometry of the IR tympanic temperature
measurement given in Figure 5, it is clear that the reading of the IR ear

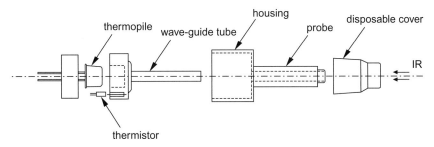

Figure 4 Simplified diagram of a thermopile IR ear thermometers using a thermistor for thermopile temperature measurement [10].

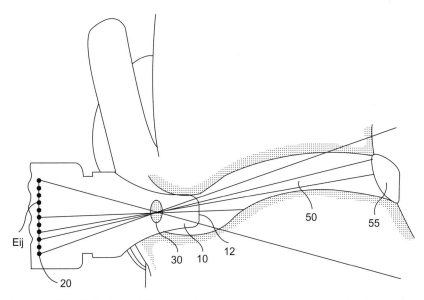

Figure 5 Detailed view of IR tympanic temperature measurement, clearly showing the curved ear canal and the different parts of the ear canal seen by the sensor [12].

thermometer can be affected by several perturbing factors. First of all, due to the curved form of the ear canal a proper alignment of the field of view by pulling the outer ear in the manner described above must be performed. Additionally, the probe of the IR ear thermometer should be placed straight and tight into the ear canal and any tilting must be avoided [13]. Otherwise, only part of the field of view includes the tympanic membrane, while the remainder will be facing the colder sidewalls of the ear canal. Furthermore, earwax may be present in the ear canal blocking the direct sight on to the tympanic membrane. As noncontact temperature measurement of human body temperature with an IR ear thermometer is not always common practice, some clinical staff may lack training in performing these measurements. It is important that clinical staff be made aware of false readings obtained with an improperly applied IR ear thermometer. Contact thermometers and their usual locations, such as the rectum and the mouth, bear a much lower risk of improper use and resulting erroneous readings and there is a greater risk of error in the use of IR ear thermometers [14]. However, if an IR ear thermometer is properly used the results obtained are excellent [12,15].

To demonstrate traceability to ITS-90 of the temperature measurements with an IR ear thermometer, specially designed blackbody radiators have been developed. Typical examples of such blackbodies are shown in Chapter 6 of the companion volume (Vol. 42 of this serial). The traceability of IR ear thermometers has been the topic of several publications and good worldwide agreement in traceability of measurement has been obtained [16−18]. In summary, the IR ear thermometer is a delicate optical instrument, which gives reliable human body temperature reading, but requires more skill and care in its use by medical staff than the contact thermometers traditionally used in medicine.

3. INFRARED THERMAL IMAGING IN MEDICINE

3.1. Background

The association between temperature and disease is almost as old as medicine itself. After Galileo introduced the thermoscope, thermometry evolved slowly and became established in medicine only by the work of Carl Wunderlich in the 19th century. Thermal imaging began in the early 19th century with William and John Herschel [19], but was not developed commercially until after World War II. Today, modern IR imaging systems offer high-resolution images of human body temperature, and can be used to quantify sensitive changes in skin temperature in relation to certain diseases, and their response to medication. Computer processing has dramatically improved the power of thermal imaging, and reliable imaging procedures have been established for medical use of this technique.

Heat transfer is principally affected by three main modes. The first is conduction, requiring contact between the object and the sensor. The second mode of heat transfer is convection, and the third is radiation. Both of the latter had led to remote detection methods of the heat. As early as in 1698, Della Porta, an Italian monk, observed the effects of reflected heat (see Figure 6). This observation was subsequently studied by William Herschel in 1800 when he identified IR radiation. John Herschel, his son, took the first "thermal image" in 1840, that of the sun, and named it a thermogram; see Figure 7.

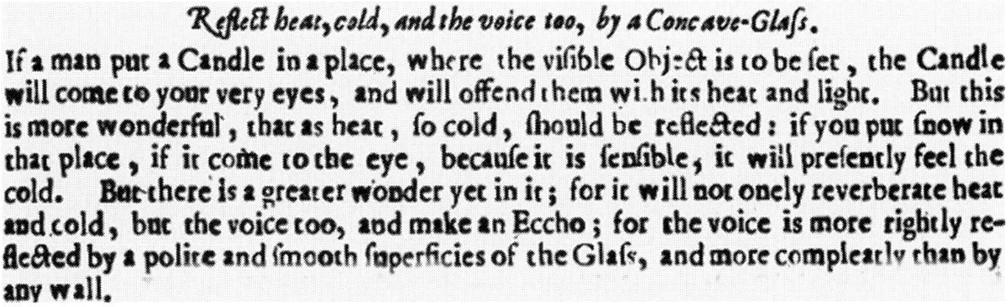

Reflect heat, cold, and the voice too, by a Concave-Glass.

If a man put a Candle in a place, where the visible Object is to be set, the Candle will come to your very eyes, and will offend them with its heat and light. But this is more wonderful, that as heat, so cold, should be reflected: if you put snow in that place, if it come to the eye, because it is sensible, it will presently feel the cold. But there is a greater wonder yet in it; for it will not onely reverberate heat and cold, but the voice too, and make an Eccho; for the voice is more rightly reflected by a polite and smooth superficies of the Glass, and more compleatly than by any wall.

Figure 6 Observation of reflected heat by Della Porta [19].

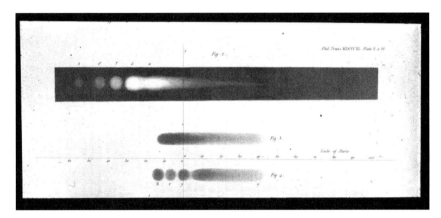

Figure 7 First thermogram of solar heat over the course of a day made by John Herschel in 1840 [19].

The German physician Carl Wunderlich established the value of studying human body temperature in 1868, and developed the clinical thermometer. This was essentially a maximum temperature device with a limited scale around the normal internal body temperature of 37°C. Wunderlich charted the temperature of all his patients daily, and sometimes two or three times during the day. An example of his temperature records are shown in Figure 8 [20].

Convection currents of heat emitted by the human body have been imaged by the technique of Schlieren Photography. The change in refractive index with density in the air around the body is made visible by special illumination. This method has been used to monitor heat loss in experimental subjects [21], especially in the design of protective clothing for people working in extreme physical environments and also in the flow evaluation of exhaled air [22].

IR thermal imaging, often referred to as thermography, is increasing used in clinical medicine, as an objective and noninvasive tool for monitoring human body temperature. This is greatly facilitated by the

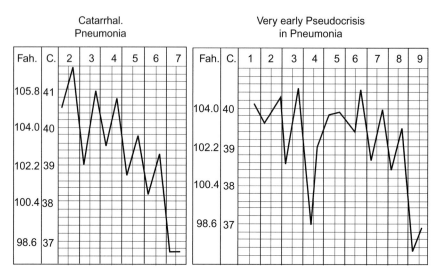

Figure 8   Early serial temperature charts by Dr Carl Wunderlich 1870 [20].

fact that human skin, regardless of color, has an emissivity of about 0.98 [23].

The human body surface requires variable degrees of heat exchange with the environment as part of the normal thermo-regulatory process and most of this heat transfer occurs in the IR [24].

Unlike a thermometer reading, which is a record of a localized temperature such as the mouth or armpit, thermal imaging registers the temperature patterns of an area of skin in a dynamic way, and can instantly show changes induced by movement, disease, or effect of a treatment. One image may contain thousands of temperature points that represent the distribution of temperature over the selected skin surface.

Skin temperature distribution shows a distinctive pattern, and under normal healthy conditions, thermal features of the body surface are predictable. For example, hair insulates the body, so little or no heat can be seen from the head unless it is bald. The inner canthus, or corner of the eye is warmer than the eye itself, while the nose may be cooler [25]. When there is localized inflammation such as that found in rheumatoid arthritis, the affected joints are several degrees warmer than those of the surrounding tissues [26,27]. In Raynaud's phenomenon (a condition first described by Maurice Raynaud), fingers and toes are abnormally cold. Here thermal imaging is used to document the temperature gradient to the fingertips before and after a thermal challenge to the hand, such as brief immersion in cold water [28]. The effects of treatment with drugs [29–31], physical therapy [32,33], and surgery, [34] which have an effect on the blood flow or skin temperature can all be documented with this technique.

In the early days of clinical use of thermal imaging many different practices were followed. This led to discrepant and sometimes contradictory conclusions. To counter this, the European Thermographic Association (formed in 1972) established study groups to formulate the guidelines for good practice. These guidelines included the requirements for patient preparation, conditions for thermal imaging and criteria for the use of thermal imaging in medicine and pharmacology [35−37].

A thermal index devised in Bath[1], UK provided a simplified measure of inflammation for clinicians [29]. A normal range of values was established for ankle, elbows, hands, and knees, with raised values obtained in osteoarthritic joints and higher values in rheumatoid arthritis. A series of clinical trials with nonsteroid antiinflammatory oral drugs and steroid analogues for joint injection were published using the index to document the course of treatment [37]. These studies showed that quantitative thermography could be a valuable tool in the objective measure of response to treatment, and was reliable enough for double blind clinical drug trials.

3.2. Thermal imaging systems for medicine

The first thermograms used in medical research were made from experimental imaging equipment constructed in the early 1940s using a liquid nitrogen cooled indium antimonide detector. These were undertaken with the prototype "Pyroscan" thermal imager (Figure 9) at The Middlesex Hospital in London, and in Bath[1] in 1959−1961. By modern standards, these thermograms were very crude. The single element detector required mechanical scanning with the result that each image required 2−5 min to record. The final image was written line by line on electrosensitive paper.

In the USA, images were made by Bowling Barnes, and early work on breast cancer studies were carried out by Dr. R. Lawson in Canada [38]. Thermal imager development progressed until the images were no longer hard printed but instead displayed on a cathode ray screens (e.g., by the equipment shown in Figure 10).

Major improvements in thermal imagers continued up to the present day with huge improvements in image quality and speed of image capture. Real-time imaging became possible using multielement detector arrays that had on-board signal processing. Figure 11 shows a thermogram of a finger taken using an early array-based thermal imager where cooling from sweat evaporation can be seen. This illustrates the advantage of high-speed and high-resolution imaging for medicine and human physiology [39].

---

[1] The Royal National Hospital for Rheumatic Diseases, Bath, UK.

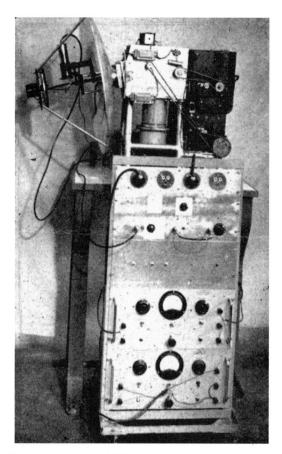

Figure 9   The Pyroscan prototype 1942.

After the introduction of multielement detectors, focal plane arrays were developed, with increasing numbers of pixels, yielding high resolution at video frame rates. These still had the disadvantage that a mechanical cooler was deployed to lower the temperature of the detector. Recently, however, uncooled bolometric detectors have also been shown to be suitable for many medical applications. The use of bolometric systems is rapidly supplanting the more traditional cooled types. The removal of electronic cooling systems for the detector means that imagers are now low maintenance, providing that regular quality assurance procedures, particularly to maintain temperature performance, are adopted [40].

Good software for image enhancement and analysis is now normal in modern thermal imaging systems. Many commercial systems use general imaging software, which is primarily designed for industrial users of the technique. A few dedicated medical software packages have been produced,

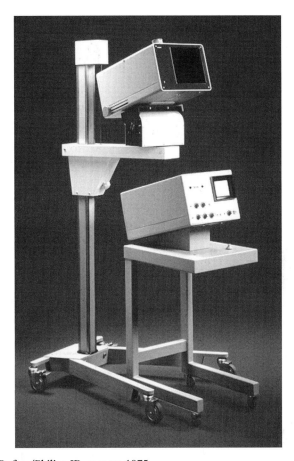

Figure 10   Bofors/Philips IR camera 1975.

which have also been used to enhance the images from older optical–mechanical IR imagers. For example, CTHERM is a simple robust and almost universally usable program for medical thermography [41].

Thermal imaging is a technique for temperature measurement based on the IR radiation emitted from the body. Unlike images created by X-rays or proton activation through magnetic resonance, thermal imaging is not related to morphology. The technique gives a map of the distribution of temperatures on the surface of the object imaged. Useful terminology relating to thermal physiology has been published [42]. Temperature values measured from a thermogram are derived from the IR radiant energy, the mean temperature of the region studied may not be the most significant statistical feature for diagnosis, for example, second-order statistical measures conferred a high-detection accuracy for breast cancer [43] and the homogeneity of temperature distribution was more successful

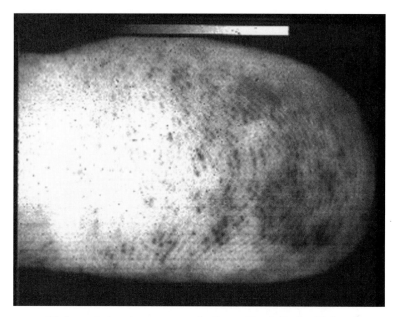

Figure 11   High-resolution thermogram of a finger (Rank Taylor Hobson).

in diagnosing complex regional pain syndrome than mean temperature values [44].

3.3. Standardization of the imaging process and digital image processing

Computerization has solved many problems in clinical thermography, providing image archives in digital form, standard regions of interest (ROI) selection, and temperature measurements obtained from the images. Manufacturers of thermal imaging equipment have adapted to the call for quantification, and small *in situ* blackbody sources now assist in the proper quantification of the technique.

Standardization of the whole process of performing medical thermography is essential. An Anglo-Polish study group have investigated the sources of error in the normal clinical thermography technique [45] and developed robust good practice guidelines for measuring skin temperatures [46]. These guidelines are given in outline below:

3.3.1. The patient
Patient information is needed to ensure full co-operation before and during the investigation, and adequate stabilization before imaging.

On arrival at the department, the patient should be informed of the examination procedure, instructed to remove appropriate clothing and jewelry, and asked to sit or rest in the preparation cubicle for a fixed time. The time required for reaching adequate stability in blood pressure and skin temperature is generally 15 min, with 10 min as a minimum.

Contact between body parts with the environment or with other body parts alters the surface temperature therefore; during the preparation time the patient must avoid folding or crossing arms and legs, or placing bare feet on a cold surface. If the lower extremities are to be examined, a stool or leg rest should be provided to avoid direct contact with the floor. If these requirements are not met, misleading measurements may result.

3.3.2. The environment

The environment refers to the examination room, with cubicles at a constant controlled temperature and draught free.

Heat generated in the investigation room affects its temperature. Possible heat sources are electronic equipment such as the imager and its computer, but also human bodies. For this reason, the air-conditioning unit should be capable of compensating for the maximum number of patients, staff, and the equipment likely to be in the room at any one time. These effects will be greater in a small room of 2×3 m or less.

Convection is a very effective method of skin cooling, therefore, air-conditioning equipment should be located so that draughts are not directed at the patient, and that overall air speed is kept as low as possible. A suspended perforated ceiling with ducts diffusing the air distribution evenly over the room is ideal.

Ambient temperature control is a primary requirement for most clinical applications of thermal imaging. In countries with a temperate climate, a range of temperatures from $18°C$ to $25°C$ should be attainable, typically $22°C$ being used and should remain stable to better than $1°C/h$. Due to the nature of human thermoregulation, stability of the room temperature is essential. When inflammatory processes are under investigation, a cooler ambient of $20°C$ can be used, which heightens contrast in skin temperature, since many inflammatory lesions resist superficial cooling. At lower ambient temperatures, the subject is likely to shiver, and over $25°C$ sweating may occur. Relative humidity should be kept to 50% or less, where there would be minimal effect on the human body in terms of thermal stabilization.

Techniques for cooling particular regions of the body have been developed [47,48]. Immersion of the hands in water at various temperatures is a common challenge for the assessment of Raynaud's phenomenon, and other vasospastic disease states [49]. A mechanical provocation may also be used, as in the case of investigating Thoracic Outlet Syndrome [50].

Figure 12 Pillar stand, an important aid to standardization and a focal plane array IR camera.

3.3.3. The imaging system

The imaging system must be stable in output, with its temperature performance checked, where possible, with an external traceably calibrated radiance source.

The imager should, if possible, be mounted on a pillar stand (such as a studio camera stand, rather than a tripod). This facilitates rapid positioning, and a reduction in the variable angles between the camera and the subject. Figure 12 shows a suitable pillar stand mount for a thermal camera in a medical installation.

Regular tests of imager temperature readout stability, and offset drift using a constant temperature reference source are strongly desirable, to monitor the system performance and indicate when the manufacturers specification may be exceeded.

3.3.4. Subject positioning

The position of the patient under study can be a large variable [51], but its effect can be readily reduced by specialized medical thermal imaging

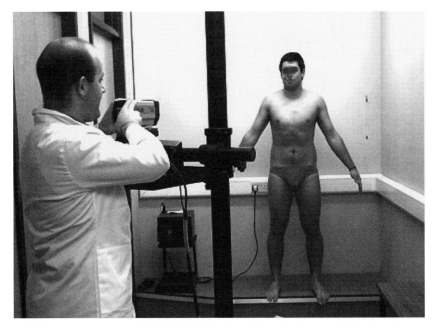

Figure 13   Recording an IR image from clinical thermography.

software. Outline masks created for a series of standard positions relating to different regions of the body under examination improve the technique for image capture. The subject's image is set to fill the outline mask, for each stage of image capture. This ensures that the size of the image remains constant each time a thermogram is recorded on a given subject (Figure 13).

3.3.5. Image analysis

The effect of image analysis can be a large source of variability, but greatly reduced by the specific anatomical definition of every region of interest used in the thermogram analysis. Almost all thermal imagers now provide software for image processing and basic quantification of the image. In some cases, this may be operated directly on the camera, or may be carried out through a computer. It is important to emphasize that false color coding of IR images may misrepresent the temperature measurement. For example if colors are separated by 1°C steps, the temperature difference between two points situated in adjacent colors may be between 0.1°C and 1.9°C. False colored images provide an indication of temperature, but not a measurement.

Reliable temperature measurements in thermal images are based on the definition of ROI. However, standards for shape, size, and placement of these regions are not yet fully established. Although a close correlation exists for ROI of different section in the same region [52], the precision of the temperature measurement *is* affected when an ROI of different size and location are used for repeated measurements. For this reason, images should be stored with the ROI previously used, so that true comparisons can be made on subsequent examinations. This process is simplified when image capture masks have been used; ensuring regular image sizes for each anatomical location are taken.

3.3.6. Reporting of results

The clinical report should carry all relevant demographic and environmental data, the images should be shown only with the temperature scale, and the thermal data extracted from the illustrated ROI for each image. For diagnosis, a good clinical knowledge of human body temperature, skin temperature patterns, and relevance to disease or injury are essential. The images and the thermal data extracted from defined ROI should be part of the clinical report, together with details of the examination room conditions and hardware used.

All details that enable the referring physician to understand the conclusions, and for any other investigator to be able to repeat the examination on the same patient should be reported.

In specialist applications such as forensic medicine in order for IR thermal imaging to be acceptable, particularly when used in legal evidence, standardization, and repeatability of the technique are essential features [53]. The ultimate aim of standardization is to achieve high accuracy and repeatability of this method for temperature measurement. It has been shown that when the procedural standards described above are applied, quantitative thermal imaging can be recommended as a responsive outcome measure for clinical trials. There are many such studies reported in peer-reviewed journals applied to rheumatology, vascular diseases and angiopathies, neuromuscular disorders, surgery. In pediatrics, the value of a noninvasive procedure for fever detection, and a number of diseases where surface temperature can be of clinical value, will be more widely applied with modern and reliable thermal imagers in conjunction with imaging protocols [54,55].

3.4. Summary

Many improvements have been made to IR thermal imaging technology over the years, both in the reliability of IR detectors, and to the technique, which have impacted on medicine. The new computer-driven technology is more complex than in the days of the pioneers, but the clinical value of human body temperature studies using thermal imaging is firmly established. *Critical* use of the technique in clinical medicine is essential to achieve reliable and useable data on human body temperature. Improved standardization of technique will enhance the uptake of the technology in medicine, and facilitate its integration into the new digital imaging systems in hospital medicine. In addition, it will pave the way to better reference data for normal and diseased patients, and in particular develop normal control images, essential for the efficient training of end users of the technique.

4. PULSED PHOTO-THERMAL RADIOMETRY (PPTR)

4.1. Background

Pulsed photo-thermal radiometry (PPTR) relies on the detection of IR radiation emitted from the surface of skin tissue in which the temperature distribution has been disturbed by the application of a fast laser pulse. A temperature rise above normal body temperature is required for effective therapy but this must be controlled carefully to prevent secondary damage, reduce pain, and ensure optimum recovery times. The use of laser therapy has recently benefited from the introduction of thermal imaging, which has led to the development of protocols for optimizing the therapy.

Lasers have been used in dermatology for over 40 years [56] but more recently there have been a number of key developments regarding the improved treatment of various skin disorders such as the removal of vascular lesions using dye lasers [57−61] and depilation using ruby lasers [62,63]. These are summarized in Table 1.

During PPTR the resultant laser−tissue interaction normally occurs subsurface and is dependent on a number of factors including the efficiency of the laser and the heat transfer effects within the tissue. Example applications include selective photo-thermolysis as in laser therapy for skin resurfacing and tissue welding [64].

The application of laser radiation to the skin results in a number of differing interactions that depend on the wavelength, interaction time, and power density. Physiological characteristics of laser therapy include selective photo-thermolysis, photo-disruption, photo-ablation, vaporization, and coagulation as illustrated in Figure 14.

An important requirement in these applications is to be able to predict subsurface temperatures and in particular to correlate the photo-thermal signal with the subsurface distribution of optical parameters and so determine the efficacy of interaction. The photo-thermal signal results from a rapid temperature increase within the subsurface target, which is due to the absorption of a pulse of optical energy. The spatial distribution of temperature in the target is determined by its optical and thermal properties, its structure and illumination geometry. Prahl et al. [66,67] derived a theory for the understanding of photo-thermal signals detected.

Table 1   Characteristics of laser therapy during and after treatment.

General indicators	Dye laser vascular lesions	Ruby laser depilation
During treatment	Varying output parameters	Varying output parameters
	Portable	Portable
	Manual and scanned	Manual and scanned
	Selective destruction of target chromophore (hemoglobin)	Selective destruction of target chromophore (melanin)
After treatment (desired effect)	Slight bruising (purpura)	No bruising (skin returns to normal coloring)
	Skin retains its elasticity	Skin retains surface markings
	Skin initially needs to be protected from UV and scratching	Skin retains its ability to tan after exposure to ultraviolet light
	Hair follicles are removed	Hair removed

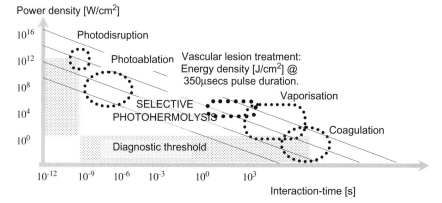

Figure 14   Physiological characteristics of laser therapy [65].

4.2. Typical applications

An outline of typical applications of PPTR is given below. Common to all these applications is that the laser energy should be uniformly applied to the skin without overlapping of the laser spot to a subdermal target region, such as a blood vessel, with the minimum of secondary damage to the surrounding tissue. Being able to measure and verify the temperature rise and distribution at the skin surface, with a firm threshold to avoid

Table 2 Interaction effects of laser light and tissue.

Effect	Interaction
Photo-thermal	
Photo-hyperthermia	Reversible damage of normal tissue $(37-42°C)$
Photo-thermolysis	Loosening of membranes (edema), tissue welding $(45-60°C)$
Photo-coagulation	Thermal-dynamic effects, microscale overheating coagulation, necrosis $(60-100°C)$
Photo-carbonization	Drying out, vaporization of water, carbonization $(100-300°C)$
Photo-vaporization	Pyrolysis, vaporization of solid tissue matrix $(>300°C)$
Photo-chemical	
Photo-chemotherapy	Photo-dynamic therapy, black light therapy
Photo-induction	Biostimulation
Photo-ionization	
Photo-ablation	Fast thermal explosion, optical breakdown, mechanical shockwave

burning/scarring is of critical importance for therapeutic efficacy. Recent advances in laser technology have resulted in new portable laser therapy devices examples of which include the removal of vascular lesions (e.g., port-wine stains (PWSs)), excess hair-removal (depilation) and wrinkle reduction [68].

There are three common types of laser–tissue interaction, namely, photo-thermal, photo-chemical, and photo-ionization (see Table 2), resulting in a number of differing interactions including photo-disruption, photo-ablation, and coagulation.

4.3. Laser–tissue interaction

The mechanisms involved in the interaction between light and tissue critically depend upon the characteristics of the impinging light and the optical properties of the target tissue such as reflectance, absorption, and scattering properties as a function of wavelength. A simplified model of laser light interaction with the skin is given in Figure 15.

Laser radiation can penetrate through the epidermis and basal structure to be preferentially absorbed within the blood layers located in the lower dermis and subcutis. This, selective photo-thermolysis, is the mechanism employed to eliminate the target without damaging surrounding tissue. For example, in the treatment of PWSs, a dye laser of wavelength 585 nm has

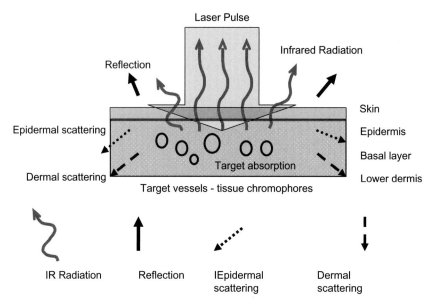

Figure 15 Passage of laser light within skin layers.

been widely used [69] where the profusion of small blood vessels (target) that comprise the PWS are preferentially targeted at this wavelength.

The spectral absorption characteristics of light through human skin has been well established [70,71] and highlighted in Figure 16, the two dominant chromophores being melanin and hemoglobin.

The application of appropriate lasers to medical problems depends critically on matching the optimum laser wavelength to the desired outcome [72]. Some typical applications and the desired wavelengths for usage are given in Table 3, where for example at 694 nm the target is melanin and at 585 nm, hemoglobin.

The therapeutic effect can be controlled by close monitoring of the temperatures arising from the skin surface post laser illumination. Typical effects of temperature rises within the tissue are indicated in Table 4.

The efficacy of active therapies has been greatly enhanced by the introduction of the latest generation of portable, high spatial, and temperature resolution thermal imaging systems. Their specific use for this application is described in the next subsection and provides a case study of where the deployment of thermal imaging in medicine has had strong beneficial consequences.

4.4. Optimizing laser therapies using thermal imaging

There are number of challenges in optimizing pulsed photo-thermal radiometry especially with reference to laser therapy; these include

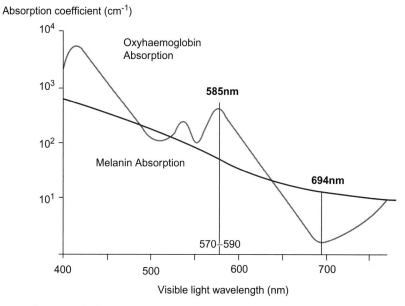

Figure 16   Spectral absorption curves for human blood and melanin.

Table 3   Laser application in dermatology.

Laser	Wavelength (nm)	Treatment
Flashlamp short-pulsed dye	510	Pigmented lesions, for example, freckles, tattoos
Flashlamp long-pulsed dye	585	Port wine stain (PWS) in children, warts, hypertrophic scars
Ruby single-pulse or Q-switched	694	Depilation of hair
Alexandrite Q-switched	755	Multicolored tattoos, viral warts, depilation
Diode Variable	805	Multicolored tattoos, viral warts
Neodymium Yttrium Aluminum (Nd—YAG) Q-switched	1,064	Pigmented lesions, adult PWSs, black/blue tattoos
Carbon dioxide continuous pulsed	10,600	Tissue destruction, warts, tumors

Table 4 Effect of temperature on tissue.

Temperature (°C)	Effect
43–45	Conformational changes, hyperthermic cell killing
50	Reduced enzyme activity
60	Protein denaturation, coagulation
80	Collagen denaturation, membrane permeabilization
100	Vaporization and ablation
300	Charring

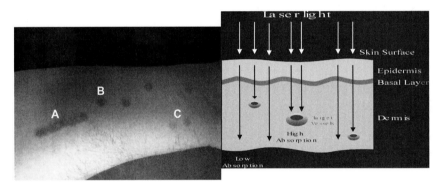

Figure 17 Some challenges to optimized laser therapy.

establishing the optimum laser parameters such as wavelength, energy density, and spot size. Combined with these are difficulties associated with handheld laser positioning. These difficulties can sometimes result in pain, excessive treatment times, and possibly psychological trauma. Figure 17 (to the left of figure) illustrates some of these difficulties as: (A) overlapping spots, (B) excessive energy density, (C) missed treatment area.

From Figure 17 (to the right of figure), the laser light (dependent on wavelength) travels through the epidermis and basal layer to the upper dermis where the interaction occurs and illustrates the true nature of the challenge; to penetrate the upper skin layers without damaging any surrounding, including surface, tissue.

By analyzing the heating and cooling of the target at the surface of the tissue, information on the optical or thermal properties of the skin can be obtained. Results have revealed that using a thermal imager is particularly useful in determining the following key treatment parameters:

- Repeatability of laser spots on tissue, that is, temperature rise.
- Identify thermal effects of different laser wavelengths with similar spot size.

- Thermal effects of varying spot size.
- Improved laser positioning resulting in:
  - Elimination of overlapping and stippling.
  - Establishment of minimum gaps between individual laser spots.
  - Optimum treatment leading to reduction in frequency of treatment.

The measurement, accuracy, and repeatability of temperature is a critical factor in optimizing therapeutic effect [65]. Figure 18 illustrates the reproducibility of repeated laser spots at different wavelengths on the same skin. A 585 nm dye laser and a 694 nm ruby laser were used to place a number of spots manually on tissue. The energy emitted by the laser was highly repeatable. Care was taken to ensure that both the laser and thermal imager position are kept constant and that the anatomical location used for the test had uniform tissue pigmentation.

It is clear, despite the complex measurement situation, that a thermal imager can be used to repeatedly and accurately measure surface tissue temperatures during active therapy. The thermal imager can be used to inform the operator when the temperature has subsided allowing further treatment without exceeding a predetermined skin damage threshold,

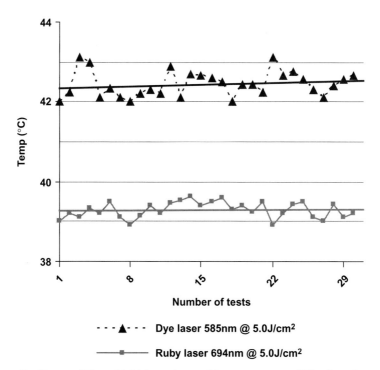

Figure 18 Repeatability of initial maximum skin temperatures (°C) of two lasers with similar energy density but different wavelengths.

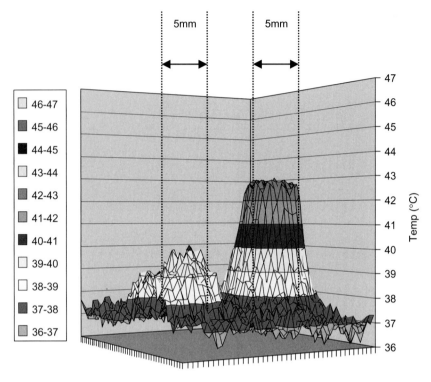

Figure 19 Skin thermal distributions (°C) from a dye laser (to right of figure 5.0 J/cm² at 585 nm, 5 mm spot) and ruby laser (5.0 J/cm² at 694 nm, 5 mm spot).

improve laser positioning (can visualize any missed areas known as stippling), and reduce pain.

Through determining the temperature rise at the skin surface the significance of different laser irradiation wavelengths can be clearly illustrated. As can be seen in Figure 19, application of a dye laser to the skin surface results in a larger increase in the skin temperature (43°C) compared a ruby laser of (40°C) with similar illumination conditions. This dramatically illustrates the strongly varying absorption properties of the skin with respect to laser wavelength.

4.4.1. Thermal effects of varying spot size

The effect of varying the laser spot size will alter the efficacy of the treatment. Larger spot sizes can reduce treatment time and reduce the period of discomfort for the patient. The purpura (bruising) threshold and skin temperature vary with spot size. A thermogram is given in

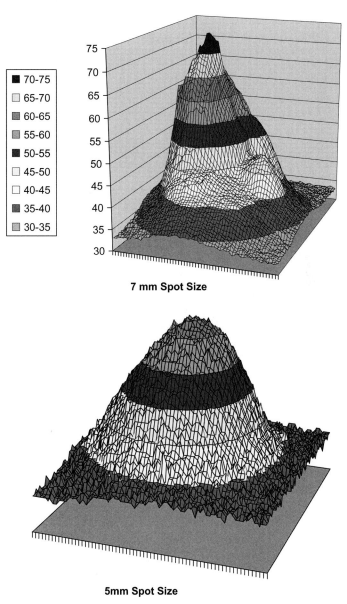

70-75
65-70
60-65
55-60
50-55
45-50
40-45
35-40
30-35

7 mm Spot Size

5mm Spot Size

Figure 20 Dye laser spot size of 7 and 5 mm, at $\lambda = 585\,\text{nm}$ and $5\,\text{J/cm}^2$.

Figure 20 with the laser having constant energy density but with varying spot size. It is clear that the same energy densities can in fact lead to different maximum tissue temperatures. It is also clear that there is a balance between laser spot size and clinical efficacy and that thermal imaging has a crucial role to play in determining where the balance lies. This effect was

Table 5 Comparison between average maximum temperatures and spot size when illuminated with laser spots of the same power density.

Laser spot size	Male A		Male B	
	Forearm (°C)	Underforearm (°C)	Forearm (°C)	Underforearm (°C)
5 mm	$45_{\text{ave. max}}$	$41_{\text{ave. max}}$	$43_{\text{ave. max}}$	$40_{\text{ave. max}}$
7 mm	$64_{\text{ave. max}}$	$56_{\text{ave. max}}$	$61_{\text{ave. max}}$	$55_{\text{ave. max}}$

investigated using two different subjects but at two similar anatomical locations with laser spot sizes of 5 and 7 mm, respectively. The outcome is given in Table 5.

It is clear that laser spot size has a significant impact on the temperature rise and therefore indirectly the efficacy of the therapy. The laser operator would assume that 5 J/cm^2 and a 5 mm spot would give the same clinical outcome as 5 J/cm^2 and a 7 mm spot, this is clearly not the case. The scattering of outer photons from the beam by the tissue leads to a greater "dilution" of the smaller spot and the larger spot attains higher temperatures. This results in a lowering of purpura and also lowering of the threshold for other unwanted side effects, which may lead to pigmentary or textural changes, indicating that a smaller spot size would lead to an enhanced clinical outcome. The treatments of vascular lesions generally utilize an energy density of $5 - 10 \text{ J/cm}^2$ [73]. It is essential to use a thermal imager to characterize the effect of the laser spot to reduce the possibility of excessive energy density and hence possible collateral damage.

4.4.2. Thermal imaging results of laser positioning

During laser therapy, the skin is treated with a number of spots, depending on the anatomical location and required treatment. For example, the type and severity of vascular lesion determines the treatment required as does its color severity (dark to light) and its position on skin (raised to level). The necessary treatment may require a number of passes of the laser over the skin. It is important that there is a physical separation between individual spots so that:

- The area is not subject to overlapping spots that could result in local excess heating effects from adjacent spots resulting in skin damage.
- The area is not under treated leaving stippled skin.
- The skin has cooled sufficiently before second or subsequent passes of the laser.

0.5 mm/div

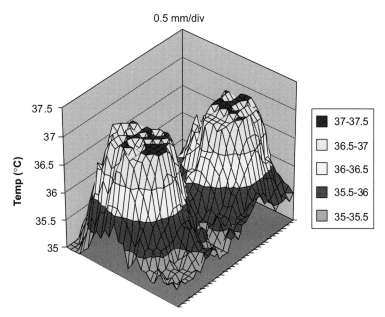

Figure 21 The temperature rise induced by two dye laser spots with a minimum of 5 mm separation (4.5 J/cm^2 at 585 nm, with 5 mm spot).

Figure 21 shows the effect of two laser pulses placed next to each other some 5 mm apart. The time between the pulses is 1 s. There are no excessive temperatures evident and no apparent temperature buildup in the gap. This suggests a minimum physical separation of 5 mm between all individual spot sizes would be sufficient to achieve uniform therapeutic and aesthetic results without either missing areas of tissue or causing excessive buildup of potentially damaging temperatures.

4.4.3. Computerized laser scanning
Having established the parameters relating to laser spot positioning, the possibility of achieving reproducible laser coverage of a lesion by computerized laser position scanning becomes possible. This has significant advantages, which include: accurate positioning of the spot with the correct spacing from the adjacent spots and accurate timing allowing the placement of the laser at a certain location at the appropriate lapsed time.

Thomas et al. [74] carried out a trial of the computer scanning method compared to the manual scanning method. It was clear that the automated scanning method was far superior to manual positioning, however this could only be demonstrated through the use of a thermal imager. Example thermal images taken during this therapy are shown in Figure 22, which captures the various stages of laser scanning of the hand using a dye laser at 5.7 J/cm^2.

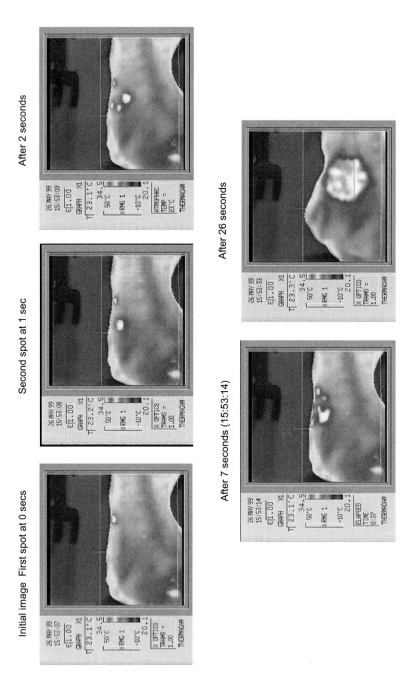

Figure 22 Sequences during computer laser scanning using a thermal imager.

Thermography confirms that due to thermal diffusion the spot temperature from individual laser spots will spread out and merge and that the efficacy of the treatment depends upon laser positioning, time duration between spots, power density, and wavelength.

The following two subsections describe examples where thermal imaging and mathematical Monte-Carlo modeling are used to validate the therapeutic process.

4.4.4. Case study 1: Port wine stain

Vascular nevi are common and are present at birth or develop soon after. Superficial lesions are due to capillary networks in the upper or mid dermis, but larger angiomas can be located in the lower dermis and subcutis. A typical example of vascular nevi is the PWS often present at birth, and is an irregular red or purple macule usually affecting one side of the face. The treatment of PWS can be carried out a number of ways often dependent on the nature, type, anatomical location, and severity of lesion location, as outlined in Table 6.

PPTR can be used very effectively for this condition. A laser wavelength of 585 nm is preferentially absorbed by hemoglobin within the blood, but there is partial absorption in the melanin-rich basal layer in the epidermis. The objective is to thermally damage the blood vessel, by elevating its temperature, while ensuring that the skin surface temperature is kept low. For a typical blood vessel, the temperature—time graph appears similar to that shown in Figure 23.

This suggests that in principle it is possible to selectively destroy the PWS blood vessels, by elevating them to a temperature in the region of 70°C, causing disruption to the small blood vessels, whilst maintaining a safe skin surface temperature. In order to realistically model laser—tissue interactions, it is necessary to consider the following:

- Transport of the incident laser radiation through the tissue media.
- Scattering and absorption of the photons as they propagate through the tissue.
- Local energy deposition, due to absorbed photon energy being converted into thermal energy and thermal transport mechanisms, with conduction being dominant within the tissue.
- Appropriate surface boundary conditions that permit convection losses to be incorporated.

To address this issue a two-stage approach is adopted, where the radiation transport is modeled using the well-established Monte-Carlo method [75—77]. Simple exponential attenuation with penetration depth (the well-known Beer—Lambert law) is inappropriate where the scattering process is significant compared to photon absorption. The two-dimensional

Table 6 Vasculature treatment types.

Treatment type	Process	Possible concerns
Camouflage	Applying skin colored pigments to the surface of the skin. Enhancement to this technique is to tattoo skin colored inks into the upper layer of the lesion.	Only a temporary measure and is very time consuming. Efficacy dependant on flatter lesions.
Cryosurgery	Applying super-cooled liquid nitrogen to the lesion to destroy abnormal vasculature.	May require several treatments.
Excision	Commonplace where the lesion is endangering vital body functions.	Not considered appropriate for purely cosmetic reasons. Complex operation resulting in a scar. Generally, only applicable to the proliferating hemangioma lesion.
Radiation therapy	Bombarding the lesion with radiation to destroy vasculature.	Induced possibility of skin cancer in a small number of cases.
Drug therapy	Widely used administering steroids.	Risk of secondary complications affecting bodily organs.

Cartesian thermal transport equation is:

$$\nabla^2 T + \frac{Q(x, y)}{k} = \frac{1}{\alpha} \frac{\partial T(x, y, t)}{\partial t} \tag{1}$$

where $T(x, y, t)$ is the local instantaneous temperature, k the thermal conductivity of the tissue layers, α the thermal diffusivity (defined as $k/\rho C$, where ρ is the layer density and C is the specific heat of that layer). The volumetric source term $Q(x, y)$ is obtained from the solution of the Monte-Carlo radiation transport problem.

The thermographic data was particularly useful in validating the computational model. Figure 24 illustrates a typical problem with multiple blood vessels buried at different depths within the dermis. A laser wavelength of 585 nm is preferentially absorbed by hemoglobin within the blood, but there is partial absorption in the melanin-rich basal layer in the epidermis. The objective is to thermally damage the blood vessel, by

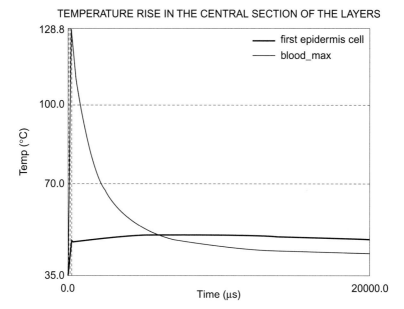

max blood above 100°C for 670.00 μs

TEMPERATURE RISE IN THE CENTRAL SECTION OF THE LAYERS

— first epidermis cell
— blood_max

Figure 23 Typical temperatures for port wine stain (PWS) problem, indicating thermal disruption of blood vessels, while skin surface temperature remains low.

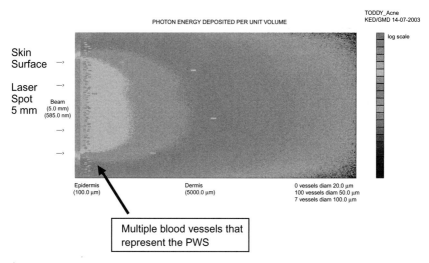

PHOTON ENERGY DEPOSITED PER UNIT VOLUME

TODDY_Acne
KED/GMD 14-07-2003

log scale

Skin Surface →

Laser Spot 5 mm

Beam
(5.0 mm)
(585.0 nm)

→

→

Epidermis
(100.0 μm)

Dermis
(5000.0 μm)

0 vessels diam 20.0 μm
100 vessels diam 50.0 μm
7 vessels diam 100.0 μm

Multiple blood vessels that represent the PWS

Figure 24 Monte-Carlo radiation transport model of laser diffusion into tissue layers.

elevating its temperature, while ensuring that the skin surface temperature is kept low. Flashlamp-pumped pulsed dye lasers (FLPPDL) are often used in clinical practice for this type of treatment, though optimum treatment regimes still need to be established.

4.4.5. Case study 2: Laser depilation

Excess hair growth is a significant problem for some individuals. Laser treatment using 694 nm wavelength laser radiation is possible due to the preferential absorption of the optical radiation by melanin, which occurs in the basal layer and particularly in the hair follicle base. A Monte-Carlo analysis was performed in a similar manner to Case Study 1 above, where the target region in the dermis is the melanin-rich base of the hair follicle.

These calculations showed that it was possible to thermally damage the melanin-rich follicle base whilst restricting the skin surface temperature to values that cause no superficial damage. This has been confirmed by clinical trials indicated that there is indeed a beneficial effect, but the choice of laser parameters still require optimizing. Results of a clinical trial are shown in Figure 25. Detailed thermometric analysis is shown in Figure 26.

Analysis of the data showed that the surface temperature was raised to about 50°C. The thermal image also clearly shows the selective absorption in the melanin-dense hair. The temperature of the hair was raised to over 200°C. This shows that selective wavelength absorption by the melanin leads to cell necrosis yielding the desired clinical outcome.

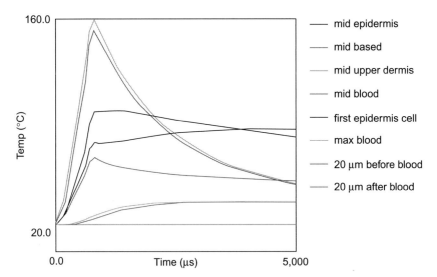

Figure 25   Temperature–time profiles for $20\,\text{J/cm}^2$ ruby (694 nm), $800\,\mu\text{s}$ laser pulse on Caucasian skin type III.

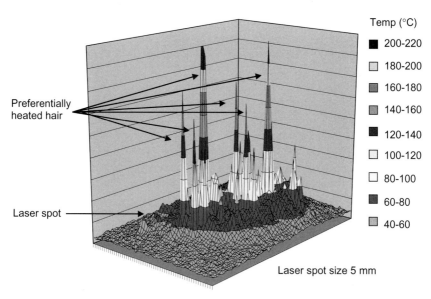

Temp (°C)
■ 200-220
□ 180-200
■ 160-180
■ 140-160
■ 120-140
□ 100-120
□ 80-100
■ 60-80
■ 40-60

Preferentially
heated hair

Laser spot

Laser spot size 5 mm

Figure 26   Post-processed results of 5 mm diameter 694 nm 20 J/cm^2 800 μs ruby pulse.

4.5. Conclusion

Laser therapy, when used in conjunction with thermal imaging is proving to be an effective therapy for differing medical conditions. This would have not been possible without the advent of high-performance thermal imagers, which have facilitated the reliable use of this therapy in medicine.

5. MICROWAVE RADIOMETRY FOR MEDICAL APPLICATIONS

5.1. Principles of medical microwave radiometry

Microwave radiometry provides a noninvasive and inherently safe method of body tissue temperature measurement for a range of medical applications. In clinical medicine, microwave radiometry is used to obtain information about internal body temperature patterns by the measurement of part of the centimetric wavelength component of the natural thermal radiation from the tissues of the body. In appropriate circumstances, knowledge of such thermal patterns can assist disease detection and diagnosis and may have a role in the monitoring of therapeutic processes [78−86]. The radiometric temperature of a volume of material is measured by coupling to its microwave thermal radiation field using an antenna, and determining the equivalent temperature of the coupled signal with a microwave radiometer receiver. The radiation measured is in the long-wavelength

Rayleigh–Jeans region of the Planck spectrum where an instrument of frequency bandwidth B will measure a signal power kTB from a source at temperature T with k being Boltzmann's constant.

At frequencies below about 6 GHz, significantly below the water molecular resonance of 26 GHz, the dielectric properties of body tissues are such that they are partially transparent to electromagnetic radiation. The generation and transmission of microwave thermal radiation then occurs over distances of approximately the radiation wavelength in the tissue material. This then allows microwave radiation coupling to temperature patterns over depths of up to a few centimeters in human tissues. It is this significant, strong coupling into tissues that distinguishes microwave radiometry temperature measurement from surface radiant or contact thermometric methods.

5.2. Microwave properties of tissues

At microwave frequencies, body tissues behave as lossy dielectrics with the dielectric properties largely determined by the water content of the tissue [87,88] see Figure 27. Tissue water content varies considerably, from about

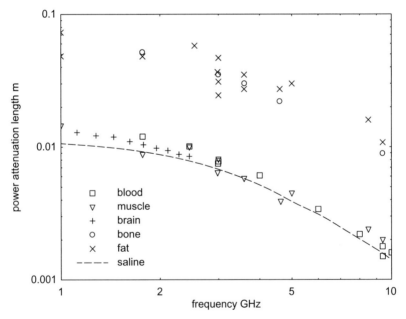

Figure 27 Attenuation lengths for microwave transmission through major human body tissues and 0.9% sodium chloride electrolyte (saline) at 37°C. This shows the dominance of the electrolyte behavior in the high water content tissues (blood, muscle, brain) and the proportionately reduced effect in low water content tissues (bone, fat) [88–90].

8–26% for fat and bone to about 62–80% for skeletal muscle, skin, brain, organ tissues, and blood, with some malignant tumors having particularly high water contents of 80–90%. The major part of the tissue water is present as 0.9% (0.15 molar) sodium chloride electrolyte. The dielectric properties of tissues and water are usually expressed through the complex permittivity $(\varepsilon'_r - j\varepsilon''_r)\varepsilon_0$, with ε'_r the relative permittivity and ε''_r the loss factor of the material and ε_0 the vacuum permittivity, and with $\sigma = \omega\varepsilon''_r\varepsilon_0$ the equivalent loss conductivity at angular frequency ω [83]. As well as being tissue type dependent these properties are also frequency and temperature dependent.

For the lower microwave frequencies of 1–6 GHz tissue propagation can be described with sufficient accuracy for practical measurements by a wavelength in tissue $\lambda_t = \lambda_0/\sqrt{\varepsilon'_r}$ (λ_0 vacuum wavelength) and power attenuation constant $2\alpha = Z_0\sigma/\sqrt{\varepsilon'_r}$ (Z_0 vacuum impedance). These quantities determine, respectively, the possible spatial resolution of tissue temperature variations and the depth over which the radiometric temperature signal is generated. In tissues at 35–37°C, the ionic conductivity and the equivalent conductivity for water molecular resonance loss are similar at about 3 GHz, giving a pronounced minimum in the product of the spatial resolution and attenuation constant factors and indicating the region of generally optimal measurement performance [84,90] (see Figure 28). Medical microwave radiometry equipment is then usually designed for operation in the 1–6 GHz region where thermal pattern resolutions of about 6–12 mm and signal generation characteristic lengths of about 10–40 mm can be obtained in tissues.

The tissue impedance for microwaves, Z_0/ε'_r, is typically 50–110 Ω, much lower than the 377 Ω impedance of air. The power reflection for a microwave signal at a tissue-to-air interface is then large and variable in the range 30–60%. For proper signal measurement, the signal coupling antennas for medical microwave radiometry must be designed to have low impedance, typically 70–100 Ω, and be operated in direct contact with the skin. This allows the remaining tissue-to-antenna reflection to be reduced, but it can remain significant at above 20%, and the radiometer receiver must correct for this to provide a standard radiation temperature measure.

5.3. Microwave radiometry measurement technique

The essential elements of a medical microwave radiometry system are a microwave antenna designed for operation in direct contact with the body so as to couple efficiently to the low tissue impedance, a reflection correcting radiometer receiver for thermal signal power measurement, and signal processing to provide a Celsius calibrated equivalent temperature output for display and recording (Figure 29) [84,85,91]. Depending on the application, the measured microwave temperature profiles may be displayed

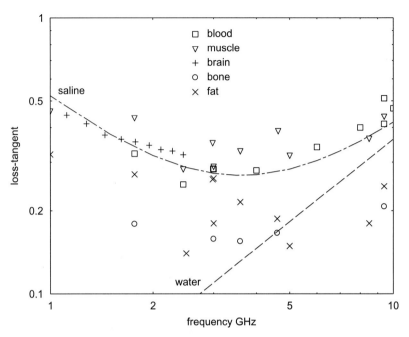

Figure 28 Tissue wavelength and attenuation constant product (loss-tangent) behavior at the lower microwave frequencies for major human body tissues at 37°C. The dominance of the permittivity contribution of the 0.9% saline electrolyte with its loss-tangent minimum around 3 GHz produces a similar minimum in the tissue values.

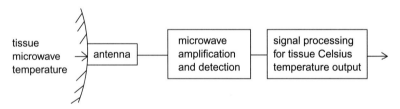

Figure 29 Basic microwave thermometry system.

for direct analysis and interpretation, or several profiles assimilated to construct a temperature pattern for the scanned area. The equipment can be easy to use, reasonably portable and compact, have accurate, stable calibration, and be suitable for routine use in a wide variety of clinical applications.

5.3.1. Measurement signal coupling
The radiometric temperature of a volume of material is determined by a convolution of the material spatial temperature pattern $T(r)$ with a coupling

spatial response or weighting function $w(r)$ for the receiving antenna within the material to give

$$T_{\text{rad}} = \int_v w(r)T(r)dv \qquad (2)$$

with $w(r)$ integrating to unity over the whole coupled volume. The form of this weighting function depends on both the antenna coupling response and on the geometry and dielectric properties of the coupled material across the measurement frequency range. Through the reciprocity principle, the weighting function is identical to the normalized power dissipation distribution when the coupling antenna is actively excited at its measurement port. The weighting function can thus be found by microwave modeling, or by measurements of the active excitation case [92−95]. When known, the weighting function can be applied at one or several frequencies to interpret measured radiometric temperatures in terms of the actual material temperature pattern. The most efficient radiometric body-contact antennas provide thermal signal information that is generated from both the antenna near-field induction zone and the far-field propagation zone in approximately equal measure (Figure 30). The spatial resolution for thermal pattern variation measurement is determined by the lateral extent of the near-field zone of about half of the maximum dimension of the antenna aperture (Figure 31).

The most efficient form of antenna for most biomedical applications is the dielectric loaded waveguide. For TE-mode propagation in the guide, loading by low-loss dielectric material reduces the impedance at the antenna aperture to close to that of the body tissues, giving a low coupling reflection coefficient and reducing the guide cross-section to give an improved near-field spatial resolution. The antenna length is typically about one guide wavelength to allow the principal mode to be established and to allow use of a wide bandwidth, low-loss transition to coaxial-cable coupling to the radiometer receiver (Figure 32). Typical operating band-widths are $> 500\,\text{MHz}$ at 3 GHz and signal losses $0.2 − 0.3\,\text{dB}$ (5 − 7%). The dimensions of these antennas make them very suitable for simple handheld scanning over body regions [84].

5.3.2. Transmission-loss temperature shift

Practical microwave signal transmission components are constructed from materials that cause some signal power loss through resistive or dielectric power absorption. These material parts will also generate thermal microwave power according to their loss and temperature and contribute to the total signal transmitted. For a power transmission α through a part of

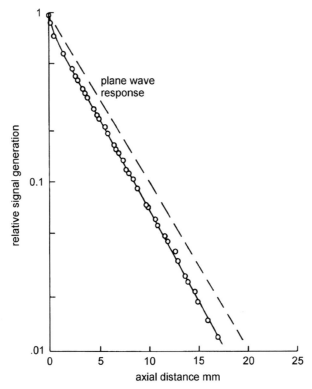

Figure 30 Signal generation along the axis of a TE$_{11}$ mode antenna (axial weighting function $w(z)$ measured in water at 3.26 GHz by nonresonant perturbation. The near-field zone extends to about 3 mm from the antenna face and contributes approximately 40% of the signal. The far-field response is essentially that of a slightly divergent wave propagating in the measurement medium [93].

a system at temperature T_{loss} the resultant effective temperature of an input signal of T_{in} is

$$T_{\text{out}} = \alpha T_{\text{in}} + (1 - \alpha) T_{\text{loss}} \tag{3}$$

and there is an error $T_{\text{out}} - T_{\text{in}}$ introduced in the measurement. Practical microwave antennas, cables, and components having signal to ambient temperature differences of $10-15\,^{\circ}$C can induce significant measurement errors of $0.5-1\,^{\circ}$C and component temperature measurement and control is essential to allow corrections to be applied to the apparent signal temperature [96,97].

5.4. Signal temperature measurement requirements

The matched impedance noise power from a source, measured by a radiometer, is kTB [98]. Since the factor kB is common to all noise signal

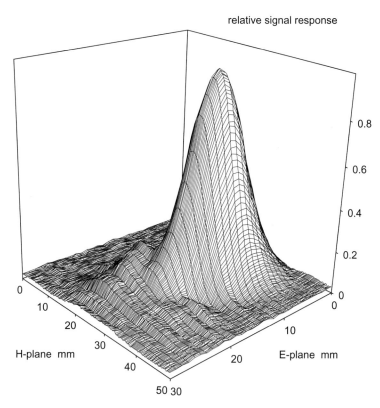

relative signal response

Figure 31 Lateral spatial response for signal generation for the TE_{11} mode antenna measured in a muscle simulating 40% sucrose solution ($\varepsilon_r = 55-j20$) at 3 mm depth at 3.26 GHz by nonresonant perturbation [93].

terms measured by a particular radiometer receiver all system signals can expressed in terms of their equivalent temperature. The receiver microwave detector is taken to have a square-law response so that the predetection high-frequency noise power signals are properly converted to low-frequency postdetection voltage signals proportional to the signal equivalent temperatures.

There are significant practical problems, which prevent the use of a simple microwave receiver for measurement of the thermal signal generated in body tissues:

- The impedance change between body tissues and an antenna will cause impedance mismatch reflection at the antenna to receiver measurement port. This mismatch must be accurately known and compensated for.
- The signal gain of microwave amplifiers and the sensitivity of microwave detectors are directly dependent on the gain of the active devices used in

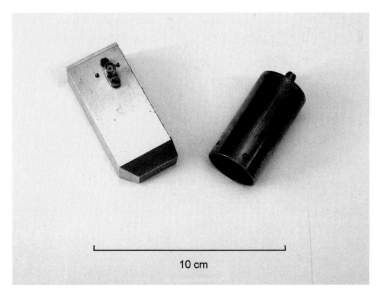

Figure 32 Examples of dielectric loaded waveguide body-contact radiometer antennas for operation at 3.0–3.5 GHz [84,85].

them and are thus device temperature dependent. Typical gain and sensitivity variations are $50-100\%$ for temperature variations of $\sim 40^\circ$C.

- The input noise temperature of a microwave amplifier is dependent on the temperature and gain of its active devices so that the receiver input noise temperature and the temperature of the noise signal emitted back out of the receiver input to the antenna-tissue reflection are not controlled or stable quantities.
- The input impedance of a microwave amplifier depends on its input active devices and may not be well matched to an antenna and coaxial line impedance. The amplifier gain and noise temperature may then vary with source impedance variations through source impedance modulation of the receiver equivalent input noise temperature.

If microwave radiometric temperatures are to be compared between different measurement systems or compared with contact thermometry measurements, it is essential that the matched-impedance, maximum power transfer temperature is measured.

5.5. Radiometer configurations

In the total power configuration radiometer, as shown in Figure 33, the microwave amplification-detection chain is continuously connected to the source to be measured. Figure 34 shows the signal components generated by a general reflection ρ between a source at equivalent temperature T_S and

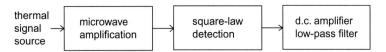

Figure 33 General arrangement of the total power configuration microwave radiometer.

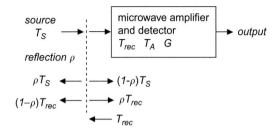

Figure 34 Signal components generated by a general reflection ρ between a source at equivalent temperature T_S and the input of a total power radiometer receiver of equivalent temperature T_{rec}, noise temperature T_A, and gain G.

the input of a total power radiometer receiver of equivalent temperature T_{rec}, noise temperature T_A, and gain G.

The radiometer output $G[(1 - \rho)T_S + \rho T_{\text{rec}} + T_A]$ has direct dependence on the receiver parameters ρ, T_{rec}, T_A and G. All of these quantities can have considerable variation with operating conditions making this an impractical radiometer configuration for medical applications. Some of these receiver- and source-dependent measurement variations are removed in the comparator or Dicke radiometer configuration [99], where the amplifier input is connected alternately to the source to be measured and to a known noise temperature reference. For medical application microwave radiometers, a comparator input circuit of the form of Figure 35 is commonly used, combining the source-reference switching with amplifier noise isolation through the use of a circulator [84,85,100].

With the switch closed the source and reflected input noise signals are passed through the radiometer and the circulator and the radiometer output signal is $G[(1 - \rho)T_S + \rho T_R + T_A]$. With the switch open the reference signal is reflected into the receiver through the circulator and the output signal is $G(T_R + T_A)$. The difference between the two input signal temperatures for the two switch positions then gives an output signal change of $G[(1 - \rho)T_S - T_R]$. With the input comparator switch operating continuously this difference signal can be efficiently extracted by synchronous demodulation of the postdetection signal. This signal is independent of T_{rec} and T_A but still dependent on gain G and reflectance ρ. If the reference noise temperature is known from calibration and can be varied to null the postdetection switching frequency signal so that $T_R = (1 - \rho)T_S$ the effect

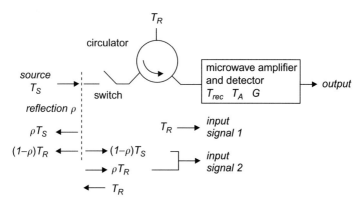

Figure 35 Comparator configuration radiometer using a switch-circulator input circuit to provide by reflection from the open switch a reference temperature signal T_R to the amplifier and the same signal to the source reflection ρ when the switch is closed, and at the same time provide isolation of the amplifier noise T_{rec} from the source reflection.

of gain dependence can be removed. Various arrangements have been used to vary the effective reference temperature and null the difference signal, but since a noise signal has to be nulled in the presence of the reflection and gain variations, there can be problems in obtaining adequate feedback stability and measurement response time may have to be increased compared to a nonfeedback radiometer configuration [91]. It may also be difficult to vary rapidly the effective noise temperature of the reference source and know accurately its calibration.

These problems of the single reference Dicke radiometer are removed in the two-reference comparator radiometer, used in conjunction with a circulator-switch input configuration [97,101]. Here the reference noise temperature is switched between two different values T_{R1} and T_{R2} during a comparator-type switching cycle (see Figure 36).

Four receiver signals are produced by the four combinations of the positions of each switch, and temperatures for each switching phase can be extracted from the postdetection signal. By using an appropriate switching pattern or a combination the four temperatures, a measured ratio can be formed that is independent of the radiometer gain, the input reflection, and the amplifier noise. This ratio with calibrated but nonmicrowave measured values for the reference temperatures gives a source temperature

$$T_S = R\left(\frac{T_{R1} - T_{R2}}{2}\right) + \left(\frac{T_{R1} + T_{R2}}{2}\right) \tag{4}$$

with the ratio R zero when the source equals the average reference temperature.

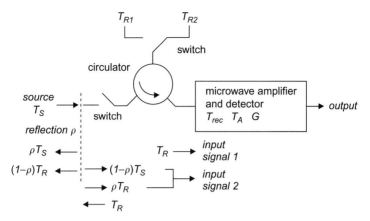

Figure 36 Two reference radiometer input circuit using a circulator for amplifier to source noise isolation and for single-pole switch reference switching.

5.6. Equivalent temperature fluctuation of radiometer measurements

All radiometers measure a randomly fluctuating noise signal. Postdetection time averaging of this signal reduces, but does not eliminate, the fluctuation inherent in the measurement. The fluctuation remaining on the output signal determines the temperature resolution that can be achieved with the system [102,103].

The detected signal fluctuation, expressed as an equivalent temperature fluctuation, depends on the total noise power measured by the radiometer and, through the detection process, on the ratio of the pre- to post-detection signal bandwidths [104–106]. The measured noise power is that due to both the source (T_S) and the radiometer (T_R), together having the system equivalent noise temperature (T_{sys}). The minimum measurement RMS temperature fluctuation, the "Gabor limit," is then of the following general form:

$$\Delta T = T_{sys}\sqrt{\frac{2B_2}{B_1}} = (T_S + T_R)\sqrt{\frac{2B_2}{B_1}} \tag{5}$$

where B_1 and B_2 are the pre- and post-detection noise power bandwidths. This fluctuation is increased by a factor of 2 or slightly more for comparator configuration radiometers where the source noise power is only measured for half of the total measurement time [97].

The postdetection signal bandwidth and frequency response determines the measurement time of the radiometer. With an optimized time response, which must include all postdetection signal processing, the measurement time is about $0.35/B_2$ [103]. This gives a temperature fluctuation for a comparator radiometer of response time τ of $\Delta T \approx 1.67 T_{sys}/\sqrt{B_1\tau}$.

5.7. Performance of practical clinical radiometers

Clinical microwave radiometry systems operate at frequencies between 1 and 6 GHz, the lower frequency being at the limit of useful tissue thermal pattern resolution and the upper frequency being in the region of significantly decreasing depth of coupling into tissues. Many systems have been designed to operate around 3 GHz where the tissue dielectric properties allow clinically useful resolution of thermal pattern features while still coupling strongly to subcutaneous tissue over the main surface temperature gradient region of the body.

The radiometric dependence of temperature resolution shows the predetection bandwidth should be the maximum possible within technical or operational constraints. Microwave circuit components can usually provide good performance over bandwidths of 20–50% of the operating frequency. Antenna coupling structures of simple form for low loss are usually limited to about a 20% bandwidth for acceptable performance. Predetection bandwidths are then typically from 200 MHz at 1 GHz, to 1 GHz at 5 GHz. For practical microwave components operating at room temperature, the noise contribution can usually be made comparable to that of the measured body tissue signal to give system noise temperatures in the region of 500–1,000 K [107]. Measurement temperature resolution is then, typically, about 0.1 K for response times of a few seconds (see Figure 37).

5.8. Multifrequency radiometry for temperature profile estimation

Microwave radiometry using a single predetection frequency response as considered above can provide only a measure of the apparent microwave temperature of the thermal radiation at the skin surface for that frequency response. This single temperature measure can provide surface temperature pattern information that is sufficient for several clinical applications but it does not provide information about internal temperature profiles that is essential for some potential applications. Attempts have been made to estimate internal temperature variations from combining surface, microwave, and core temperatures, usually combined with thermal modeling [97], but critical applications require the profile estimation that can only be provided by multifrequency microwave measurements [108–111].

The basis of multifrequency radiometry is the variation of the measured temperature with the variation of the weighting function of the signal coupling antenna and tissue system with frequency [112,113]. The antenna form, dimensions, and operating mode particularly determine the near-field zone weighting and have some effect on the far-field zone. The tissue attenuation factor has a significant effect on the near-field zone weighting

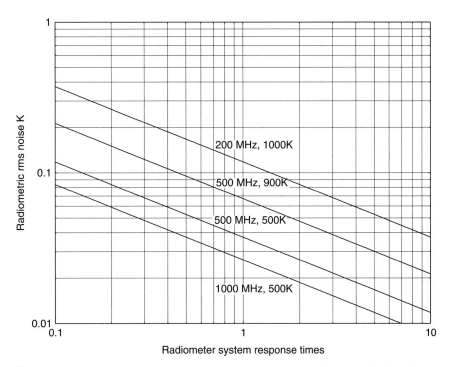

Figure 37 Radiometric noise limited temperature resolution for practical clinical microwave radiometer operating bandwidths and system noise temperatures.

but particularly determines the variation of the far-field zone contribution. For practical applications, the weighting function behavior is usually numerically modeled for an assumed tissue region [114]. To interpret the radiometric temperature measurements some form of the region temperature profile must be assumed, with the number of individual frequency range temperature measurements taken at any position determining the form of profile that can be estimated. A two-frequency measurement allows estimation of a symmetric parabolic profile; three-frequency measurement allows estimation of an asymmetric quadratic profile or, commonly, an exponential form [113].

A basic problem for the clinical application of multifrequency radiometry is the very limited range of radiometric temperature available for measurement. Taking the typical subcutaneous temperature gradient of 0.3°C/mm with the dielectric behavior of high water content tissues, even with the ideal antenna response of only a far-field zone the microwave temperatures between 1.5 and 5 GHz vary by only approximately 1°C. With measurement noise of ~0.05°C per measurement channel and interchannel calibration uncertainties of at least a similar order, a three-frequency measurement will give a good estimate of an average tissue temperature but an estimate of a core temperature that may be in error by $\pm 1°C$ or more [111,115].

5.9. Clinical applications

In clinical applications of microwave radiometry, the receiving antenna is usually in direct contact with the skin [100], although techniques using noncontacting antennas have also been reported [116]. In many earlier applications of microwave radiometry only an estimate of the relative temperature was sought. Examples are the search for local "hot spots" while scanning an area with a single-frequency device [84] and microwave radiometric imaging using a multiprobe radiometer. However, absolute temperatures have also been retrieved [112,115], and more recently the inhomogeneous dielectric nature of the body has been accounted for [114,117].

The use of microwave radiometry to detect tumors and circulatory disorders has been envisaged within early proposals that it could be a diagnostic tool in medicine [78,81,82] used alone or in combination with IR thermography [86]. For breast cancer detection, the method is based on the detection of the local temperature elevations in the breast that may arise due to factors associated with tumors, such as greater metabolic rate and differing vascularity, with respect to normal breast tissue so that thermal asymmetries may be observed between normal and tumor-bearing breasts. In a study involving 120 patients, analysis of thermal images of the breasts indicated that microwave radiometric imaging alone had a sensitivity of 75% and a specificity of 73% in detecting the presence of tumor. These figures are comparable with other noninvasive diagnostic techniques but suggest that when used alone, radiometry leads to an unacceptable rate of false diagnoses. However, when used in combination with mammography, it may lead to more accurate positive prediction rates [83,118] but as yet there has been no widespread use of microwave radiometry for cancer screening.

Another application of microwave radiometry in the treatment of cancer is the noninvasive monitoring and/or controlling of various thermal therapies. For example, there are several reports of its use to monitor superficial localized hyperthermia [92,101], interstitial hyperthermia, and ablation treatments [119]. When using both single and multifrequency radiometry, there will be uncertainty in the predicted tissue temperature due to incomplete tissue composition information, though in practice the hyperthermia bolus depth and its water temperature are found to be of greater importance [117].

Hypothermal neural rescue therapy is the clinical application of cooling of infants who suffer hypoxic-ischemic problems at birth. However, full evaluation of brain cooling requires that temperatures be measured, especially in the deep brain where cell loss leads to the most severe long-term neurological impairments. Multifrequency microwave radiometry has been investigated for this application since this method (unlike magnetic resonance imaging methods) could be made available over the prolonged periods of up to 72 h that mild hypothermia may be maintained in neural protection. A feasibility study has been completed and a method for retrieving the physical temperature profile within the baby's head developed [114]. Tractable solutions to the temperature retrieval problem in a realistic three-dimensional model of the head involving extensive numerical modeling of weighting functions and heat transfer have been developed for use with a five frequency-band radiometer system [111,117].

Microwave radiometry has been used to assess joint inflammation in patients with rheumatoid arthritis. Reports suggest that the technique is reproducible, quick, and simple to use at the bedside without the need for a controlled environment. The resulting data from studies of knees have been shown to correlate with relevant clinical and laboratory measures [120]. The technique was employed in a subsequent study to investigate the effectiveness of three drug regimes in improving pain relief associated with rheumatoid knees [121].

Microwave radiometry has also been used in the *in vivo* detection of extravasation of fluids. For example, a canine study [122] has shown that extravasation of small volumes of the chemotherapeutic agent adriamycin could be detected and the authors suggested that radiometry could be used as an alarm to detect such extravasation. Subsequent studies demonstrated that radiometry has clinical potential for the early detection of extravasation of contrast media administered with power injectors and for reliably detecting small (0.005 ml) emboli in both ionic and nonionic low-osmolar contrast media [123].

As indicated above, there have been numerous assessments of potential applications of microwave radiometry in medicine, and investigations into new potential applications continue to be reported. Since the diagnostic basis of microwave radiometry is the change in tissue temperature with pathology, it can provide useful information, noninvasively and with complete safety, when there is an inflammatory response to disease or injury. However, this information appears to be of limited clinical assistance for several of the conditions considered, particularly when more specific, more reliable or more established modalities are available. Thus, whilst many of the reported studies have been successful in demonstrating that the technique is capable of providing appropriate and useful clinical information and of supporting specialized studies, there has, however, been no general take-up of microwave radiometry to date. The inherent advantages of the technique, though, continue to attract investigation and development which, with the ongoing reduction in equipment costs and increase in computational analytical capabilities, should make it more accessible for clinical application and evaluation as well as for research.

6. SUMMARY

The measurement of the temperature of the human body has been of clinical importance for the last 200 years. However, during the last 50 years technology and physical science have so advanced that noncontact measurement of body temperature is now a reality. Radiometry of the ear, perhaps the most recent, has become a common technique, replacing the mercury-in-glass thermometer. Microwave imaging research continues, with its potential for less superficial temperature detection than in IR thermography. In this last area, the detectors and imaging systems have developed far beyond the pioneer devices of the 1950s and 1960s. A wide demand for thermal imaging in industry, military, and civil engineering has ensured a substantial development in technology, assisted by the now commonplace computer image processing. This chapter brings together the research and applications of thermal radiometry to medicine.

REFERENCES

[1]   P. E. Daneman, "Fever Thermometry — a Review," in *Temperature, Its Measurement and Control in Science and Industry*, Vol. 6, Part 2, American Institute of Physics, New York, pp. 1179–1183 (1992).
[2]   J. M. S. Pierce, "A Brief History of the Clinical Thermometer," QJM: Int. J. Med. **95**, 251–252 (2002).
[3]   P. A. Mackowiak, "Concepts of Fever," Arch. Intern. Med. **158**, 1870–1881 (1998).
[4]   M. Sund-Levander, C. Forsber, and L. K. Wahren, "Normal Oral, Rectal, Tympanic and Axillary Body Temperature in Adult Men and Women: a Systematic Literature Review," Scand. J. Caring Sci. **16**, 122–128 (2002).
[5]   T. Togawa, "Body Temperature Measurement," Clin. Phys. Physiol. Meas. **6**, 83–108 (1985).
[6]   M. Schünke, E. Schulte, and U. Schumacher, *Prometheus: Lernatlas der Anatomie*, Georg Thieme Verlag, Berlin (2006).
[7]   P. W. McCarthy and A. I. Heusch, "The Vagaries of Ear Temperature Assessment," J. Med. Eng. Tech. **30**, 242–251 (2006).
[8]   R. L. Pullen, "Using an Ear Thermometer," Nursing **33**, 54 (2003).
[9]   J. Fraden, "Medical Infrared Thermometry (Review of Modern Methods)," in *Temperature, Its Measurement and Control in Science and Industry*, Vol. 6, Part 2, American Institute of Physics, New York, pp. 825–830 (1992).
[10] V. Betta, F. Cascetta, and D. Sepe, "An Assessment of Infrared Tympanic Thermometers for Body Temperature Measurement," Physiol. Meas. **18**, 215–225 (1997).
[11] I. Pušnik and J. Drnovšek, "Infrared Ear Thermometers — Parameters Influencing Their Reading and Accuracy," Physiol. Meas. **26**, 1075–1084 (2005).

[12] B. Kraus and F. Beerwerth, "Method for Determining Temperature; Radiation Thermometer with Several Infrared Sensor Elements," US Patent 6,898,457 (2005).

[13] J. V. Craig, G. A. Lancaster, S. Taylor, P. R. Williamson, and R. L. Smyth, "Infrared Ear Thermometry Compared with Rectal Thermometry in Children: A Systematic Review," The Lancet 360, 603–609 (2002).

[14] J. L. Robinson, R. F. Seal, D. W. Spady, and M. R. Joffres, "Comparison of Esophageal, Rectal, Axillary, Bladder, Tympanic and Pulmonary Artery Temperatures in Children," J. Pediatr. 133, 553–556 (1998).

[15] C. Childs, R. Harrison, and C. Hodkinson, "Tympanic Membrane Temperature as a Measure of Core Temperature," Arch. Dis. Child 80, 262–266 (1999).

[16] I. Pušnik, E. van der Ham, and J. Drnovšek, "IR Ear Thermometers: What Do They Measure and How Do They Comply with the EU Technical Regulation?," Physiol. Meas. 25, 699–708 (2004).

[17] R. Simpson, G. Machin, R. L. Rusby, and H. C. McEvoy, "Traceability and Calibration in Temperature Measurement: A Clinical Necessity," J. Med. Eng. Tech. 30, 212–217 (2006).

[18] J. Ishii, T. Fukuzaki, and H. C. McEvoy, et al., "A Comparison of the Blackbody Cavities for Infrared Ear Thermometers of NMIJ, NPL, and PTB," in Proceedings of Tempmeko 2004, edited by D. Zvizdic, LPM/FSB, Zagreb, pp. 1093–1098 (2005).

[19] E. F. J. Ring, "The Historical Development of Temperature Measurement in Medicine," Infrared Phys. Technol. 49, 297–301 (2007).

[20] C. A. Wunderlich, "On the Temperature in Diseases, A Manual of Medical Thermometry," Translated from the second German edition by W. Bathurst Woodman, The New Sydenham Society, London (1871).

[21] B. A. Craven and G. S. Settles, "A Computational and Experimental Investigation of the Human Thermal Plume," J. Fluids Eng. 128, 1251–1258 (2006).

[22] E. Bjørn and P. V. Nielsen, "Dispersal of Exhaled Air and Personal Exposure in Displacement Ventilated Rooms," Indoor Air 12(3), 147–164 (2002).

[23] J. D. Hardy, "The Radiation of Heat from the Human Body – I, II, III," J. Clin. Invest. 13, 593–620 (1934).

[24] Y. Houdas and E. F. J. Ring, Human Body Temperature, Its Measurement and Regulation, Plenum Press, New York (1982).

[25] A. Khallaf, R. W. Williams, E. F. J. Ring, et al., "Thermographic Study of Heat Loss from the Face," Thermol. Österreich 4, 49–54 (1994).

[26] A. J. Collins, E. F. J. Ring, J. A. Cosh, et al., "Quantitation of Thermography in Arthritis Using Multi-Isothermal Analysis," Ann. Rheum. Dis. 33, 113–115 (1974).

[27] E. F. J. Ring, A. J. Collins, P. A. Bacon, et al., "Quantization of Thermography in Arthritis Using Multi-Isothermal Analysis. II. Effect of Nonsteroidal Anti-Inflammatory Therapy on the Thermographic Index," Ann. Rheum. Dis. 33, 353–356 (1974).

[28] E. F. J. Ring, N. J. M. Aarts, C. M. Black, et al., "Raynaud's Phenomenon: Assessment by Thermography (EAT Report)," Thermology 3, 69–732 (1988).

[29] H. A. Bird, E. F. J. Ring, and P. A. Bacon, "A Thermographic and Clinical Comparison of Three Intra-Articular Steroid Preparations in Rheumatoid Arthritis," Ann Rheum Dis 38(1), 36–39 (1979).

[30] M. V. Kyle, G. Belcher, and B. L. Hazleman, "Placebo Controlled Study Showing Therapeutic Benefit of Iloprost in the Treatment of Raynaud's Phenomenon," J. Rheumatol 19, 1403–1406 (1992).

[31] E. F. J. Ring, "Objective Assessments of Physical Diagnosis and Therapy," Thermol. Österreich 4, 5–9 (1994).

[32] K. Ammer, "Temperature Effects of Thermotherapy Determined by Infrared Measurements," Phys. Med. 20(Suppl. 1), 64–66 (2004).

[33]   J. M. Engel, J. A. Cosh, E. F. J. Ring, et al., "Thermography in Locomotor Diseases: Recommended Procedure," Eur. J. Rheumatol. Inflamm. **2**, 299–306 (1979).

[34]   L. de Weerd, J. B. Mercer, and L. B. Setsa, "Intraoperative Dynamic Infrared Thermography and Free-Flap Surgery," Ann. Plast. Surg. **57**, 279–284 (2006).

[35]   E. F. J. Ring, J. M. Engel, and D. P. Page-Thomas, "Thermological Methods in Clinical Pharmacology," Int. J. Clin. Pharmacol. **22**, 20–24 (1984).

[36]   K. Ammer, "European Congress of Thermology 1974–2006: A Historical Review," Thermol. Int. **16**, 85–95 (2006).

[37]   P. A. Bacon, E. F. J. Ring, and A. J. Collins, "Thermography in the Assessment of Anti-Rheumatic Agents," in *Rheumatoid Arthritis*, edited by J. L. Gordon and B. L. Hazleman, Elsevier, Amsterdam, pp. 105–110 (1977).

[38]   R. N. Lawson and M. S. Chughtai, "Breast Cancer and Body Temperatures," Can. Med. Assoc. J. **88**, 68–70 (1963).

[39]   J. K. A. Alderson and E. F. J. Ring, "'Sprite' High Resolution Thermal Imaging System," Thermology **1**, 110–114 (1985).

[40]   P. Plassmann, E. F. J. Ring, and C. D. Jones, "Quality Assurance of Thermal Imaging Systems in Medicine," Thermol. Int. **16**, 10–15 (2006).

[41]   B. F. Jones and P. Plassmann, "Digital Thermal Imaging of Human Skin," IEEE Eng. Med. Biol. Mag. **21**(6), 41–48 (2002).

[42]   The Commission for Thermal Physiology of the International Union of Physiological Sciences (IUPS Thermal Commission), "Glossary of Terms for Thermal Physiology – Third Edition," Jpn. J. Physiol. **51**, 245–280 (2001).

[43]   T. Jakubowska, B. Wiecek, M. Wysocki, et al., "Thermal Signatures for Breast Cancer Screening Comparative Study," in *Engineering in Medicine and Biology Society, Proceedings of the 25th Annual International Conference of IEEE*, Vol. 2, pp. 1117–1120 (2003).

[44]   F. J. P. M. Huygen, S. Niehof, J. Klein, et al., "Computer-Assisted Skin Videothermography is a Highly Sensitive Quality Tool in the Diagnosis and Monitoring of Complex Regional Pain Syndrome Type," I. Eur. J. Appl. Physiol. **91**, 516–524 (2004).

[45]   E. F. J. Ring, K. Ammer, A. Jung, et al., "Standardization of Infrared Imaging," Conf. Proc. IEEE Eng. Med. Biol. Soc. **2**, 1183–1185 (2004).

[46]   E. F. J. Ring and K. Ammer, "The Technique of Infrared Imaging in Medicine," Thermol. Int. **10**, 7–14 (2000).

[47]   R. Schuberr, J. V. d Haute, J. Hassenbürger, et al., "Directed Dynamic Cooling, a Methodic Contribution in Telethermography," Acta Thermograph. **2**, 94–99 (1977).

[48]   A. Di Carlo, "Thermography in Patients with Systemic Sclerosis," Thermol. Österreich **4**, 18–24 (1994).

[49]   E. F. J. Ring, "Cold Stress Test for the Hands," in *Thermal Image in Medicine and Biology*, edited by K. Ammer and E. F. J. Ring, Uhlen Verlag, Vienna, pp. 237–240 (1995).

[50]   K. Ammer, "Nerve Entrapment and Skin Temperature of the Human Hand," in *A Casebook of Infrared Imaging in Clinical Medicine*, edited by A. Jung, J. Zuber, and F. Ring, Medpress, Warsaw, pp. 74–76 (2003).

[51]   K. Ammer, "The Glamorgan Protocol for Recording and Evaluation of Thermal Images of the Human Body," Thermol. Int. **18**, 125–144 (2008).

[52]   K. Ammer and E. F. J. Ring, "Influence of the Field of View on Temperature Readings from Thermal Images," Thermol. Int. **15**, 99–103 (2005).

[53]   G. E. Sella, "Forensic Criteria of Acceptability of Thermography," Eur. J. Thermol. **7**, 208–212 (1997).

[54] K. Siniewicz, B. Więcek, J. Baszczyński, and S. Zwolenik, "Thermal Imaging before and after Physical Exercises in Children with Orthostatic Disorders of the Cardiovascular System," Thermol. Int. **12**, 139–146 (2002).

[55] K. Wojaczynska-Stanek, E. Marszal, A. Krzemien-Gabriel, et al., "Biostimulating Laser Treatment of Chronic Sinusitis in Children — Monitoring by Thermal Imaging," Thermol. Int. **15**, 140–145 (2005).

[56] R. G. Wheeland, "Clinical Uses of Lasers in Dermatology," Lasers Surg. Med. **16**(1), 2–23 (1995).

[57] R. J. Barlow, N. P. J. Walker, and A. C. Markey, "Treatment of Proliferative Haemangiomas with 585 nm Pulsed Dye Laser," Br. J. Dermatol. **134**, 700–704 (1996).

[58] J. M. Garden, L. L. Polla, and O. T. Tan, "Treatment of Port Wine Stains by Pulsed Dye Laser — Analysis of Pulse Duration and Long Term Therapy," Arch. Dermatol. **124**, 889–896 (1988).

[59] E. Glassberg, G. Lask, L. G. Rabinowitz, and W. W. Tunnessen, "Capillary Haemangiomas: Case Study of a Novel Laser Treatment and a Review of Therapeutic Options," J. Dermatol. Surg. Oncol. **15**, 1214–1223 (1989).

[60] S. W. Lanigan, "Port Wine Stains on the Lower Limb: Response to Pulsed Dye Laser Therapy," Clin. Exp. Dermatol. **21**, 88–92 (1996).

[61] R. J. Motley, G. Katugampola, and S. W. Lanigan, "Mircovascular Abnormalities in Port Wine Stains and Response to 585 nm Pulsed Dye Laser Treatment," Br. J. Dermatol. **135**(Suppl. 47), 13–14 (1996).

[62] M. C. Grossman, C. Dierickx, W. Farinelli, T. Flotte, and R. R. Anderson, "Damage to Hair Follicle by Normal Mode Ruby Laser Pulse," J. Am. Acad. Dermatol. **35**, 889–894 (1996).

[63] D. Gault, R. M. Clement, R. B. Trow, and M. N. Kiernan, "Removing Unwanted Hairs by Laser," Face **6**(2), 129–130 (1998).

[64] R. M. Clement, M. N. Kiernan, R. A. Thomas, K. E. Donne, and P. J. Bjerring, "The Use of Thermal Imaging to Optimise Automated Laser Irradiation of Tissue," Skin Res. Technol. **5**(2), *6th Congress of the International Society for Skin Imaging*, July 4–6, Royal Society London (1999).

[65] R. A. Thomas, "An Integrated Approach to Condition Monitoring of the Skin," PhD Thesis, Swansea Institute, UK (2002).

[66] S. A. Prahl, I. A. Vitkin, U. Bruggemann, B. C. Wilson, and R. R. Anderson, "Determination of Optical Properties of Turbid Media Using Pulsed Photothermal Radiometry," Phys. Med. Biol. **37**, 1203–1217 (1992).

[67] U. S. Sathyam and S. A. Prahl, "Limitations in Measurement of Subsurface Temperatures Using Pulsed Photothermal Radiometry," J. Biomed. Opt. **2**, 251–261 (1997).

[68] R. M. Clement, K. D. Donne, R. A. Thomas, and M. N. Kiernan, "Thermographic Condition Monitoring of Human Skin during Laser Therapy," *Quality Reliability Maintenance, the 3rd International Conference*, St. Edmund Hall, University of Oxford, March, pp. 30–31 (2000).

[69] M. N. Kiernan, "An Analysis of the Optimal Laser Parameters Necessary for the Treatment of Vascular Lesions," PhD Thesis, the University of West of England (1997).

[70] R. R. Andersen and J. A. Parrish, "Microvasculature Can Be Selectively Damaged Using Dye Lasers," Lasers Surg. Med. **1**, 263–270 (1981).

[71] A. J. Welsh and M. V. C. van Gemert, *Optical-Thermal Response of Laser-Irradiated Tissue*, Plenum Press, New York (1995).

[72] R. A. Thomas, *Thermography*, Coxmoor Publishers, Oxford, pp. 79–103 (1999).

[73]   J. M. Garden and W. Bakus, "Clinical Efficacy of the Pulsed Dye Laser in the Treatment of Vascular Lesions," J. Dermatol. Surg. Oncol. **19**, 321–326 (1996).

[74]   R. A. Thomas, K. E. Donne, R. M. Clement, and M. Kiernan, "Optimised Laser Application in Dermatology Using Infrared Thermography," in *Proc. SPIE Thermosense XXIV*, Orlando, FL (2002).

[75]   E. D. Cashwell and C. J. Everett, *The Monte Carlo Method for Random Walk Problems*, Pergamon Press, London (1959).

[76]   G. Daniel, "An Investigation of Thermal Radiation and Thermal Transport in Laser-Tissue Interaction," PhD Thesis, Swansea Institute, UK (2002).

[77]   B. C. Wilson and G. A. Adam, "Monte Carlo Model for the Absorption and Flux Distributions of Light in Tissue," Med. Phys. Biol. **10**, 824–830 (1983).

[78]   A. H. Barrett and P. C. Myers, "Subcutaneous Temperatures: A Method of Noninvasive Sensing," Science **190**, 669–671 (1975).

[79]   A. H. Barrett, P. C. Myers, and N. L. Sadowsky, "Microwave Thermography in the Detection of Breast Cancer," Am. J. Roentgenol. **134**, 365–368 (1980).

[80]   K. L. Carr, I. A. El-Mahd, and J. Shaeffer, "Passive Microwave Thermography Coupled with Microwave Heating to Enhance Early Detection of Cancer," Microwave J. **25**, 125–136 (1982).

[81]   J. Edrich and P. C. Hardee, "Thermography at Millimeter Wavelengths," Proc. IEEE **62**, 1391–1392 (1974).

[82]   B. Enander and G. Larson, "Microwave Radiometric Measurements of the Temperature Inside a Body," Electron. Lett. **10**, 317 (1974).

[83]   K. R. Foster and E. A. Cheever, "Microwave Radiometry in Biomedicine: A Reappraisal," Bioelectromagnetics **13**, 567–579 (1992).

[84]   D. V. Land, "A Clinical Microwave Thermography System," Proc. IEE **134A**, 193–200 (1987).

[85]   Y. Leroy, A. Mamouni, J. C. Van De Velde, B. Bocquet, and B. Duardin, "Microwave Radiometry for Non-Invasive Thermometry," Automedica **8**, 181–202 (1987).

[86]   P. C. Myers, N. L. Sadowsky, and A. H. Barrett, "Microwave Thermography: Principles, Methods and Clinical Applications," J. Microwave Power **14**, 105–114 (1979).

[87]   J. L. Schepps and K. R. Foster, "The U.H.F. and Microwave Properties of Normal and Tumour Tissues: Variation in Dielectric Properties with Water Content," Phys. Med Biol. **25**, 1149–1159 (1980).

[88]   H. P. Schwan and K. R. Foster, "RF-field Interactions with Biological Systems: Electrical Properties and Biophysical Mechanisms," Proc. IEEE **68**, 104–113 (1980).

[89]   J. M. Allison and R. J. Sheppard, "Dielectric Properties of Human Blood at Microwave Frequencies," Phys. Med. Biol. **38**, 971–978 (1993).

[90]   J. B. Hasted, *Aqueous Dielectrics*, Chapman and Hall, London (1973).

[91]   K. M. Ludeke, B. Schiek, and J. Koehler, "Radiation Balance Microwave Thermograph for Industrial and Medical Applications," Electron. Lett. **14**, 194–195 (1978).

[92]   S. Jacobsen, P. R. Stauffer, and D. G. Neuman, "Dual-Mode Antenna Design for Microwave Heating and Noninvasive Thermometry of Superficial Tissue Disease," IEEE Trans. Biomed. Eng. **47**, 1500–1509 (2000).

[93]   D. V. Land, "Measurement of VHF, UHF and Microwave Biomedical Antenna Behaviour by Nonresonant Field Perturbation," IEEE-MTT/URSI Microwaves Med. 233–236 (1993).

[94]   G. P. Rine, T. V. Samulski, W. Grant, and C. A. Wallen, "Comparison of Two-Dimensional Numerical Approximation and Measurement of SAR in a Muscle

Equivalent Phantom Exposed to a 915 MHz Slab-Loaded Waveguide," Int. J. Hyperthermia **6**, 213−225 (1990).

[95] M. Robillard, M. Chive, Y. Leroy, J. Audet, Ch. Pichot, and J. Ch. Bolomey, "Characteristics of Waveguide Applicators and Signatures of Thermal Structures," J. Microwave Power **17**, 97−105 (1982).

[96] D. V. Land, "Radiometer Input Circuit Requirements for Microwave Thermography," Electron. Lett. **19**, 1040−1042 (1983).

[97] D. V. Land, "An Efficient, Accurate and Robust Radiometer Configuration for Microwave Temperature Measurement for Industrial and Medical Applications," J. Microwave Power Electromagn. Energy **36**, 139−154 (2001).

[98] P. Kittel, "Comment on the Equivalent Noise Bandwidth Approximation," Rev. Sci. Instrum. **48**, 1214−1215 (1977).

[99] R. H. Dicke, "The Measurement of Thermal Radiation at Microwave Frequencies," Rev. Sci. Instrum. **17**, 268−275 (1946).

[100] Y. Leroy, B. Bocquet, and A. Mamouni, "Non-Invasive Microwave Radiometry Thermometry," Physiol. Meas. **19**, 127−148 (1998).

[101] M. Chive, M. Plancot, G. Giaux, and B. Prevost, "Microwave Hyperthermia Controlled by Microwave Radiometry: Technical Aspects and First Clinical Results," J. Microwave Power **19**, 233−241 (1984).

[102] D. Gabor, "Communication Theory and Physics," Philos. Mag. **41**, 1161−1187 (1950).

[103] D. V. Land, "Radiometer Receivers for Microwave Thermography," Microwave J. **26**, 196−201 (1983).

[104] D. V. Land, A. P. Levick, and J. W. Hand, "The Use of the Allan Deviation for the Measurement of the Noise and Drift Performance of Microwave Radiometers," Meas. Sci. Technol. **18**, 1917−1928 (2007).

[105] R. Meredith, F. L. Warner, Q. V. Davies, and J. L. Clark, "Superheterodyne Radiometers for Short Millimetre Wavelengths," Proc. IEE **111**, 241−256 (1964).

[106] A. Van der Ziel, *Noise*, Chapman and Hall, New York, Chapter 13 (1955).

[107] R. Adler, R. S. Engelbrecht, S. W. Harrison, H. A. Haus, M. T. Lebenbaum, and W. W. Munford, "Description of the Noise Performance of Amplifiers and Receiving Systems," Proc. IEEE **51**, 436−442 (1963).

[108] M. El-Shenawee, "Numerical Assessment of Multifrequency Microwave Radiometry for Sensing Malignant Breast Cancer Tumors," Microwave Opt. Tech. Lett. **36**, 394−398 (2003).

[109] B. Stec, A. Dobrowolski, and W. Susek, "Multifrequency Microwave Thermograph for Biomedical Applications," IEEE **BME-51**, 548−551 (2004).

[110] T. Sugiura, Y. Kouno, A. Hashizume, H. Hirata, J. W. Hand, Y. Okita, and S. Mizushina, "Five-Band Microwave Radiometer System for Non-Invasive Measurement of Brain Temperature in New-Born Infants: System Calibration and Its Feasibility," in *Proceedings of IEEE 26th Annual Conference of the Engineering in Medicine and Biology Society*, pp. 2292−2295 (2004).

[111] G. M. J. Van Leeuwen, J. W. Hand, J. B. Van de Kamer, and S. Mizushina, "Temperature Retrieval Algorithm for Brain Temperature Monitoring Using Microwave Brightness Temperatures," Electron. Lett. **37**, 341−342 (2001).

[112] Y. Hamamura, S. Mizushina, and T. Sugiura, "Non-Invasive Measurement of Temperature-versus-Depth Profile in Biological Systems Using a Multiple Frequency-Band Microwave Radiometer System," Automedica **8**, 213−232 (1987).

[113] S. Mizushina, Y. Hamamamura, M. Matsuda, and T. Sugiura, "A Method of Solution for a Class of Inverse Problems Involving Measurement Errors and Its Application to Medical Microwave Radiometry," IEEE MTT-S Int. Microwave Symp. Digest **1**, 171−174 (1989).

[114] S. Mizushina, K. Maruyma, T. Sugiura, et al., "Algorithm for Retrieval of Deep Brain Temperature in New-Born Infant from Microwave Radiometric Data," IEEE MTT-S Int. Microwave Symp. Digest **2**, 1033–1036 (2000).

[115] F. Bardati, V. J. Brown, and P. Tognolatti, "Temperature Reconstructions in a Dielectric Cylinder by Multi-Frequency Microwave Radiometry," J. Electromagn. Waves Appl. **7**, 1549–1571 (1993).

[116] J. Montreuil and M. Nachman, "Multiangle Method for Temperature Measurement of Biological Tissues by Microwave Radiometry," IEEE Trans. **MTT-39**, 1235–1239 (1991).

[117] J. W. Hand, G. M. J. Van Leeuwen, S. Mizushina, et al., "Monitoring of Deep Brain Temperature in Infants Using Multi-Frequency Microwave Radiometry and Thermal modeling," Phys. Med. Biol. **46**, 1885–1904 (2001).

[118] S. Mouty, B. Bocquet, R. Ringot, N. Rocourt, and P. Devos, "Microwave Radiometric Imaging (MWI) for the Characterisation of Breast Tumours," Eur. Phys. J. (Appl. Phys.) **10**, 73–78 (2000).

[119] S. S. Wang, B. A. Van der Brink, J. Regan, et al., "Microwave Radiometric Thermometry and Its Potential Applicability to Ablative Therapy," J. Interv. Card. Electrophysiol. **4**, 295–300 (2000).

[120] A. G. MacDonald, D. V. Land, and R. D. Sturrock, "Microwave Thermography as a Noninvasive Assessment of Disease Activity in Inflammatory Arthritis," Clin. Rheumatol. **13**, 589–592 (1994).

[121] T. Blyth, A. Stirling, J. Coote, D. Land, and J. A. Hunter, "Injection of the Rheumatoid Knee: Does Intra-Articular Methotrexate or Rifampicin Add to the Benefits of Triamcinolone Hexacetonide?," Br. J. Rheumatol. **37**, 770–772 (1998).

[122] J. Shaeffer, A. M. El-Mahdi, A. E. Hamwey, and K. L. Carr, "Detection of Extravasation of Antineoplastic Drugs by Microwave Radiometry," Cancer Lett. **31**, 285–291 (1986).

[123] J. Shaeffer, S. V. Sigfred, S. Leslie, P. Kolm, L. A. Rogus, R. S. Grabowy, and K. L. Carr, "Detection of Air Emboli in Radiographic Contrast Media by Microwave Radiometry," Eur. Radiol. **6**, 570–573 (1996).

APPENDIX A: FUNDAMENTAL AND OTHER PHYSICAL CONSTANTS

Quantity	Symbol	Value	Uncertainty
Avogadro constant	N_A	$6.02214179 \times 10^{23} \, \mathrm{mol}^{-1}$	$3.0 \times 10^{16} \, \mathrm{mol}^{-1}$
Boltzmann constant	k	$1.3806504 \times 10^{-23} \, \mathrm{J/K}$	$2.4 \times 10^{-29} \, \mathrm{J/K}$
Electrical permittivity (vacuum)	ε_0	$8.854187817 \times 10^{-12} \, \mathrm{F/m}$	Exact
Electron mass	m_e	$9.10938215 \times 10^{-31} \, \mathrm{kg}$	$4.5 \times 10^{-38} \, \mathrm{kg}$
Elementary charge	e	$1.602176487 \times 10^{-19} \, \mathrm{C}$	$4.0 \times 10^{-27} \, \mathrm{C}$
First radiation constant	c_1	$3.74177118 \times 10^{-16} \, \mathrm{W \, m^2}$	$1.9 \times 10^{-23} \, \mathrm{W \, m^2}$
First radiation constant for spectral radiance	c_{1L}	$1.191042759 \times 10^{-16} \, \mathrm{W \, m^2/sr}$	$5.9 \times 10^{-24} \, \mathrm{W \, m^2/sr}$
Magnetic permeability (vacuum)	μ_0	$4\pi \times 10^{-7} \, \mathrm{N/A^2}$	Exact
Molar gas constant	R	$8.314472 \, \mathrm{J/(mol \, K)}$	$1.5 \times 10^{-5} \, \mathrm{J/(mol \, K)}$
Planck constant	h	$6.62606896 \times 10^{-34} \, \mathrm{J \, s}$	$3.3 \times 10^{-41} \, \mathrm{J \, s}$
Planck constant over 2π	$\hbar$	$1.054571628 \times 10^{-34} \, \mathrm{J \, s}$	$5.3 \times 10^{-42} \, \mathrm{J \, s}$
Second radiation constant	c_2	$1.4387752 \times 10^{-2} \, \mathrm{m \, K}$	$2.5 \times 10^{-8} \, \mathrm{m \, K}$
Second radiation constant (ITS-90)	c_2	$1.4388 \times 10^{-2} \, \mathrm{m \, K}$	Exact
Speed of light in vacuum	c_0	$299{,}792{,}458 \, \mathrm{m/s}$	Exact
Stefan–Boltzmann constant	σ	$5.670400 \times 10^{-8} \, \mathrm{W/(m^2 \, K^4)}$	$4.0 \times 10^{-13} \, \mathrm{W/(m^2 \, K^4)}$
Third radiation constant	c_3	$2.8977685 \times 10^{-3} \, \mathrm{m \, K}$	$5.1 \times 10^{-9} \, \mathrm{m \, K}$

Source: http://physics.nist.gov/cuu/index.html.

Subject Index

Aberration, 23, 27, 106, 282
Absorber, 191, 337
Absorptance, absorptivity, 235
Absorption, 7, 11, 12, 14, 30, 44, 68, 154, 235, 280, 339, 411
Aerial survey, 308–309
Aghulas Current, 375
Air leakage, 303–304
AIRI facility, 76, 78, 79
Airtightness, 303, 304
Albedo, 335
Alumina (cement), 171, 174, 178
Aluminum extrusion, 41–42
Aluminum industry, 37–42
Amorphous silicon (ASi), 159, 284
Amplifier, 81–82, 119, 146, 151, 165, 285–286
Angle of acceptance, 146
Annealing, 222
Annealing line, 229–230
ANSI (American National Standards Institute), 327–328
Antenna, 360, 383, 427, 429–434, 438–440
Arc flash, 299
Argon, 244
Arsenic, 169, 203
Ashing, 152
ASTM (American Society for Testing and Methods), 306, 327
Atmospheric absorption, 7, 192
Atmospheric attenuation, 348
Atmospheric correction algorithm, 347–351
Atmospheric effects, 347
Atmospheric emission, 7
Atmospheric scattering, 7
Atmospheric (temperature) profile, 381
Atmospheric transmission, 345
Atmospheric transmission spectrum (or spectra), 345, 348
Atmospheric window, 348–349
Atomic layer deposition (ALD), 142, 152

Background radiation, 247–248
Background temperature, 52, 74–79, 113, 291–292
Band(s), 346, 347, 349
Bandgap, 156–157, 192–193, 200
Bandpass, 157, 162, 166, 192–193, 195, 200
Barium strontium titanate (BST), 283, 322
Baseline, 311
Batch annealing, 222

Batch process, 224
Batch production, 218
Batch rolling process, 221
Bead blasting, 189
Bias current, 284
Blackbody
   heat pipe, 76, 100–103
   Na heat-pipe, 179
   pseudo-, 231
   quasi-, 253, 395
Blackbody radiation source
   fixed point, 62, 64-65, 76, 86
   high temperature, 87–93, 99
   radiometric measurements, 74–97
   reflectometric measurements, 60–74
   temperature distribution, 99–103
Blackbody simulator, 166–168
Bolometer, 3, 4
Boltzmann's constant, 346, 428
Boron, 169, 203
BRDF (bidirectional reflectance distribution function), 60
Breast cancer, 403, 440
Brewster's angle, 207
Brightness temperature, 345, 346, 347
BSRIA (Building Services Research and Information Association), 306
BSRIA Guidelines, 306
Buoys
   drifting, 349, 365, 366
   moored, 366
Busbar, 293, 296
Business risk, 299, 315

Calibration wafer, 178–184
Calibration, 164–169
Cast or caster or casting, 46–47
CCD camera, 116, 241, 244
Cellulose, 298
CGL (continuous galvanizing line), 222
Channel(s), 346–347
Charge-coupled device (CCD), 286, 244, 286
Chemical vapor deposition (CVD), 142, 152
Chemical-mechanical planarization (CMP), 152
Chemical-mechanical polishing (CMP), 142, 152
Cladding, 148, 239
CLARREO mission 67, 74
Climate change, 378–379
Climate Data Records (CDRs), 362

CMOS (complementary metal-oxide semiconductor), 286
Coalescence, 172
Coastlines, 374–376
Coating, 31–32
Cold-rolled process, 219–224
Cold-rolled strip, 222
Component failure, 299
Compound semiconductor, 142, 152, 162, 163
Coral bleaching, 379
CoSi phase, 159
Crystal growth
    Czochralski (CZ) method, 142
    float-zone method, 142

Defect, 147–148, 189–190, 308, 312
Degree of activation, 156
Delamination, 172
Depth profiling, 172
Design guidelines, 321–327
Design robustness, 324
Detectablility, 303–304
Detectivity, 4–5
Detector
    indium antimonide, 5–6, 403
    InGaAs, 5–7, 151, 165
    mercury cadmium telluride, 5–6
    photo-, 411–427
    quantum, 151, 282–283, 396
    sensitivity of, 283, 285
    thermal, 3, 5–6, 282–283, 396
Detector array, 282, 284
Detector responsivity, 202
Dichroic mirror, 207
Dielectric constant, 142
Dielectric properties of tissues, 428–429
Dielectrics
    high-k, 162
    low-k, 428
Diffraction grating, 157, 204
Diffuse surface, 98–99, 247
Diffusely reflecting, 229
Diffusion barrier, 172
Diurnal (heating), 335, 340f, 342, 344, 369
Diurnal thermocline, 342–345
Diurnal warm layer, 344

Ear canal, 395, 396, 399
Ear IR thermometer, 394–400
Earth Observing System (EOS), 96, 359–361
E-beam evaporation, 142
Edge-defined film feed growth (EFG), 148
Edge exclusion, 202
Effective temperature range, 324, 325
Effective wavelength, 9, 11–12
Electromotive force (EMF), 141, 174, 175–177
Electron microscope, 187
Electroplating, 222

Ellipsometry, 169, 198, 205
El Niño, 378
Emissivity or emittance
    atmospheric, 7
    cavity, 146, 187, 194
    human skin, 396, 402, 414
    microwave, 353–354
    silicon, 154, 156–159
    water, 154, 175
Emissivity compensation, 266–270
Emissivity-free, 234, 236
Emissometer, 195–199
Epi or epitaxial growth, 142, 162
ESA (European Space Agency), 335, 352
Etch or etching
    caustic, 189
    dry-plasma, 152
    plasma, 152
    wet-plasma, 152
European Global Monitoring for Environment and Security (GMES), 382
European Organisation for the Exploitation of Meteorological Satellites (EUMETSAT), 356
Explorer of the Seas (cruise ship), 365
Exposure time, 27
Extinction length, 185
Extrusion
    blown film, 44–46
    cast film, 46–47
    sheet, 47
Extrusion coating, 47–48
Extrusion of aluminum, 41–42

Fab (fabrication facility), 141
Far-field propagation zone, 431
Ferroelectric metal, 283
Fiber optics, 141, 143–151
Figures of merit, 287
Finite-element model, 180, 182
Fire detection, 380
Firefighting, 280–328
Fire service, 319–321, 324, 326–327
Flash lamp, 207
Flashpoint, 298
FOT (flouroptic thermometer), 141
FOV (field-of-view), 4, 21t, 22, 25t, 26–27
FPA (focal plane array), 4, 283, 285, 404
FTIR (Fourier transform infrared spectrometer), 64, 68, 79–81, 96, 106
Furnace
    anealing, 223
    basic oxygen (BOF), 219–220
    blast (BF), 218, 219–220, 239, 242
    continuous annealing (CAF), 222, 248–250, 252, 258, 260
    electric (EF), 220, 221
    heat treatment, 218, 245

metal freezing point, 76, 87, 88
reheating, 221, 224
tunnel, 51–53

Gallium arsenide, 3, 159
Galvanizing or galvanized, 220, 222–223, 261, 266
Galvannealing, 223, 261
Germanium (Ge), 282
Glass
  automotive, 23, 30, 34–35
  cobalt, 237
  container, 23, 30, 32–34
  flat, 30–32
  soda-lime, 292f–293f
Glass industry, 28–36
Glass wool, 35–36
Global warming, 67, 336, 378
Gold-cup, 38, 253, 260–261
Grain size, 172, 221
Growth chamber, 162
Gulf Stream, 374–375

Heat capacity, 287
Heat flux
  latent, 338
  radiant, 294
  sensible, 338
  turbulent, 338
Heat loss, 26, 86, 100, 103, 240, 246, 293, 302, 310, 340, 342, 401
Heat transfer modes
  conduction, 306–307, 335–341
  convection, 292, 296, 306–307, 341–344
  evaporation, 338–339
  radiation, 306–307
Helmholtz reciprocity principle, 59, 105
Hemiellipsoidal mirror, 60
Hot dipping or hot-dip process, 222
Hot rolling of aluminum, 40–41
Hot strip mill, 220, 221
Hot strip rolling, 221–222
Humidity, 6–7, 27, 36, 169, 322, 350, 407
Hybrid system, 153, 285–286
Hyperthermia, 440
Hysteresis, 172, 180

Ice bath, 175
Ignition temperature, 297–298
Image processing (or processor), 241–245, 406–410
Image quality, 286, 309, 320, 322, 324–327, 403
Immersion thermocouple, 237–240, 243–244
Immersion-type optical fiber, 238–241
Implant, 142
In-chamber calibration, 170
Incandescent lamp, 70–71

Index of refraction, 148
Industries
  aluminum, 37–42
  glass, 28–36
  plastic, 43–49
  semiconductor, 50–51
  steel, 36–37
Infrared (IR)
  longwave or long wave, 84–85, 335
  shortwave or short wave, 49, 292, 335
Infrared window, 348, 381–382
Insolation, 337, 339
Integrated circuit (IC), 138, 142, 152, 161
Integrated pyrometer-emissometer, 196–199
Integrating sphere, 61–64, 70–71, 109–112
Interferometer, -try
  Marine-Atmospheric Emitted Radiance interferometer (M-AERI), 364–365
  Michelson interferometer, 96, 204
  Speckle interferometry, 204
Ion implanting, 152, 169
IR camera or IR imager, 84, 405f, 408f
IR optics, 282
IR spectrum, 394
Irradiance
  solar, 301, 302, 335, 337
  spectral, 90
IRT (infrared thermography), 288
ISO (International Organization for Standardization), 327–328
Isothermal cavity, 58, 194
Istituto di Metrologia "G. Colonnetti" (IMGC), 171
ITS-90, 2, 15, 54, 88, 90, 225

Japan Aerospace Exploration Agency (JAXA), 382

Kirchhoff's Law, 58, 98, 191

LABB (large-area blackbody), 103
Lambertian, 59, 98
Lamp
  halogen, 111
  infrared, 261
  quartz halogen, 207
LART (Laser absorption radiation thermometry), 235–237
Laser (types of)
  Ar-ion, 114
  CO_2, 114
  dye, 411, 413, 417–418
  excimer, 174, 185
  He-Ne, 114
  Nd:YAG, 114
  pulsed, 185, 205
  Q-switched, 414

quantum cascade, 67–68
ruby, 411, 417–418
solid state, 235–236
Laser cutting, 49–50
Laser depilation, 426–427
Laser energy, 412
Laser spot size, 418–420
Laser therapy, 411–412, 420
Laser-tissue interaction, 413–414
Laser welding, 49–50
Lattice vibration mode, 156
LBIR facility, 93–94
LCD (liquid crystal display), 281
LEDs (light-emitting diodes), 142, 167f
Lesions
    flatter, 424t
    inflammatory, 407
    pigmented, 415t
    superficial, 423
    vascular, 411, 412t, 420
Linear light source, 266–270
Line-scanner, 19–24
Liquid nitrogen, 82, 364, 403
Lithography, 142, 152
LNE (Laboratoire National de Metrologie et
    d'Essais), 81
LPRT (lightpipe radiation thermometer), 141,
    165–166
LST (land surface temperature), 338–339

Mammography, 440
Marine atmosphere, 365
Marine-Atmospheric Emitted Radiance
    Interferometer (M-AERI), 364–365
MBF facility, 96–97
Melting point of silicon, 138
Metal organic chemical vapor deposition
    (MOCVD), 140, 142, 152, 162
Michelson interferometer, 204
Microbolometer, 284–285, 287, 322
Microprocessor, 152
Microwave radiometer or radiometry, 353–354,
    359–360
Microwave radiometry (medical), 427–442
Microwave thermometry, 430f
MIRCO facility, 81
MODIS UCSB Emissivity Library, 352
Molecular beam epitaxial (MBE), 140, 142
Molten iron, 237–238
Molten metal, 225, 237–238
Molten steel, 220–221, 237–238
Monolithic, 285–286
Monte Carlo modeling or ray tracing, 99
Moore's Law, 152
Multi-channel, 349, 353, 359, 369, 382
Multi-junction PV, 161–162
Multiple reflections, 250–252

NASA (National Aeronautics and Space
    Administration), 96, 356, 360, 382
NASA EOS program, 367
National Research Council (NRC), 336
Near-field induction zone, 431
NETA (International Electrical Testing
    Association), 297
NFPA (National Fire Protection Agency), 320,
    327–328
NiSi phase, 159
NIST (National Institute of Standards and
    Technology), 79, 155, 367
NIST Transfer Radiometer (TXR), 82–84
NOAA (National Oceanic and Atmospheric
    Administration), 336, 355
Noise-equivalent-temperature difference
    (NETD), 287, 326
Non-distructive testing (NDT), 288
Nonuniformity, 93
Numerical aperture (NA), 141, 149, 150
NWP (numerical weather prediction), 377

OFT (optical fiber thermometry), 141, 143–151
Opacity, 192, 200, 339
Optical fiber
    graded-index, 239
    metal sheath, 239
Optical micrometer, 204
Optical microscopy, 172
Optical properties, 23, 37, 99, 153, 156–159,
    182, 193, 232, 358, 413
Optical scattering, 117
Outer-diameter (OD) ground, 147
Oxidation resistance, 223
Oxide layer, 157
Oxide thickness, 169
Oxidized steel, 234

Personal protection equipment (PPE), 308
Personal safety, 308
Phosphor or Phosphorus, 141, 153, 169
Photon flux, 5
Photosensitive resist, 142
Photothermal (or photo-thermal) effect, 237
Photovoltaic (PV), 161, 163
Physical vapor deposition (PVD), 142
Pig iron, 220, 239
Pixel, 283–286, 291
Planck's constant, 346
Planck's law or Planck's function, 9, 345
Plasma processing, 208
Plastics industry, 43–49
Polarization technique, 234
Polysilicon (poli-silicon), 157
Polytetrafluoroethylene (PTFE), 63
Polyvinyl chloride (PVC), 296
Port wine stain, 413–414, 423–426

PPTR (pulsed photo-thermal radiometry), 411–427

PRT (platinum resistance thermometer), 76, 185–188

PTB (Physikalisch-Technische Bundesanstalt), 114

Pyroelectric sensor, 398f

Pyromark paint, 79, 108

Pyrometry or pyrometer
blow-tube, 237, 243
dual wavelength, 192, 256–261
dynamic surface anneal (DSA), 206–207
flash rapid thermal annealing (fRTA), 207
laser spike annealing (LSA), 207
near-infrared (NIR), 138
optical, 138, 207f, 237, 238
ripple, 202–203
two-color, 192, 235, 256–261

Pyrophoric carbon, 298

Quality assurance, 327

Quality control, 327

Quartz, 148

Quartz lightpipe, 143, 145, 146, 147

Quartz plate, 179

RADCAL, 317

Radiance, 317

Radiance temperature, 11, 58, 59, 74, 75–79, 82, 178, 256, 260, 301, 311, 317

Radiant energy, 281

Radiant power, 9, 11, 13, 192

Radiation
IR (infrared), 280, 292, 318, 322, 400, 405, 411
solar, 301, 302, 337, 338, 339
thermal, 2–6, 8, 23–27, 59, 68, 70, 76, 80, 96, 138, 141, 191

Radiation pyrometry, see pyrometry

Radiation therapy, 424t

Radiation thermometer (RT) or thermometry
active method, 227–228
broadband, 10–13
dual wavelength method, 256-261
emissivity-compensated, 261
industrial, 4–24
infrared ear thermometry, 394–400
laser absorption, see LART
lightpipe, see LPRT
multi-wavelength, 15
passive method, 226–227
polarization, 232–234
ratio, 13–14
scanning, 9–24
specifications of, 24–27
spectral range of, 7, 52
total, 13

Radiation thermometry, see radiation thermometer

Radiative transfer, 348, 349, 358

Radiometers (satellite-borne or spacecraft)
AATSR (Advanced Along-Track Scanning Radiometer), 336, 357–359, 371–372
AIRS (Atmospheric InfraRed Sounder), 336
AMSE-E (Advanced Microwave Scanning Radiometer), 336, 359–361
AMSR-E (Advanced Microwave Scanning Radiometer), 359–361, 372–373
ATSR (Along-Track Scanning Radiometers), 339, 351, 358–359
AVHRR (Advanced Very High Resolution Radiometer), 355–356, 368–369
MODIS (MODerate resolution Imaging Spectroradiometer), 335, 356–357, 369–371
SGLI (Second-Generation Global Imager), 381, 382
SLSTR (Sea and Land Surface Temperature Radiometer), 381, 382
VIIRS (Visible Infrared Imager / Radiometer Suite), 336, 381–382

Radiometers (ship-based)
CIRIMS (Calibrated InfraRed in situ Measurement System), 363, 364
ISAR (Infrared Scanning Autonomous Radiometer), 363, 364
SISTeR (Scanning Infrared Sea Surface Temperature Radiometer), 363, 364

Radiometry (or radiometer)
cryogenic, 87, 89, 94, 96
infrared, 73, 76, 77f, 96, 334–337, 339, 347–352, 367, 380–383
microwave, 353–354, 359–360
multifrequency, 438–440
pulsed photo-thermal, see PPTR
satellite-borne, 354–361
scanning, 334, 336, 339, 357, 359, 383
ship-based, 363
spacecraft, 354–361

Radiosity, 200–201

Raynaud's phenomenon, 402

Reactor, 142, 162

Readout integration circuit (ROIC), 285, 286

Reference junction, 175, 182

Reflection
diffuse, 8, 106, 115, 267
specular, 106, 110, 267

Reflectance
directional-hemispherical, 60, 105, 106–108
hemispherical, 98, 105
spectral, 70, 72, 106, 122

Reflectometry
broadband sources, 70–74
laser source, 60–70

Relative spectral response function, 345, 346–347
Repeatability, 27, 152–153, 164, 165, 168–169
Reproducibility, 183, 417
Resistance thermal device (RTD), 141, 154
Resolution
    spatial, 324–325
    temporal, 201
Response time, 16–18t, 25t, 26, 27, 437, 438
Ripple technique, 141, 200–203
Risk assessment, 299, 308
ROI (region of interest), 166, 406, 409–410
Rolling mill, 40–41
Rolling process, 40–41
Rolling temperature, 40–41
Roll-strip wedge, 250–253
Rosenstiel School of Marine and Atmospheric
    Science, 367
Roughness, 8, 37–38, 110, 121, 156, 157, 191,
    204, 225, 228, 266, 267, 339, 351, 360
RTA (rapid thermal annealing), 143, 169, 198,
    199f
RTO (rapid thermal oxidation), 143, 169, 198,
    199f
RTP (rapid thermal processing), 50, 143
RTP test bed, 178–179
Rutherford Appleton Laboratory, 358

Sapphire, 146–147, 148, 149
Sapphire lightpipe, 143–144, 145f, 146
Satellites
    Earth-observing, 354
    European MetOp, 381
    GCOM (Global Change Observation
        Mission), 382–383
    Geostationary meteorological, 359, 372
    NPOESS, 381–382
    polar-orbiting, 359, 372, 382
Satellites (list of)
    Aqua, 335–336, 352, 356, 357, 361, 366, 369,
        370t, 371, 373t
    Envisat, 336, 371(t), 373
    ERS-l, ERS-2, 351,
    GOES (Geostationary Operational
        Environmental Satellite), 359, 372, 380
    Terra, 335, 352, 356, 357, 365, 369, 370t, 371,
        374f
Scanning electron microscopy, 172
Schlieren photography, 401
Secondary ion mass spectrometry (SIMS), 172
Seebeck coefficient or Seebeck effect, 172, 176,
    178, 179, 182, 183
Semiconductor industry, 50–51, 138–209
Semiconductor manufacturing technology
    (SEMATECH), 173
Shielding flange, 229, 245–250
Signal-to-noise ratio, 192, 357, 365
Silicide formation, 152, 160f
Silicon carbide, 106, 142

Silicon dioxide, 160f, 221, 239
Silicon nitride, 142, 157, 158f, 284–285
Size-of-source effect, 10, 25t, 27
Slag, 225, 237, 239–240, 242, 243
Smelting and refining process, 220–221
Soak time, 153
Soaking, 222
Solar cell, 142, 161, 162
Solar energy, 335
Specific heat, 424
Speckle interferometry, 204
Sputtering, 142
SST (sea surface temperature)
    bulk, 339
    Multi-Channel .. (MCSST), 349
    Non-Linear .. (NLSST), 349
    Pathfinder .. (PFSST), 349
    skin, 339
Stability, 27, 67
Stainless steel, 223–224
Standardization, 327–328
STARR facility, 115
Steel
    cold-rolled, 222
    electrical or silicon, 223–224
    hot-rolled, 221–222
Steel industry, 36–37
Steel plant, 218, 237, 250, 253, 258
Steel process, 223–224
Steel production, 219–224
Stefan-Boltzmann constant, 337–338
Stefan-Boltzmann equation or law, 337–338
Stirling cycle, 364
Stray light, 62, 145, 161, 208
Substrate, 141–142, 156–159, 174, 185, 206–207f
Surface renewal (theories), 341–342
Surface temperature
    global, 334, 372
    ice, 353
    land, see LST, 338–339
    of the earth, 334–384
    sea, see SST, 339–341
    snow, 353
Susceptor, 139t, 146, 154, 162

TAAS facility, 119
Taphole, 220, 239–240, 241f
Temperature
    air, 294, 301–302
    ambient, 291–292, 307
    apparent, 317
    background, 291, 294
    brightness, 345, 346, 347
    bulk, 341, 344
    core (body), 394–395
    environmental, 300
    human body, 395–396
    ignition, 297, 298

land surface temperature, *see* LST
radiance, 317
sea surface temperature, *see* SST
skin, 410
sky, 302
surface, *see* surface temperature
Temperature coefficient of resistance (TCR), 284
Temperature control, 3, 28, 141, 153, 161, 165, 221–224, 407
Temperature deficit, 347
Temperature scale, 2, 54, 76, 87, 153, 155, 166, 170, 410, 410
TFTC (thin-film thermocouple)
  Pt/Pd wire, 170, 171–172
  Rh/Pt, 170, 171–172
Thermal anomalies, 300, 306, 310f
Thermal background, 228
Thermal budget, 155, 199
Thermal cycle, 176, 187–188
Thermal environment, 316–319
Thermal equilibrium, 191
Thermal evaporation, 142
Thermal expansion, 180, 204
Thermal image or imaging, 280, 296, 302f, 316–321
Thermal imaging camera, *see* TIC
Thermal imaging (medical), 400–411
Thermal impulse, 206
Thermal index, 307
Thermal insulation, 36, 285f, 302–303, 304–305, 306
Thermal map, 309
Thermal mass, 185
Thermal resistance, 300
Thermal scene, 282, 322, 323f
Thermal sensitivity, 286, 325–326
Thermal skin layer, 340, 342
Thermocouple (TC)
  type-E, 185, 186, 188f
  type-K, 170–171
Thermoelectric (TE) cooler, 151, 322
Thermoforming, 49
Thermogram, 400–401, 403, 406f
Thermographic survey, 295, 296, 300, 315
Thermography
  building (fabric), 299–310
  clinical, 400–411
  comparative, 288–293
  electrical, 293–299
  engineering, 286–316
  infrared, *see* IRT
  mechancial plant, 310–316
  medical, 400–411
  quanlitative, 289–290
  quantitative, 288–289
Thermometer or thermometry
  acoustic, 205

clinical, 440–442
contact, 394
ear, 394–400
flouroptic (FOT), 141
infrared or IR, 32–34, 37, 43–45
infrared (or IR) ear, 394–400
laser, 49–50
lightpipe (LPRT), 141, 166–167
mercury-in-glass, 395
microwave, 430f
optical fiber (OFT), 141, 143–151
platinum resistance (PRT), 76, 185–188
radiation (RT), 2–54, 138–209
resistance, *see* RTD
vacuum infrared standard, *see* VIRST
wedge, 250–253
Throughput, 122, 202
TIC (thermal imaging camera), 279–287
Time of flight, 205
Tissue, 402, 411–420, 423, 428–430, 438–441
Tissue temperature, 417, 419, 427, 429, 440
Tool-to-tool variability, 169
Traceability, 367
Transistor, 152, 206
Transmission, 292–293
Transmittance, 7, 24, 201, 294
Tumor, 429, 440
TXR, 82–84
Tympanic membrane, 396–397, 399
Tympanic temperature, 397–399

Uncertainty
  expanded, 72, 91f, 94, 110–111, 174–175
  measurement, 13, 25t, 186, 192, 260, 344, 363–366, 371, 382
  relative, 59, 111
  standard, 183, 319t
  temperature, 41, 182–184, 356, 359, 361, 440
Underwriters Laboratory (UL), 327

Validation, 361–367
Vanadium oxide (VOx), 4, 284, 322
Video Electronics Standards Association (VESA), 327
Video signal, 280
VIRST (vacuum infrared standard thermometer), 97

Wafer
  GaAs, 142
  InP, 142
  proof, 170–174

sapphire, 146–147
Si, 187
TC-instrumented, 170, 171
temperature-monitor, 163
test, 163, 186
thin-film instrumented, 155
Wafer back side, 146, 153, 189
Wafer temperature, 50, 139, 146, 151f, 161, 163, 170, 181, 183t, 205
Water content of tissue, 428–429

Wedge thermometer, 252
Wien's approximation or Wien's law, 193, 247, 257, 260

Yttrium aluminum garnet (YAG), 147

Zinc-coated, 222, 223
Zinc selenide (ZnSe), 5, 282
Zirconia, 185